This introductory textbook describes modern cosmo for advanced undergraduates who are familiar with mathematical and basic theoretical physics. It is intended for use on courses in theoretical physics or applied mathematics that include modern cosmology. An introductory survey of the large scale structure of the universe is followed by an outline of general relativity. This is then used to construct the standard models of the universe. The early and very early stages of the big bang are described, and this includes primordial nucleosynthesis, grand unified theories, primordial black holes, and the era of quantum cosmology. The problem of the formation of structure in the universe is then addressed. This textbook also gives brief outlines of alternative cosmologies. The theory is complemented by two chapters on observations of the local universe and the distant universe that are significant to cosmology. The book includes 400 problems for students to solve, and is accompanied by numerous worked examples.

INTRODUCTION
TO
COSMOLOGY

INTRODUCTION
TO
COSMOLOGY

Second edition

JAYANT V. NARLIKAR

Inter-University Centre for Astronomy and Astrophysics, Pune, India

CAMBRIDGE
UNIVERSITY PRESS

Published by the Press Syndicate of the University of Cambridge
The Pitt Building, Trumpington Street, Cambridge CB2 1RP
40 West 20th Street, New York, NY 10011-4211, USA
10 Stamford Road, Oakleigh, Victoria 3166, Australia

First published by Jones and Bartlett Publishers Inc. 1983
Second edition published by Cambridge University Press 1993

Printed in Great Britain at the University Press, Cambridge

A catalogue record for this book is available from the British Library

Library of Congress cataloguing in publication data

Narlikar, Jayant V.
Introduction to cosmology / Jayant V. Narlikar. — 2nd ed.
p.   cm.
Includes bibliographical references and index.
ISBN 0 521 41250 1 (hc) — ISBN 0 521 42352 X (pb)
1. Cosmology.   I. Title.
QB981.N3 1993
523.1–dc20   92–32390   CIP

ISBN 0 521 41250 1 hardback
ISBN 0 521 42352 X paperback

Contents

Foreword

This is an important book, which I hope will be studied by everybody concerned with physics and astronomy. I can guarantee that the student who works steadfastly through the many splendid examples will end by knowing a very great deal about relativity and cosmology. I can also guarantee that the practised expert will find much that is a surprise and a delight.

Backed by many years of distinguished research, the book is a masterpiece of clarity. From his earliest days as a graduate student, Jayant Narlikar has always been an incisive writer and lecturer. The mystery about a lecture by Narlikar is to understand how he manages to go at such an apparently leisurely pace, to write on a blackboard with extreme precision and without haste, and yet at the end of an hour to have covered an immense amount of ground. The solution to the mystery has to be that he wastes less time than most of us on irrelevancies, which is just what the reader of this book will find from the first page to the last. Author, publisher, and reader are all to be congratulated.

I wrote the above at the time of the first edition of this book. No word needs changing but a few need adding. This is not only an important book. It is the best book, and I believe by a considerable margin. Pity the student who doesn't work from it.

Fred Hoyle

Preface to the first edition

The progress of modern cosmology has been guided by both observational and theoretical advances. The subject really took off in 1917 with a paper by Albert Einstein that attempted the ambitious task of describing the universe by means of a simplified mathematical model. Five years later Alexander Friedmann constructed models of the expanding universe that had their origin in a big bang. These theoretical investigations were followed in 1929 by the pioneering work on nebular redshifts by Edwin Hubble and Milton Humason, which provided the observational foundations of present-day cosmology. In 1948 the steady state theory of Hermann Bondi, Thomas Gold, and Fred Hoye added a spice of controversy that led to many observational tests, essential for the healthy growth of the subject as a branch of science. Then in 1965 Arno Penzias and Robert Wilson discovered the microwave background, which not only revived George Gamow's concept of the hot big bang proposed nearly two decades before, but also prompted even more daring speculations about the early history of the universe.

The landmarks mentioned above have led to many popular and technical books on cosmology. In particular, the rapid growth of interest in the areas of general relativity and cosmology during the 1970s was reflected in a number of classic textbooks that came out in the early 1970s. The purpose of the present textbook is to introduce the reader to the state of the subject in the early 1980s. However, the approach adopted here is different from that found in most other texts on the subject, and it is perhaps desirable to state what the differences are and why they have been introduced.

For example, it is usual to find cosmology appearing at the end of a text on general relativity, introduced more as an appendage than as a subject in its own right. Perhaps this is one reason why cosmology still stands

apart from the rest of astronomy, to which it really belongs. The
astronomer tends to regard cosmology as a playground for general
relativists rather than as a logical extension of extragalactic astronomy. To
correct this tendency, the relative importance of cosmology and general
relativity has been inverted in this text. Chapter 2 introduces general
relativity more as a tool for studying cosmology than as a subject in its own
right. Thus the relativist may find many topics dealt with at a superficial
level or not at all. This chapter covers only those topics that are really
necessary for understanding the large-scale geometrical properties of the
universe. I have taken this approach in the hope that the relatively
elementary treatment of general relativity will not put a newcomer off, as
a more exhaustive treatment might well do. The expert relativist may skip
this chapter and refer to it only for fixing the notation.

Chapters 3 and 4 introduce the standard models of cosmology as
solutions of Einstein's equations. The tools developed in Chapter 2 will be
found applicable here, and the reader will find the pace more relaxed than
in Chapter 2.

Chapters 5, 6, and 7 concentrate on the physical aspects of standard
cosmology. Gamow's idea of primordial nucleosynthesis, the current state
of ignorance on galaxy formation, the properties of the microwave
background, and the various recent contributions of particle physics to our
understanding of the early universe are discussed here.

Perhaps this would have been the appropriate stage to move on to
observational cosmology. However, I felt that the reader should also be
taken on a short excursion into nonstandard cosmology. Contrary to the
view propagated (unfortunately) by many experts in cosmology today, the
subject is not a closed book, nor is standard cosmology the only answer to
the problem of the origin and the evolution of the universe. Part III of this
book introduces some alternatives to the standard models.

Although some readers may prefer to see an observational test
discussed immediately after the theoretical prediction, I have left observations to the last part of the book. This approach has made an overall
assessment of the various models possible. A survey of cosmological
observations shows how better techniques and a better appreciation of
errors and uncertainties have led to frequent reassessments (a classic
example being the value of Hubble's constant, which is still uncertain!). I
have therefore not gone into very many observational details, but have
emphasized how the observations are made and the likely sources of
errors. In any case it would be unwise to go into too many details in an
introductory text.

In spite of many remarkable advances, cosmology is still very much an open subject. On the observational side, the launching of the space telescope in the mid-1980s is likely to revolutionize our view of the universe. On the theoretical side, the Grand Unified Theories (GUTs) are still grappling with the problem of the early universe, while quantum cosmology is in a rudimentary state. Cosmologists have yet to appreciate the problems posed by life in the universe. How did life come into existence? Is it confined to the Earth or is it widespread in the universe? A text of the future may well devote a large part of its discussion on cosmology to contributions from biology.

It is assumed that the reader is familiar with standard mathematical methods like differential equations, vector analysis, Fourier series and transforms, the calculus of variations, and so on. A knowledge of basic physics including mechanics, elementary thermodynamics, electromagnetic theory, atomic structure, and fluid dynamics is also assumed. Similarly, basic knowledge of elementary astronomy will be useful. The text is intended for advanced undergraduates, graduate students, and teachers of astronomy and cosmology.

This book contains over 400 exercises, of which over 80 per cent are of a computational nature. Many of them are designed to illustrate or amplify the material described in the text. It is hoped that they serve their intended purpose.

I thank Art Bartlett for encouraging me to write the book. Comments received from Bob Gould, Bob Wagoner, Dimitri Mihalas, Richard Bowers, and Geoff Burbidge were of great help during the preparation of the manuscript. Last, but not least, it was Fred Hoyle who introduced me to the fascinating field of cosmology as a graduate student, and I am indebted to him for agreeing to write the Foreword.

I began writing this book while visiting the Department of Applied Mathematics and Astronomy at the University College, Cardiff, Wales. I am grateful to the head of the department, Chandra Wickramasinghe, for the facilities extended to me at Cardiff. For the prompt typing of the manuscript I am indebted to Ms Suzanne Ball and Mr P. Joseph. It is also a pleasure to acknowledge the help I received from the Drawing Office and Xerox Facility of the Tata Institute of Fundamental Research.

Bombay, India   *Jayant Narlikar*

Preface to the second edition

I am happy that the revised second edition of *Introduction to Cosmology* is seeing the light of the day. The motivation and format of this edition continue to be the same as for the earlier edition and hence this preface only supplements the more detailed preface of the first edition given above.

The changes incorporated in this edition broadly reflect the new developments in cosmology that came in the 1980s, e.g. inputs from particle physics including the inflationary universe, new attempts at structure formation, recent observations of the large-scale structure and the improved (more sensitive) limits on the intensity fluctuations of the microwave background. The observational sections have been updated although no text book can really keep pace with the rapid advances in cosmological observations.

A comparison of the two editions will reveal a slight rearrangement of the chapters including a streamlining of the part devoted to alternative cosmologies. The final chapter is perhaps more critical of standard cosmology than before. This is necessary, in my opinion, in order to correct the prevailing impression that the standard hot big bang model describes the universe so well that no significant new or alternative inputs are required.

I thank Simon Mitton for encouraging me to proceed with the job of revising the book for Cambridge University Press. Thanks to speedy typing by Santosh Khadilkar and help with artwork by Arvind Paranjpye, the job could be completed within the time frame set by Simon. I also thank the numerous reviewers of the first edition whose constructive comments helped in preparing the revised manuscript.

Inter-University Centre for
Astronomy and Astrophysics
Pune, India

Jayant V. Narlikar

1

The large-scale structure of the universe

1.1 Astronomy and cosmology

No branch of science can claim to have a bigger area of interest than cosmology, for cosmology is the study of the universe, and the universe by definition contains everything. Although, because of its profound implications, cosmology had traditionally excited the imaginations of poets, philosophers and religious thinkers, our approach to the subject will be through the science of astronomy. Astronomy started as a study of the properties of planets and stars, and gradually reached out to include the limits of the Milky Way System, which is our Galaxy. Modern astronomical techniques have taken the subject beyond the Galaxy to distant objects from which light may take billions of years to reach us.

Cosmology is concerned mainly with this extragalactic world. It is a study of the large-scale structure of the universe extending to distances of billions of light-years – a study of the overall dynamic and physical behaviour of billions of galaxies spread across vast distances and of the evolution of this enormous system over several billion years.

At first such a study may appear an ambitious task. Are our tools of observation good enough to provide sufficient scientific information about the large-scale structure of the universe? Is our knowledge of the laws of nature sufficiently advanced and mature to interpret this information? We may answer these questions with a remark of Albert Einstein: 'The most incomprehensible thing about the universe is that it is comprehensible.' Although our observing techniques are far from perfect and our knowledge of physical laws still leaves considerable room for improvement, we are now in position to make some sense out of what we observe about the universe. We can begin to study cosmology as a branch of science, just as we study the structure of the universe. This is what this book is all about.

We will begin with a brief survey of some of the features of the universe that are pertinent to the subject of cosmology

1.2 Our Galaxy

Figure 1.1 shows a schematic representation of the Milky Way. In Figure 1.1(a) we see it face-on and in Figure 1.1(b) edge-on. The bright parts are made of light from many stars, while the dark parts are the observations produced by absorbing gas and dust clouds. The face-on picture shows the spiral structure of the galaxy, while the edge-on picture demonstrates that it is a disc with a central bulge. The disc is also referred to as the galactic plane.

Although the physicist would prefer the light-year as a unit of astronomical distance, the astronomer (for historical reasons) has grown accustomed to using the parsec (pc), the kiloparsec (kpc), and the megaparsec (Mpc) as distance units. $1\,\mathrm{pc} \approx 3.26$ light-years $\approx 3.0856 \times 10^{18}$ cm. Using the kiloparsec as the unit for galactic dimensions, the diameter of the disc is estimated to be $\sim 30\,\mathrm{kpc}$, and its thickness $\sim 1\,\mathrm{kpc}$. The Sun along with all its planets is located $\sim 10\,\mathrm{kpc}$ from the centre. The galaxy rotates about its polar axis as shown in

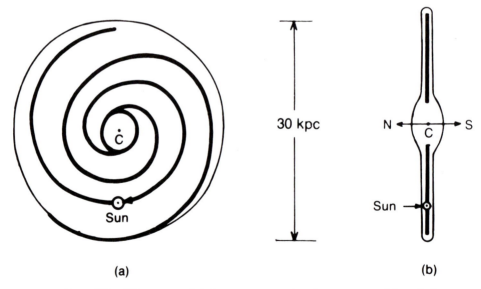

(a)   (b)

Fig. 1.1 The Milky Way, seen (a) face-on as a circular system with spiral arms, and (b) edge-on as a disc with a central bulge. We (that is, the Sun and its planets) are located about two-thirds of the way out from the centre. The Galaxy rotates about a central axis, with N and S the galactic North and South poles. C is the centre of the Galaxy.

Figure 1.1, although not as a rigid body. The Sun, for example, takes ~ 200 million years to make one complete orbit. Other stars have highly eccentric orbits that take them out of the galactic plane and also to the galactic centre. The former type of stars (like our Sun) with nearly circular orbits in the disc are called Population I stars, while the latter type of stars are called Population II stars. From the metal contents of the two types of stars and the theory of nucleosynthesis it is possible to argue that Population II stars are older than Population I stars. Astronomers also refer to an even earlier generation of stars called the Population III stars, which were very massive and burnt out quickly.

The mass of our Galaxy is estimated at ~ $1.4 \times 10^{11} M_{\odot}$, where $M_{\odot} \equiv$ mass of the Sun $\approx 2 \times 10^{33}$ g (a convenient mass unit in astronomy.) It is estimated that there are upwards of 10^{11} stars in the galaxy. However, stars alone do not make up the whole of the galaxy. The dark lanes in Figure 1.1 show that obscuring matter is also present.

Absorption lines in the spectra of galactic stars show that absorbing gases are present in the interstellar medium. Gas appears in various forms – atomic and molecular, hot and cold. Emission nebulae around stars are made of gas that absorbs the ultraviolet radiation from stars and radiates it as visible light in spectacular colours. The so-called H II regions are hot regions near stars and contain hydrogen gas that has been ionized by the ultraviolet light of the stars. By contrast, the H I regions are cool regions of atomic hydrogen. The 21-cm observations in radio astronomy were largely responsible for detecting neutral hydrogen in the galaxy. Moreover, since the 1960s radio and microwave studies have revealed the existence of several complex molecules in the interstellar gas clouds.

Dark nebulae in the Galaxy are, by contrast, due to the presence of dust (see Figure 1.2). Interstellar dust may exist in several forms, such as graphite, silicates, or solid hydrogen. The effect of dust is to reduce the intensity of light from distant stars in the Galaxy. In the early days astronomers overestimated stellar distances in the Galaxy because they failed to correct for interstellar absorption. (Without correction, the faintness of a star was assumed to be wholly due to its distance from us.) The early astronomers also mistook dark regions for 'holes' or empty regions in the Galaxy.

The distances between stars in the Galaxy were determined in the early days by the trigonometric method. Unfortunately, this method loses accuracy beyond ~ 50 to 100 pc. A more reliable method that made use of the variable stars called Cepheids became available in 1912. H. Shapley used this method to measure the distances of remote stars in our galaxy

Fig. 1.2 The Horsehead Nebula in Orion. The dark shape arises from interstellar dust. (Courtesy of Kitt Peak National Observatory.)

and showed that our galaxy was much larger than it was previously thought to be.

A few years later, Hubble discovered that certain bright nebulae previously considered part of the Galaxy were actually remote objects lying well beyond it. Hubble's discovery finally laid to rest the belief that the whole of the observable universe was contained in our Milky Way, an island floating in infinite space. The nebulae that Hubble had proved to be extragalactic turned out to be galaxies in their own right. Today the astronomer has a much better perspective on the vastness of the extragalactic world. The following section describes broad features of various types of galaxies known today. There we shall also see that the galaxies appear to contain dark matter that extends substantially beyond their visible boundaries.

1.3 Galaxy types

The spiral structure of our Galaxy shown in Figure 1.1(*a*) was difficult to establish observationally, since we view it from within. It is easier to see this structure in other galaxies, unless we are viewing them edge-on. Our nearest large galaxy, labelled M31 (see section 1.7 for the meaning of this label), in the Andromeda constellation, has a similar spiral structure (see Figure 1.3). *Spiral galaxies*, as such galaxies are called, are probably the

Fig. 1.3 The Great Galaxy in Andromeda, a spiral galaxy of type Sb. (Courtesy of Kitt Peak National Observatory.)

most numerous among the various bright galaxy types (see Figures 1.4 and 1.5). Like our Galaxy they show rotation, flattening with a central bulge, and dark lanes of absorbing matter.

In 1926 Hubble classified the various galaxy types in the following way.

Fig. 1.4 Galaxy of type Sb in Ursa Major, M81. (Courtesy of Kitt Peak National Observatory.)

Fig. 1.5 Galaxy of type Sc in Pisces, M74. (Courtesy of Kitt Peak National Observatory.)

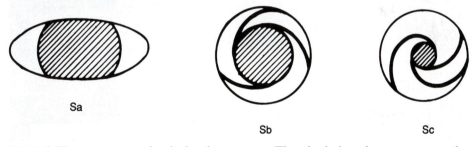

Fig. 1.6 The sequence of spiral galaxy types. The shaded region represents the nucleus.

The various classes of spiral galaxies are called Sa, Sb, Sc, and so on. The sequence is in decreasing order of the importance of the central *nucleus* or bulge in relation to the surrounding disc. Our galaxy and M31 are of type Sb. Some spirals have bars in the central region. These are called *barred spirals* and are categorized as SBa, SBb, SBc, and so on. See Figure 1.6 for schematic illustrations of these types.

While spirals are most numerous among bright galaxies, the most numerous among *all* galaxies are those classified as *ellipticals*. These are ellipsoidal in shape, show very little rotation, and have very little gas and dust (see Figures 1.7 and 1.8). The various types of ellipticals are placed in the sequence E0, E1, ..., E7. This sequence describes progressively flattened profiles of galaxies, E0 being nearly spherical and E7 markedly flattened lenticular form. These types are illustrated in Figure 1.9.

Unlike star images, which tend to be pointlike, galaxies have nebulous shapes like those described above. Astronomers can measure the distribution of light across a galaxy with great accuracy using solid-state instruments like the charge-coupled device (CCD). The light distribution is conveniently described by *isophotes*, or contours of equal intensity. In many galaxies, especially the ellipticals, increasing sensitivity of measurement shows that the boundary of a galaxy does not come to an abrupt end; rather, there is a gradual diminution of intensity of light outwards from the centre. In this connection astronomers often use the so-called *Holmberg radius*, which corresponds to the isophote at which the surface brightness drops to $26.5m_{\mathrm{pg}}$ (photographic magnitude) per square arc

Fig. 1.7 Elliptical galaxy of type E0 in Virgo, M87. Its nucleus is believed to contain a highly collapsed mass of the order of $5 \times 10^9 M_\odot$. (Courtesy of Palomar Observatory, California Institute of Technology.)

Fig. 1.8 Elliptical galaxy of type E2 in Andromeda, M32. (Courtesy of Kitt Peak National Observatory.)

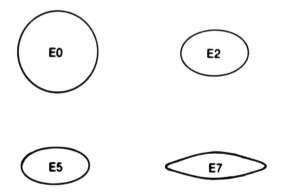

Fig. 1.9 The sequence of elliptical galaxy types. Not all types between E0 and E7 are shown.

second, as some kind of observational limit to a galaxy size. (Magnitude is a measure of the brightness of a celestial object. For quantitative details see section 3.6.)

For many decades since the discovery of galaxies it was believed that they extend as far as they are visible. Thus the Holmberg radius was taken as the extent of a typical galaxy. However, in the seventies the orbits of neutral hydrogen clouds circling around a spiral galaxy indicated that the masses enclosed within them far exceeded the visible mass of the galaxy.

Figure 1.10 shows the typical rotation curve of a spiral galaxy. At a distance r from the centre O of the galaxy, a Keplerian orbit will have velocity

$$v = \left(\frac{GM(r)}{r}\right)^{1/2},$$ (1.1)

where $M(r)$ is the galactic mass up to radius r from the centre. The point A represents the visible extent of the galaxy. If all the mass were visible

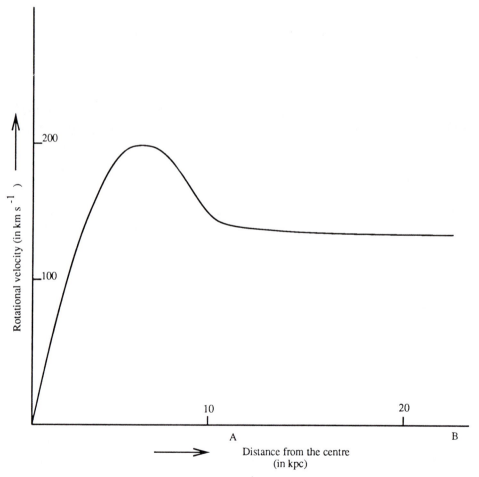

Fig. 1.10 The rotation curve of a spiral galaxy is flat right up to point B, well beyond the visible extent up to A.

then $M(r) = $ constant beyond A and v should have dropped as $r^{-1/2}$. In reality, v is more or less constant as far as point B, which may be two or three times farther away from O than A.

If Newtonian laws of gravity and mechanics hold then we have to conclude that $M(r)$ keeps on increasing beyond A; in other words, there is unseen matter present well beyond the visible radius of the galaxy. This dark matter poses many problems for cosmological theories which we shall encounter later in this book.

Another type of galaxy, called S0, is intermediate between the ellipticals and the spirals (see Figure 1.11). Like the ellipticals, the S0 galaxies have little gas and dust, while their isophotes are more like those of the spirals (see Figure 1.12). These galaxies may have formed from collisions of spirals and ellipticals. Galactic collisions are not uncommon, especially in rich clusters of galaxies. Stars may go through a collision relatively unscathed, since they are widely spaced, but interstellar gas and dust may be spewed out into intergalactic space. In such a case the isophotes (which arise from starlight) may remain intact.

Fig. 1.11 Galaxy of type S0 in Virgo, M84. (Courtesy of Kitt Peak National Observatory.)

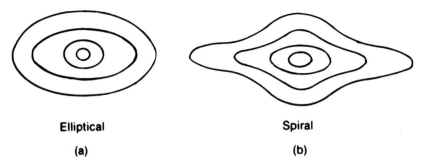

Elliptical   Spiral

(a)   (b)

Fig. 1.12 The isophotes (contours of equal brightness) of an S0 galaxy are more like those of a spiral (b) than an elliptical (a).

In addition to the types of galaxies already mentioned, there are others that are broadly classified as 'irregular'. However, some rarer species of galaxies in this group stand out because of certain special features. For example, in 1943 Seyfert investigated a class of galaxies in which the nuclei show many features common to stars, such as broad emission lines. (The spectra of galaxies as a rule show absorption lines from interstellar gas.) The Seyfert galaxies also have a large amount of infrared emission; in some cases the infrared luminosity may be as much as 100 times the visual luminosity of our Galaxy. (See Figure 1.13.)

Another group of galaxies with bright nuclei like the Seyferts are the so-called N-galaxies. These galaxies emit radio waves and have large redshifts, whereas Seyferts are radio quiet and have small redshifts. (For a discussion of redshifts, see sections 1.5 and 1.8.) There is considerable similarity between Seyferts, N-galaxies, and another class of astronomical objects, the quasars (described in section 1.5).

Apart from these morphological types, galaxies are also classified by their spectral features and luminosities. W. W. Morgan introduced the formal spectral classification, while van den Bergh introduced the luminosity classes. We will not go into details of these classifications here.

1.4 Radio sources

The advent of radio astronomy led to the discovery of strong sources of radio emission outside the Galaxy. As we shall see in Chapter 10, these radio sources also serve as useful probes of the structure of the universe. The first extragalactic radio source, Cygnus A, was discovered by J. S. Hey, S. J. Parsons, and J. W. Phillips in 1946. When the position of the radio source in the sky could be accurately specified, W. Baade and R. M.

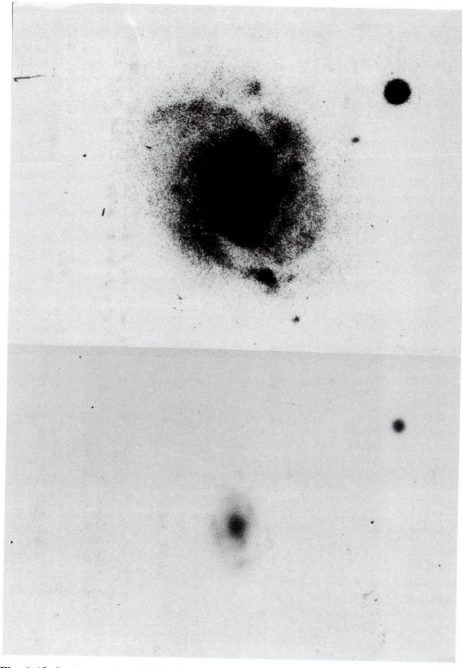

Fig. 1.13 Seyfert galaxy NGC 1068. The bottom picture is underexposed to show the nucleus only. (Courtesy of the Indian Institute of Astrophysics.)

Minkowski, at the Mt Wilson and Palomar Observatories, located what looked like a pair of colliding galaxies at the position of the radio source (see Figure 1.14). This process of identifying an object on the photographic plate at (or very close to) the position of the radio source is known as *optical identification* of the radio source. The discovery of Cygnus A led to the early view that radio sources arise from collisions of galaxies.

Eventually, however, it turned out that Baade was wrong in considering the optical object at Cygnus A a pair of colliding galaxies. In the seventies it became possible to study structures of radio sources in great detail. (Very-long-baseline interferometry can detect structures at the angular scale of less than a milliarc second.) The picture that has emerged not only for Cygnus A but for a majority of extragalactic radio sources is shown in Figure 1.15.

Fig. 1.14 The radio source Cygnus A is located around the optical object at the centre of the photograph. (Courtesy of Palomar Observatory, California Institute of Technology.)

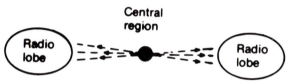

Fig. 1.15 The most common type of extragalactic radio source is a double source with two radio-emitting blobs located symmetrically on the opposite sides of a central region. The central region is believed to be the source of activity that generates fast particles moving out along the broken lines.

Here we have two radio-emitting blobs on opposite sides of a central component, usually located close to, and on opposite sides of, a galaxy or a quasar. It is believed that radio emission takes place in the blobs from the acceleration of fast-moving electrons by ambient magnetic fields, a process known as *synchrotron emission*. The particles themselves may have been fired in an explosion in the central region of the object. The source of the explosion is still a mystery. In 1963 F. Hoyle and W. A. Fowler suggested that gravitational energy in a collapsed object may somehow have been converted into the kinetic energy of the electron. In the late 1970s several scenarios were proposed involving a supermassive black hole of mass $\sim 10^8 \, M_\odot$. As first pointed out by G. Burbidge in 1958, a powerful energy machine is needed to generate energy reservoirs of 10^{58} to 10^{62} erg in these radio sources. The potential energy of two colliding galaxies falls far short of this target.

1.5 Quasars

The term *quasar* is a short form for the full name 'quasi-stellar radio source'. Quasars were first discovered in 1963 as a result of the optical identification programme. The radio position of the quasar 3C 273 (see section 1.7 for the meaning of these catalogue numbers) was accurately determined by lunar occultation. If the Moon happens to cross the line of sight to a source, the source is said to be *occulted*. The drop in the intensity of a radio source as it is blocked by the Moon and the rise when the Moon has moved out of its way give accurate indication of when it is occulted and hence where it is located on the sky. The optical identification of this object (see Figure 1.16) and of another radio source, 3C 48, revealed starlike objects with emission lines, and it was originally asumed that these were radio stars in the galaxy. When their spectra were carefully examined, however, it became clear that the wavelengths were strongly redshifted.

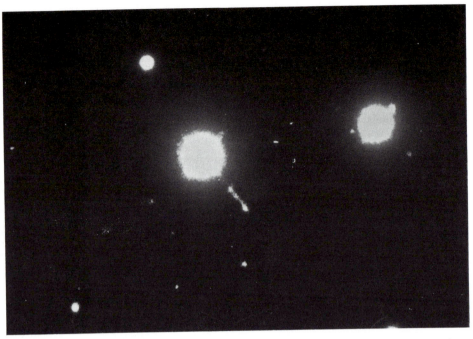

Fig. 1.16 The quasar 3C 273. (Courtesy of Kitt Peak National Observatory.)

If the wavelength of an emission line in the laboratory is λ_0 and if the observed wavelength is $\lambda > \lambda_0$, then the line is said to be redshifted by a fraction z given by

$$z = \frac{\lambda - \lambda_0}{\lambda_0}. \tag{1.2}$$

It is usual to call z the *redshift* of the object. For 3C 273, $z = 0.158$, while for 3C 48, $z = 0.367$. (The word *redshift* is used to indicate a shift to the red end of the visual spectrum.)

These were high values of z for stars in the galaxy, which have values $< 10^{-3}$. What were these objects? In 1964 Terrell suggested that they were high-velocity stars ejected from the galaxy. The more popular interpretation, however, has been that the redshifts arise from the expansion of the universe, a concept we will discuss in section 1.8.

If this latter interpretation is correct, it implies that quasars are very distant objects, and since from such large distances they look bright enough to be mistaken for stars, they must be intrinsically very powerful. Many quasars show rapid variation in their light and radio output. This fact places a limit on their physical size; for if an object shows variability on a characteristic time scale T, its size must be limited by cT, where

Table 1.1. *Some rich clusters of galaxies*

Name of cluster	Distance from Earth[a] (Mpc)
Virgo	21
Pisces	82
Perseus	100
Coma	122
Hercules	190
Gemini	430
Hydra II	1110

[a] Distances corresponding to $H_0 = 50 \, \text{km} \, \text{s}^{-1} \, \text{Mpc}^{-1}$.

c = the speed of light. This limitation, arising from the special relativistic result that no physical disturbance can propagate with a speed $> c$, makes quasars very compact indeed. We saw in section 1.2 how big our galaxy is. A quasar by comparison may emit a comparable amount of energy per unit time from a volume whose linear extent may be only a few light-hours!

By now more than 5000 quasars are known. Only a few per cent of the total quasar population emit radio waves. Thus the early qualification 'radio source' is not applicable to the bulk of the quasar population, and although the term 'quasar' is used today also for radio quiet objects, the purist may prefer the term 'quasi-stellar object' (QSO). More recently, the X-ray astronomy satellite 'Einstein Observatory' has revealed that X-ray emission is also a common feature among quasars, indeed is much more common than radio emission.

1.6 Structures on the largest scale

A galaxy that is not a member of a group of galaxies is called a *field* galaxy. Other galaxies are members of groups or *clusters* that may contain from a handful to hundreds of big galaxies. Our galaxy, for example, is a member of a group of ~ 20 galaxies, known as the Local Group, that are separated by distances of up to $\sim 1 \, \text{Mpc}$. The nearest members of the group are the Large and Small Magellanic Clouds, which are located $\sim 50 \, \text{kpc}$ from us.

Table 1.1 lists a few of the larger clusters (see Figure 1.17). The distances quoted in this table are not exact because of the uncertainty

Fig. 1.17 The Coma Cluster of galaxies. (Courtesy of Kitt Peak National Observatory.)

surrounding the measurements of extragalactic distances. The extra-galactic distance scale is related to the magnitude of Hubble's constant (see section 1.8). Currently there is disagreement among astronomers as to the true value of this constant. The ratios of these numbers should, however, give us reliable estimates of the relative distances of these clusters.

G. Abell has catalogued clusters out to distances of the order of that of Hydra II using strict criteria of what constitutes a cluster. In order to pick out a cluster one has to look for an enhancement of the number density of galaxies within a specified region compared with the overall background density. The order of 'richness' of a cluster is accordingly fixed by specifying the size, brightness range, and background. F. Zwicky has also catalogued clusters of galaxies, but with less strict criteria than those adopted by Abell.

How much matter is contained in a cluster? We will attempt to answer this question in section 9.4. For the time being we may say that a cluster may contain a mass of the order of $\sim 10^{14} \, M_\odot$. Further, if we try to estimate the mean density of matter in the universe by taking account of how much matter we see in clusters of galaxies then we come up with a figure lying between 10^{-31} and $10^{-30} \, \mathrm{g \, cm^{-3}}$. However, as we shall see in Chapters 7 and 9, even clusters may have dark matter in substantial amounts. Thus these density estimates may have to be enhanced.

The mean density of matter in a galaxy, on the other hand, is $\sim 10^{-24} \, \mathrm{g \, cm^{-3}}$. Thus the volume occupied by galaxies is $\leqslant 10^{-6}$ of the total volume of the universe. This also explains why galaxies are considered as points when cosmological models are constructed.

Apart from optical emission, clusters of galaxies also show radio and X-ray emission. These arise not only from individual sources located in the clusters, but also in a diffuse fashion throughout the clusters.

Does a structure larger than clusters exist in the universe? This can be decided by studying the distribution of galaxies across the sky and looking for nonrandomness (that is, grouping or clumping) on larger and larger scales. Such studies have revealed the existence of larger structures on the scale of $\sim 50 \, \mathrm{Mpc}$, compared to cluster scales of $\sim 5 \, \mathrm{Mpc}$. These larger units are referred to as *superclusters*.

For example, G. de Vaucouleurs has found that our Local Group is a member of a supercluster centred on the Virgo cluster of galaxies. C. D. Shane and co-workers at the Lick Observatory found similar clumpiness in other regions of the sky. Abell also found clumpiness in an analysis of the plates in the National Geographic–Palomar Sky Survey.

In the 1970s and 1980s there were considerable improvements in the techniques of observing discrete extragalactic objects. With distances determined by Hubble's law (section 1.8) it became possible to have three-dimensional perspective of matter distributions in the universe. These are beginning to indicate that discrete objects show a large-scale inhomogeneity of distribution. There are, for example, the following features revealed by such surveys:

1. *Superclusters*: As was seen above, these are on the scales of $\sim 50 \, \mathrm{Mpc}$ or more and contain several thousand galaxies. For example, the Local Supercluster containing the Local Group is shaped like a flattened ellipsoid which has a plane of symmetry called the *supergalactic plane*. It passes through the centre of the Virgo cluster and the centre of our own Galaxy.
2. *Voids*: These are gaps in the distribution of large superclusters, with sizes of the order of 100–200 Mpc. There are apparently no galaxies in these regions (see Figure 1.18)

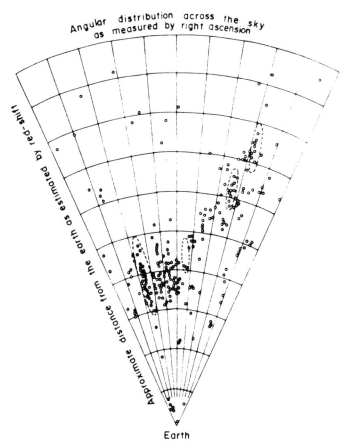

Fig. 1.18 Galaxy distributions show giant voids and long filaments. The above region contains the Perseus and Pegasus superclusters. The circles are galaxies while the closed dashed curves enclose the Abell clusters. (After J. O. Burns, *Scientific American*, July 1986, p. 40.)

3. *Filaments*: The boundaries of voids are filamentary distributions of galaxies in clusters and superclusters. Figure 1.18 indicates this feature clearly.

4. *The Great Attractor and the Great Wall*: In the late 1980s it became apparent that galaxies in and around the Local Group seem to have a large-scale streaming motion towards the Hydra–Centaurus supercluster in the Southern sky. The typical streaming velocity is around $600 \, \text{km s}^{-1}$, against the reference frame in which the cosmic microwave background (see section 1.9) is isotropic. This motion is believed to have been caused by a 'great attractor' mass of some tens of thousands of galaxies. The volume of the attractor is as large as $10^6 \, \text{Mpc}^3$. Such massive structures may be present elsewhere in the universe also.

Mapping of the universe on a large scale also indicates the presence of a large but thin sheet of mass. Known as the 'great wall', it has an extent of $60 \times 170 \, \mathrm{Mpc}^2$ (using a Hubble constant of $100 \, \mathrm{km \, s^{-1} \, Mpc^{-1}}$). Figure 1.19 shows its existence. These structural inhomogeneities therefore span distances as large as 10 per cent of the characteristic size of the universe as given in section 1.10.

In the 1920s and 1930s the general belief was that the universe is homogeneous on the large scale. The cosmological models which arose in those days make this simple assumption. In Chapters 3 and 4 we will outline these models. But it is now becoming apparent that these models were too simplistic. They face the problem of explaining how such a large-scale structure came into existence.

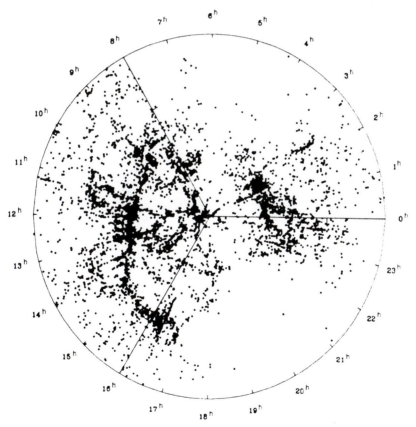

Fig. 1.19 The distribution of galaxies in a thin sheet called the 'great wall'. (Source: M. J. Geller and H. P. Huchra, *Science*, **246**, 897 (1989)).

1.7 Coordinates and catalogues of astronomical objects

Before proceeding further we will describe how the astronomer locates the
position of a heavenly body in the sky. In general the astronomer does not
know the distance of the body from us; he sees it projected on the sky, on
what is known as the *celestial sphere*. Two coordinates, akin to longitude
and latitude, are therefore needed to specify the position of the body on
the sphere.

Figure 1.20 shows two different coordinate systems, both useful to the
astronomer in different contexts. The system in Figure 1.20(a) uses *right
ascension* (RA, denoted by α) and *declination* (δ), coordinates fixed by
the geometry of the Sun–Earth system. Here the poles are the points N, S
on the celestial sphere where the Earth's axis of rotation intersects it. The
celestial equator is the great circle on the celestial sphere whose plane is
perpendicular to NS. The plane in which the Sun appears to go round (as
seen from the Earth) intersects the celestial sphere in another great circle
called the *ecliptic*. The ecliptic and the celestial equator intersect in two
points γ and Ω, corresponding to the position of the Sun on 21 March and
22 September, respectively. Now α and δ are the longitude and latitude of
a celestial object measured with respect to the celestial equator and the
great circle through N, γ, S, and Ω. This latter circle, known as the
celestial meridian, plays the role of the Greenwich meridian on the Earth,

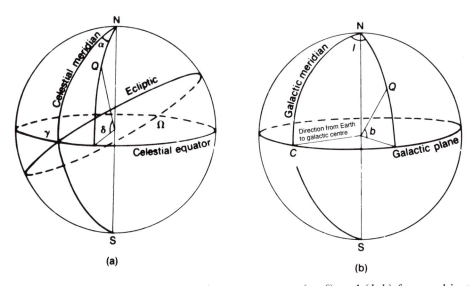

Fig. 1.20 This figure demonstrates how to measure (α, δ) and (l, b) for an object
Q in the sky using two different coordintate systems. (a) The coordinate system
based on the geometry of the Sun–Earth system. (b) The coordinate system based
on the geometry of our Galaxy.

Table 1.2. *Some catalogues of use in cosmology*

Name	Type of object	Catalogue code
Messier	Nebulae and galaxies	M followed by catalogue number.
New General	Nebulae and galaxies	NGC followed by catalogue number in increasing RA.
Abell	Clusters	A followed by catalogue number in increasing RA.
Cambridge (3rd, 4th, 5th surveys)	Radio sources	3C, 4C, 5C followed by catalogue number in increasing order of RA.
Ohio source	Radio sources	O followed by a letter (B to Z omitting O) and a number. The letter gives hours of RA, the first digit the declination in 10° intervals, and the last two digits the decimal part of the RA to two places. Thus 1443 + 101 is OQ 172.

with γ the point of zero α. It is customary to measure α in hours and minutes, with the range 360° corresponding to 24 hours. The declination is written in degrees, minutes, and seconds, with + for North, − for South.

While (α, δ) coordinates are convenient for measurements made from the Earth, the cosmologist is often interested in knowing how the object is located *vis-à-vis* the plane of the Galaxy. For such purposes the *galactic coordinates* are useful. These are illustrated in Figure 1.20(b). The *galactic equator* is the great circle where the plane of the Galaxy intersects the celestial sphere. N, S are the North and South galactic poles, while the 'zero' meridian is the one passing through the points N, S, and the point C where the direction from Earth to the centre of the Galaxy meets the celestial sphere. This meridian is also called the *galactic meridian*. The galactic longitude is denoted by l, and latitude by b. In terms of the (α, δ) system, the point C has the coordinates $\alpha \approx 17^h 42^m.4$, $\delta \approx -28° 55'$. It is possible to convert from one coordinate system to another using spherical trigonometry.

Astronomical objects are catalogued in many ways. Table 1.2 lists some of the catalogues referred to in this book and their code letters. This is not an exhaustive list, but is given as an illustration of how sources are numbered and listed. A more systematic method common in recent compilations is to list the object by its (α, δ) values in the form $\alpha(\pm)\delta$.

Thus the object 1143−245 has right ascension 11^h43^m and declination $−24° 30'$ ($\equiv −24.5°$).

1.8 Expansion of the universe

We now come to the observations that launched modern cosmology. Between 1912 and 1925, V. M. Slipher measured the shifts in the spectra of more than 20 objects that later turned out to be galaxies. Slipher was surprised that all shifts were towards the red end. Later, E. Hubble and M. Humason extended Slipher's list of observations to more galaxies and to the brightest cluster galaxies. An example of the pattern that emerged when the redshift was plotted against distance of a galaxy is shown in Figure 1.21 (see also Figure 1.22).

If all galaxies seen are equally bright, then the magnitudes are proportional to the logarithm of distances. Thus the straight line drawn through the cluster of points corresponds to the linear relation

$$V \equiv cz = H_0 D, \qquad (1.3)$$

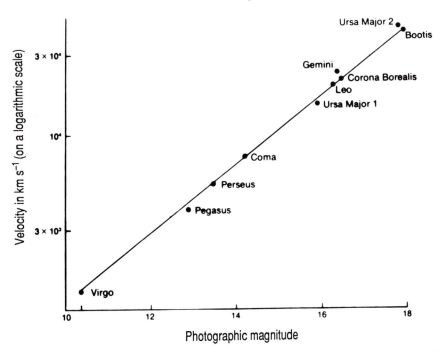

Fig. 1.21 Hubble's plot for the fifth brightest member in clusters of galaxies. The magnitudes are photographic. In Chapter 9 we will see how to convert magnitudes into distances. The velocities are obtained by multiplying the observed redshifts by c. (After E. Hubble, *The Realm of the Nebulae* (New Haven, Conn.: Yale University Press, 1936).)

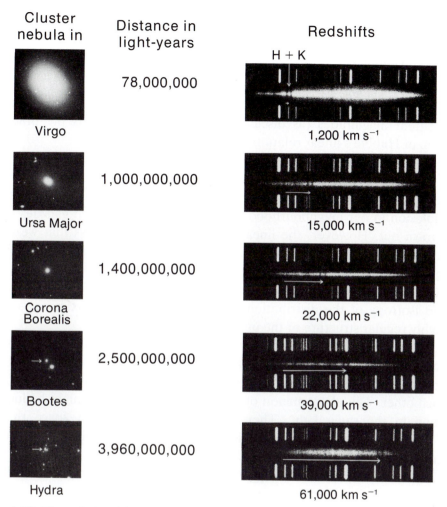

Cluster nebula in	Distance in light-years	Redshifts

Fig. 1.22 The relationship between redshift and distance for extragalactic nebulae. Redshifts are expressed as velocities, $c\,d\lambda/\lambda$. Arrows indicate shift for calcium lines H and K. Distances are based on an expansion rate of $50\,\mathrm{km\,s^{-1}\,Mpc^{-1}}$. (Courtesy of Palomar Observatory, California Institute of Technology.)

where D is the distance of the galaxy and z its redshift. If the redshift were due to the Doppler effect, then we could ascribe to the galaxy a velocity of recession V relative to us. (Since $z \ll 1$ in the observations of Hubble and Humason, and the Newtonian Doppler shift formula is valid.) The constant H_0 is now known as *Hubble's constant*.

If instead of plotting z against the distance D, $\log z$ is plotted against the apparent magnitude m of the galaxy, then another straight-line relation shows up (see section 3.6 for a definition of apparent magnitude).

For,

$$m = 5 \log D + \text{constant}, \qquad (1.4)$$

and (1.3) implies

$$m = 5 \log z + \text{constant}. \qquad (1.5)$$

Since the distances of remote galaxies are determined through their apparent magnitudes (as discussed in Chapter 9), (1.5) is the practical form of Hubble's linear relation (1.3).

The relation (1.3) is called *Hubble's law*. It was published as a linear law by Hubble in 1929, and it caused great excitement. For the prima facie interpretation of Hubble's law seemed to be that there was a great explosion in our neighbourhood of the universe from which galaxies were thrown out. However, the linearity of Hubble's law shows that we need not consider ourselves in any special position in the universe. If we viewed the population of galaxies from any other galaxy, we would notice the same Hubble's law. The combination of this fact with the homogeneity and isotropy of the distribution of the population of galaxies suggests a highly regular structure of the universe.

Imagine a piece of dough with self-raising flour being baked in the oven, and suppose we have spread caraway seeds uniformly throughout the dough. As the dough bakes it expands, and the seeds move away from each other. The phenomenon of the recession of galaxies might be looked upon in the same light. They are points embedded in space that is expanding. This notion of galaxies embedded in expanding space led to the concept of the expanding universe.

The rate of expansion is characterized by Hubble's constant. Hubble obtained a value for H_0 in the neighbourhood of $530 \, \text{km s}^{-1} \, \text{Mpc}^{-1}$. (Note that these units arise because H_0 is velocity divided by distance. The dimensions of H_0^{-1} are simply those of time.) As we will discuss in section 9.2, Hubble had grossly underestimated the galactic distances, with the result that his value of H_0 was too high. The value of H_0 is now believed to lie in the range of 50 to $100 \, \text{km s}^{-1} \, \text{Mpc}^{-1}$. We will write it as $100 h_0 \, \text{km s}^{-1} \, \text{Mpc}^{-1}$, where h_0 lies between 0.5 and 1. Notice that if we assume Hubble's law we can estimate the distance of an extragalactic object from its redshift.

1.9 The radiation backgrounds

Apart from matter in its visible form, we may look for radiation at various frequencies. In general, measurements of the electromagnetic radiation at a given frequency (or in a given range of frequencies) reveal peaks that

Table 1.3. *Radiation background at different levels*

Type of radiation	Wavelength, λ Frequency, ν Energy range, E	Energy density ($\mathrm{erg\,cm^{-3}}$)
Radio	$\nu \leqslant 4080\,\mathrm{MHz}$	$\leqslant 10^{-18}$
Microwaves	λ in 80 cm to 1 mm	$\approx 4 \times 10^{-13}$
Optical	λ in 4000 Å to 8000 Å	$\approx 3.5 \times 10^{-15}$
X-rays	E in 1 to 40 keV	$\approx 10^{-16}$
γ-rays	$E \geqslant 100\,\mathrm{MeV}$	$\leqslant 2 \times 10^{-17}$

are associated with relatively nearby discrete sources, many of which can be identified in specific directions. However, after these peaks are eliminated, there is still a residual background of radiation. This background radiation could also arise from discrete sources that are located much farther away and therefore cannot be resolved, or it could arise from processes in the intergalactic spaces. Table 1.3 gives a rough estimate of the energy densities in the various wavelength ranges. It should be remembered that the measurements in X-rays, γ-rays, and so on became possible only from the early 1960s with the advent of space astronomy.

One thing is immediately clear from Table 1.3. Compared with the estimates of matter density, the radiation energy density is less by about three orders of magnitude. This observation is often expressed by the statement that the universe is at present 'matter-dominated'.

It is also clear from Table 1.3 that the most dominant form of radiation background is in the microwaves. The spectrum of the microwave background is very nearly that of the blackbody radiation of temperature $\sim 3\,\mathrm{K}$. Moreover, the extreme homogeneity of this radiation on small angular scale seems to rule out the possibility that it could have arisen from discrete sources. As we shall see in Chapter 5, the most popular interpretation of this radiation is that it is a relic of an early hot epoch when the universe was much denser than it is now. Unlike the matter distribution, this relic radiation is extremely homogeneous. This contrast further exacerbates the difficulty of understanding the origin of discrete structures against a smooth radiation background.

1.10 Relativistic cosmology

If Hubble's observation launched modern observational cosmology, it was Einstein's general theory of relativity that laid the foundations of modern theoretical cosmology. We will discuss in Chapter 3 the details of how the

Table 1.4. *Spatial dimensions and masses of astronomical systems*

Object	Linear size	Mass
Sun	7×10^{10} cm (radius)	2×10^{33} g $\equiv M_{\odot}$
Galaxy	≈ 15 kpc	$\approx 10^{11} \, M_{\odot}$
Cluster	≈ 5 Mpc	$\approx 10^{13} - 10^{14} \, M_{\odot}$
Supercluster	≈ 50 Mpc	$\approx 10^{15} \, M_{\odot}$
Universe	≈ 3000 Mpc	$\approx 10^{21} \, M_{\odot}$

theoretical developments in cosmology actually began more than a decade before Hubble's exciting observations. We conclude the present chapter by considering the general question of why relativity is taken to be so important for cosmology.

Table 1.4 shows the orders of magnitude involved in the large-scale structure of the universe. The last entry refers to the characteristic distance scale c/H_0 that emerges from Hubble's constant and the mass contained in the 'observable' volume of radius c/H_0 if the density were that seen for visible matter in our neighbourhood. Similarly, the time scale characteristic of the universe is $H_0^{-1} \approx 10^{10}$ years.

What interaction in physics is likely to be influential over such long distances and such large masses? Of the four known interactions, only gravity and electromagnetism are of long range. Although the electromagnetic interaction is much stronger than gravity on the scale of atoms, it is ineffective in determining the large-scale structure of the universe, since all indications are that an electric charge balance is preserved in galaxies, clusters, and intergalactic space. Nor is there any evidence for large-scale electric currents that could interact with the magnetic fields in the universe to produce large forces. By contrast, the enormous masses of astronomical objects generate huge gravitational fields. Gravity is therefore the most relevant force in cosmology.

Given that we need a theory of gravity for cosmology, what is wrong with the Newtonian framework? It has worked well in the theory of stellar structure. It is even used in stellar dynamics in the Galaxy. Why not use it in cosmology? Let us try to understand the answer with the help of the entries in Table 1.4.

Newtonian gravity is a theory of instantaneous action at a distance. As such, it is inconsistent with the special theory of relativity, in particular with the limit (c) placed by that theory on the speed with which any interaction can propagate across space. In those parts of astronomy where the distances across which gravity is suppose to act are relatively small, the use of Newtonian gravity is permissible. As seen in Table 1.4, however, the distances in cosmology are so large that action at a distance with infinite speed is unrealistic. This is not so with stellar dimensions or even for galaxies.

Special relativity itself is suspect in the presence of gravity. The concepts of the inertial frame and the inertial observer (on whom no force acts), which are so basic to special relativity, are unrealizable in the presence of gravity. Gravity seems to be an ever-present force that cannot be switched off altogether. Since all matter attracts gravitationally, an inertial observer cannot exist at all! Nevertheles, it was shown in 1934 by E. A. Milne and W. H. McCrea that with suitable compromise Newtonian gravity and special relativity can describe cosmology in an adequate manner. Although Newtonian cosmology is simple to understand, it is based on insecure foundations. It is preferable instead to resort to a framework that is free from conceptual difficulties and compromises.

As we shall see in Chapter 2, general relativity provides a framework that is free from the difficulties of Newtonian gravity with respect to special relativity and of special relativity with respect to gravity. It is for these conceptual reasons, apart from the experimental successes of general relativity in the various solar system experiments (see section 2.10), that cosmologists feel at home with the use of this theory.

It is therefore appropriate that we begin our discussion of cosmology by outlining the general theory of relativity.

2

General relativity

2.1 Space, time, and gravitation

Every major scientific theory carries its own mark of distinction. The distinctive feature of Newtonian gravitation is the radial inverse square law. To those uninitiated in the laws of dynamics, the fact that a planet goes *round* the Sun under a force of attraction *towards* the Sun comes as a surprise. The major achievement of Maxwell's electromagnetic theory was the unification of electricity and magnetism and the demonstration that light itself is an electromagnetic wave. The unique place held by the speed of light characterizes Einstein's special theory of relativity, while quantum mechanics can point to the uncertainty principle as the crucial feature that sets it apart from classical mechanics.

To what distintive feature can general relativity lay its own special claim? A clue to the answer to this question is provided in the title of this section.

Let us compare gravitation with electricity. We know that two unlike electric charges attract each other through the Coulomb inverse square law, just as any two masses attract each other gravitationally by the Newtonian inverse-square law. To this extent, electricity and gravitation are similar. However, we can go no further! We also know that two like electric charges repel each other and that this property seems to have no parallel in gravitation. Every bit of matter attracts every other bit and, as yet, we do not have any instance of gravitational repulsion.

We can express this difference between electricity and gravitation in another, more practical way. The existence of repulsion as well as attraction enables us to construct a closed chamber whose interior is completely sealed from any outside electrical influence. Not so with gravitation! We cannot point to any region of space as being totally free from external gravitational influences. Gravitation is permanent: it cannot be switched off at will.

This ever-present nature of gravitation plays the key role in Einstein's general theory of relativity. Einstein argued that because of its permanence, gravitation must be related to some intrinsic feature of space and time. And, with a master stroke of genius, he identified this feature as the geometry of space and time. He suggested that any effects we ascribe to gravitation actually arise because the geometry of space and time is 'unusual'. Let us now try to understand what is meant by the word 'unusual', and how this property of space and time leads to gravitational effects – for therein lies the distinctive characteristic that sets general relativity apart from other physical theories.

The 'usual' geometry of space, the geometry that we learn at school and learn to apply in so many ways, is the geometry whose foundations were laid by the Greek mathematician Euclid around 300 BC. Euclidean geometry is a logical structure wherein theorems about triangles, parallelograms, circles, and so on are proved on the basis of postulates that are taken as self-evident. Thus the results shown in Figure 2.1 follow as theorems in Euclid's geometry, based on the original postulates of Euclid.

It was only in the last century that mathematicians realized that there is nothing sacrosanct about Euclid's postulates. Provided they are not mutually contradictory, a new set of postulates can lead to a new type of geometry. Indeed, as the work of mathematicians like Gauss (1777–1855), Bolyai (1802–60), Lobachevsky (1793–1856), and Riemann (1826–66) showed, a host of such new geometries can be constructed. These are collectively called non-Euclidean geometries. For instance, the geometry on the surface of a sphere is non-Euclidean. If we define a straight line on the surface of a sphere as the line of shortest distance between two points, it is easy to see that these lines are great circles and that any two straight lines intersect. Thus there are no parallel lines in the geometry. Figure 2.2 demonstrates how the theorems of Figure 2.1 break down when applied to the non-Euclidean geometry of the surface of the sphere.

The concept of the geometry of space can be extended to the geometry of space and time, thanks to the foundations laid by Einstein's special theory of relativity. Let us first recall a familiar result from special relativity in the following form. Let (x, y, z) denote a Cartesian coordinate system and t the time measured by an observer O at rest in an inertial frame. (That is, an observer who is acted on by no force. We will return to a discussion of such observers later.) Let two neighbouring events in space

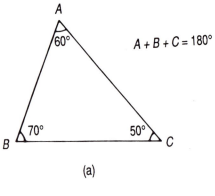

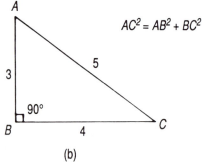

Fig. 2.1 (a) The three angles of any triangle ABC add up to 180°. (b) The well-known theorem of Pythagoras for a typical right-angled triangle ABC.

and time be labelled by the coordinates (x, y, z, t) and $(x + dx, y + dy, z + dz, t + dt)$. The resulting analogue of the Pythagorean theorem shown in Figure 2.1(b) is as follows. The square of the 'distance' between the two events is given by

$$ds^2 = c^2 dt^2 - dx^2 - dy^2 - dz^2. \tag{2.1}$$

The distance ds is invariant in the sense that another inertial observer O' using a different coordinate system (x', y', z', t') to measure this distance will find the same answer.

However, when we make a transition from special to general relativity and quantify Einstein's idea that the geometry of space and time is unusual in the presence of gravitation, we abandon the simple form of (2.1) in favour of a more complicated form. This is comparable with the transition from Figure 2.1(b) to Figure 2.2(b). The more complicated form is still quadratic, and we may state it formally as follows:

$$ds^2 = \sum_{i,k=0}^{3} g_{ik} \, dx^i \, dx^k. \tag{2.2}$$

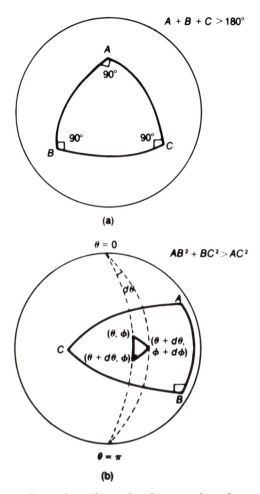

Fig. 2.2 (a) On the surface of a sphere the three angles of any triangle add up to more than 180°. For the triangle shown, the three angles add up to 270°. (b) The Pythagorean theorem breaks down for a finite spherical right triangle (shown inside $\triangle ABC$), but it looks more complicated in spherical coordinates (θ, ϕ): $ds^2 = a^2(d\theta^2 + \sin^2 \theta \, d\phi^2)$, where a = radius of the sphere.

Here we have modified the notation as follows. The coordinates are now called x^i, with $i = 1, 2, 3$ representing the three space coordinates and $i = 0$ the time coordinate. The coefficients g_{ik} are functions of x^i with the property that the matrix $\|g_{ik}\|$ has the signature -2. (This means that if the quadratic equation (2.2) is diagonalized, it has one square term with a positive coefficient and three square terms with negative coefficients. The signature equals the number of positive terms minus the number of negative terms.) It is convenient to refer to this unified structure of space and time as *spacetime*.

Clearly, the geometry of spacetime in which the basic invariant distance is given by (2.2) instead of by (2.1) is going to be 'unusual'. Its properties will depend on the function g_{ik}. But do these properties tell us about the presence of gravitation? In what way can we interpret manifestly gravitational phenomena like the motion of planets as effects of geometry? The remainder of this chapter attempts to answer these questions.

2.2 Vectors and tensors

Let us consider again the example of geometry on the surface of a sphere of radius a. If we consider the sphere as embedded in a three-dimensional space with the Cartesian coordinates x, y, z, we may write the equation of the surface of the sphere as

$$x^2 + y^2 + z^2 = a^2. \tag{2.3}$$

For describing the geometry on the surface of the sphere it is, however, more convenient to use coordinates intrinsic to the surface of the sphere. Such coordinates are available and are like the latitude and longitude used to locate a point on the Earth. More specifically,

$$x = a \sin\theta \cos\phi, \qquad y = a \sin\theta \sin\phi, \qquad z = a \cos\theta, \tag{2.4}$$

so that for any (θ, ϕ) with $0 \leqslant \theta \leqslant \pi$ and $0 \leqslant \phi < 2\pi$ we can locate a point (x, y, z) on the surface of the sphere. Spherical trigonometry tells us how to measure and relate the angles, sides, and so on of triangles drawn on this surface. The rules of Euclid's geometry do not apply to these measurements.

In our example above, the square of the distance between two neighbouring points (θ, ϕ) and $(\theta + d\theta, \phi + d\phi)$ is given by

$$d\sigma^2 = dx^2 + dy^2 + dz^2 = a^2(d\theta^2 + \sin^2\theta\, d\phi^2). \tag{2.5}$$

Thus we have examples of g_{ik} that are not constants. (The coefficient of $d\phi^2$ is $a^2 \sin^2\theta$.)

However, the nonconstancy of g_{ik} or its nondiagonal nature do not guarantee that we are dealing with a non-Euclidean geometry. For example, in three-dimensional Euclidean space, the transformation

$$x = r \sin\theta \cos\phi, \qquad y = r \sin\theta \sin\phi, \qquad z = r \cos\theta, \tag{2.6}$$

with (θ, ϕ) as defined before and $0 \leqslant r \leqslant \infty$ gives

$$d\sigma^2 = dx^2 + dy^2 + dz^2 = dr^2 + r^2(d\theta^2 + \sin^2\theta\, d\phi^2). \tag{2.7}$$

Again we have g_{ik} as functions of r and θ. But now we know that we are dealing with Euclidean geometry and that the dependence of g_{ik} on r and θ is purely a coordinate effect.

Thus we clearly have to devise a means of extracting essential geometrical information as distinct from pure coordinate effects. In a qualitative way we can see that the essential information must survive even when we change from one coordinate system to another. In order to extract such information, we must devise machinery that tells us what things remain unchanged under coordinate transformations. Such machinery is provided by the invariants, the vectors, and the tensors, which we will now study.

Let us first introduce a summation convention. We will frequently encounter sums like

$$\sum_{i=0}^{3} A_i B^i, \qquad \sum_{k=0}^{3} A_{ik} B^k, \qquad \sum_{i,k=0}^{3} P_{ik} \xi^i \xi^k, \dots$$

It is convenient in such cases to drop the summation symbol and write these quantities as

$$A_i B^i, \qquad A_{ik} B^k, \qquad P_{ik} \xi^i \xi^k, \dots,$$

the rule being that whenever an index appears once as a subscript and once as a superscript in the same expression, it is automatically summed over all the values (from 0 to 3). Thus we can rewrite (2.2) in the more compact form

$$ds^2 = g_{ik} \, dx^i \, dx^k. \tag{2.8}$$

A warning must be issued here: the summation convention does not apply under any other circumstances. Thus it does not apply to quantities like

$$A_i \, B_i, \, A_{ik} \, B_i \, C_i, \dots$$

However, such expressions do not arise in most relativistic calculations.

We will assume that the Latin indices i, j, k, ... will run over all four values 0, 1, 2, 3. On some (infrequent) occasions we may want to refer to index values 1, 2, 3 only. These values are usually reserved for space components, and we will use Greek indices μ, ν ... to represent these.

It is worth pointing out here that many other textbooks use the convention of denoting the spacetime coordinates by Greek indices λ, μ, ν, etc. and the space coordinates by Latin indices i, j, k, etc. Also, many authors prefer to write (2.1) with the opposite sign for the right-hand side. These differences are of cosmetic nature and do not affect the 'physics' being described.

2.2.1 Scalars

A *scalar* or an *invariant* does not change under any change of coordinates. Thus, if $\phi(x^i)$ is a function of coordinates, then it is invariant provided it retains its value under a transformation from x^i to new coordinates x'^i:

$$\phi(x^i) = \phi[x^i(x'^k)] = \phi'(x'^k). \tag{2.9}$$

Note that the form of the function may change, but its value does not.

2.2.2 Contravariant vectors

Suppose we are given a curve in space and time, which is parametrized by λ. Thus, the points along the curve have coordinates

$$x^i \equiv x^i(\lambda), \tag{2.10}$$

where x^i are given functions of λ. The direction of the tangent to the curve at any point on it is given by a vector with four components,

$$A^i \equiv \frac{dx^i}{d\lambda}. \tag{2.11}$$

Notice that the direction of a tangent to the curve is an invariant concept: a change of coordinates should not alter this concept, although its four components in the new coordinates will be different. Suppose the new coordinates are x'^i and the new components are A'^i. Then

$$A'^i \equiv \frac{dx'^i}{d\lambda}. \tag{2.12}$$

Unless otherwise stated, we will assume that the transformation functions

$$x^i = x^i(x'^k), \qquad x'^k = x'^k(x^i) \tag{2.13}$$

are continuous and possess at least second derivatives. It is then easy to see that A'^i and A^i are related by the linear transformation

$$A'^k = \frac{\partial x'^k}{\partial x^i} \frac{dx^i}{d\lambda} = \frac{\partial x'^k}{\partial x^i} A^i. \tag{2.14}$$

We use (2.14) as the transformation law for *any* vector A^i. Quantities that transform according to the above linear law are called *contravariant vectors*. The four components of a contravariant vector are specified by a superscript.

For example, consider the curve parametrized by

$$x^0 = \text{constant}, \qquad x^1 = \text{constant}, \qquad x^2 = \lambda, \qquad x^3 = \lambda^2.$$

The tangent to this curve is specified by the contravariant vector A^i, with components

$$A^0 = 0, \qquad A^1 = 0, \qquad A^2 = 1, \qquad A^3 = 2\lambda.$$

2.2.3 Covariant vectors

Consider next a scalar function $\phi(x^k)$. The equation

$$\phi(x^k) = \text{constant} \tag{2.15}$$

describes a hypersurface (that is, a surface of three dimensions) whose normal has the direction given by the four quantities

$$B_i = \frac{\partial \phi}{\partial x^i}. \tag{2.16}$$

Again, the concept of a normal to a hypersurface should be independent of the coordinates used. Under (2.13), the new components are

$$B_i' = \frac{\partial \phi}{\partial x'^i}. \tag{2.17}$$

It is easy to see that $B_i' \leftrightarrow B_i$ is a linear transformation:

$$B_k' = \frac{\partial x^i}{\partial x'^k} B_i. \tag{2.18}$$

Again, we generalize (2.18) as a transformation law of any vector B_i. Quantities that transform according to this rule are called *covariant vectors*.

For example, the normal to the unit sphere given by

$$\phi = (x^1)^2 + (x^2)^2 + (x^3)^2 = 1$$

has the covariant components

$$B_0 = 0, \qquad B_1 = 2x^1, \qquad B_2 = 2x^2, \qquad B_3 = 2x^3.$$

2.2.4 Tensors

The concept of a vector can be generalized to that of a tensor. Thus a contravariant tensor of rank 2 is characterized by the following transformation law:

$$T'^{ik} = \frac{\partial x'^i}{\partial x^m} \frac{\partial x'^k}{\partial x^n} T^{mn}. \tag{2.19}$$

A covariant tensor of rank 2 is similarly characterized by the transformation law

$$T_{ik}' = \frac{\partial x^m}{\partial x'^i} \frac{\partial x^n}{\partial x'^k} T_{mn}. \tag{2.20}$$

It is also possible to have mixed tensors. Thus T_k^i is a mixed tensor of rank 2, with one contravariant index and one covariant index. It transforms as

$$T'^i_k = \frac{\partial x'^i}{\partial x^m} \frac{\partial x^n}{\partial x'^k} T_n^m. \tag{2.21}$$

Again, these concepts are easily generalized to tensors of higher rank than 2. The rule is to introduce a transformation factor $\partial x'^i/\partial x^m$ for each contravariant index i and a factor $\partial x^n/\partial x'^k$ for each covariant index k.

Example 1 The quantities g_{ik} transform as a covariant tensor. This result follows from the assumption that ds^2 as given by (2.8) is invariant. For

$$ds^2 = g_{ik}\,dx^i\,dx^k$$

$$= g_{ik}\left(\frac{\partial x^i}{\partial x'^m}\,dx'^m\right)\left(\frac{\partial x^k}{\partial x'^n}\,dx'^n\right)$$

$$= \left(g_{ik}\frac{\partial x^i}{\partial x'^m}\frac{\partial x^k}{\partial x'^n}\right)dx'^m\,dx'^n$$

$$= g'_{mn}\,dx'^m\,dx'^n,$$

that is,

$$g'_{mn} = \frac{\partial x^i}{\partial x'^m}\frac{\partial x^k}{\partial x'^n}\,g_{ik}. \tag{2.22}$$

This tensor is called the *metric* tensor. The quadratic expression for ds^2 is called the *line element* of spacetime or the *spacetime metric*.

Example 2 The *Kronecker delta* defined by

$$\delta^i_k = 1 \quad \text{if} \quad i = k, \qquad \text{otherwise} \quad \delta^i_k = 0 \tag{2.23}$$

is a mixed tensor of rank 2.

Example 3 Define $\|g^{ik}\|$ to be the inverse matrix of $\|g_{ik}\|$, assuming that $g = $ determinant of $\|g_{ik}\| \neq 0$. (Since g_{ik} has signature -2, g is negative.) Thus we have

$$g_{ik}g^{kl} = \delta^l_i. \tag{2.24}$$

It can be shown that g^{ik} transforms as a contravariant tensor of rank 2. (See Exercise 6.)

2.2.5 Symmetric and antisymmetric tensors

If tensors S_{ik} and A_{ik} satisfy the relations

$$S_{ik} = S_{ki}, \qquad A_{ik} = -A_{ki}, \tag{2.25}$$

then they are respectively symmetric and antisymmetric tensors of rank 2. These ideas can be generalized to higher-rank tensors, and we will encounter specific tensors having the properties of symmetry and antisymmetry with respect to some or all indices.

Example 1 g_{ik} and g^{ik} are symmetric tensors.

Example 2 Consider the symbol ε_{ijkl} with the following properties:

$$\varepsilon_{ijkl} = +1 \qquad \text{if } (ijkl) \text{ is an even permutation of } (0123),$$
$$\varepsilon_{ijkl} = -1 \qquad \text{if } (ijkl) \text{ is an odd permutation of } (0123),$$
$$\varepsilon_{ijkl} = 0 \qquad \text{otherwise.} \tag{2.26}$$

We will show that

$$e_{ijkl} = (-g)^{1/2} \varepsilon_{ijkl} \tag{2.27}$$

transforms as a tensor. First take the determinant of (2.22). Let J denote the Jacobian $\partial x^i / \partial x'^m$. Then, using the rule that the determinant of a product of matrices is equal to the product of their determinants, we get

$$g' = J^2 g. \tag{2.28}$$

However, we have from the definition of a determinant

$$\varepsilon_{mnpq} J = \varepsilon_{ijkl} \frac{\partial x^i}{\partial x'^m} \frac{\partial x^j}{\partial x'^n} \frac{\partial x^k}{\partial x'^p} \frac{\partial x^l}{\partial x'^q}. \tag{2.29}$$

Using (2.28) and (2.29), the result follows: e_{ijkl} is a tensor that is totally antisymmetric. Strictly speaking, e_{ijkl} is a *pseudotensor*, since it changes sign under transformations involving reflection; for example, $x'^0 = -x^0$, $x'^1 = x^1$, $x'^2 = x^2$, and $x'^3 = x^3$.

Exercises 3 to 10 at the end of this chapter will help in understanding the operation of vectors and tensors. We end this section with one important operation.

2.2.6 Contraction

Contraction consists of identifying a lower index with an upper index in a mixed tensor. This procedure reduces the rank of the tensor by 2.

Thus $A^i B_k$ is a tensor of rank 2 if A^i and B_k are vectors. The identification $i = k$ gives a *scalar*:

$$A^i B_i = A^0 B_0 + A^1 B_1 + A^2 B_2 + A^3 B_3.$$

As in special relativity, we define a vector A^i to be *spacelike*, *timelike*, or *null* according to

$$g_{ik} A^i A^k < 0, \qquad g_{ik} A^i A^k > 0, \qquad g_{ik} A^i A^k = 0.$$

It is convenient to define associated tensors by the relations

$$A_i = g_{ik} A^k, \qquad A^k = g^{ik} A_i. \tag{2.30}$$

Thus $g_{ik}A^iA^k = A_kA^k$. The operations embodied in (2.30) are called *lowering* and *raising* the indices. We may frequently refer to A^i and A_i as the same object.

From the above manipulations of tensors it is clear (and can be easily proved) that the product of two tensors is a tensor. A reverse result is sometimes useful in deducing that a certain quantity is a tensor. This result is known as the *quotient law*. It states that if a relation such as

$$PQ = R$$

holds in all coordinate frames, where P is an arbitrary tensor of rank m and R a tensor of rank $m + n$, then Q is a tensor of rank n.

2.3 Covariant differentiation

A *vector field* is a vector function of position defined over a subspace of spacetime. Thus $B_i(x^k)$ is a covariant vector field whose four components transform according to the rule in (2.18) at each point (x^k) where it is defined. Suppose B_i is a differentiable function of (x^k). Do the derivatives $\partial B_i/\partial x^k$ transform as a tensor?

We have already seen that the derivatives $\partial \phi/\partial x^k$ of a scalar transform as a vector. So at first sight the answer to the above question might be 'yes'. Indeed, in special relativity we do encounter such results. For example, if A_i is the 4-potential of the electromagnetic field (described in the four-dimensional language of special relativity), then $\partial A_i/\partial x^k$, for Cartesian coordinates (x, y, z) and the time t of (2.1), do transform as a tensor. In our more general spacetime with an arbitrary coordinate system, however, the answer to the above question is in the negative.

This result is easily verified by differentiating (2.18). We get

$$\frac{\partial B'_k}{\partial x'^m} = \frac{\partial x^i}{\partial x'^k}\frac{\partial x^n}{\partial x'^m}\frac{\partial B_i}{\partial x^n} + \frac{\partial^2 x^i}{\partial x'^m \partial x'^k}B_i. \tag{2.31}$$

Thus, whereas the first term on the right-hand side does appear in the right form to make $\partial B_i/\partial x^n$ a tensor, the second term spoils the effect. It also gives a clue as to why this happens. The second derivative

$$\frac{\partial^2 x^i}{\partial x'^m \partial x'^k}$$

is in general nonzero and indicates that the transformation coefficients in equation (2.18) vary with position in spacetime. Thus when we seek to construct the derivative $\partial B_i/\partial x^n$, we are forced to define it as a limit:

$$\frac{\partial B_i}{\partial x^n} = \lim_{\delta x^n \to 0} \left[\frac{B^i(x^k + \delta x^k) - B^i(x^k)}{\delta x^n} \right].$$

However, the two terms in the numerator do not transform as vectors at the same point because of the variation of the transformation coefficients with position. Therefore their difference is not expected to be a vector.

This situation is illustrated in Figure 2.3. *P* and *Q* are the two neighbouring points (x^k) and $(x^k + \delta x^k)$, with the vectors B_i shown there with continuous arrows. In order to describe the change in the vector from *P* to *Q*, we must somehow measure this difference at *the same* point. How can this be achieved?

This is achieved by a device known as *parallel transport*. Assume that the vector B_i at *P* is moved from *P* to *Q* as if its magnitude and direction did not change. In Figure 2.3 this is shown by a dotted vector at *Q*. The difference between the vector $B_i(x^k + \delta x^k)$ and this dotted vector is a vector at *Q*. So we may after all be able to define a process of differentiation of vectors, provided we know what happens to B_i during a parallel transport from *P* to *Q*.

First we have to note that the dotted vector need not have the same components as the undotted vector at *P*. It is only with Cartesian coordinates that the components are the same. Consider, for example, the Euclidean plane with a polar coordinate system. A vector *A* at a point *P* with coordinates (r, θ) has components A_r and A_θ in the radial and transverse directions. If we now move from *P* to a neighbouring point *Q* with polar coordinates $(r + \delta r, \theta + \delta \theta)$, the radial and transverse directions at *Q* will not necessarily be parallel to those at *P*. Hence after parallel transport of *A* from *P* to *Q*, its radial and transverse components at *Q* will be different from A_r and A_θ.

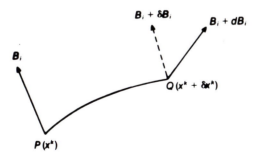

Fig. 2.3 The vector field has the components B_i at *P* and $B_i + dB_i$ at *Q*. If B_i were transported parallel to itself along an infinitesimal curve connecting *P* to *Q*, its components at *Q* would be $B_i + \delta B_i$.

A simple calculation (see Exercise 11) shows that the components of A at Q are $A_r + \delta\theta A_\theta$ and $A_\theta - \delta\theta A_r$. To return to our general case, there will be changes in B_i that are proportional to the original components and also to the displacement in position from P to Q. We may express the change in general as

$$\delta B_i = \Gamma_{ik}^l B_l \, \delta x^k, \tag{2.32}$$

where the coefficients Γ_{ik}^l are, in general, functions of space and time. These quantities are called the *three index symbols* or the *Christoffel symbols*.

Notice that the introduction of (2.32) is something new in addition to the introduction of the metric. The metric tells us how to measure distance between neighbouring points, whereas (2.32) tells us how to define parallel vectors at neighbouring points. This property of local parallelism is often called the *affine connection* of spacetime.

Returning to (2.32), we see that the difference between the continuous and the dotted vectors at Q is given by

$$B_i(x^k + \delta x^k) - [B_i(x^k) + \delta B_i] = \left(\frac{\partial B_i}{\partial x^k} - \Gamma_{ik}^l B_l \right) \delta x^k. \tag{2.33}$$

We may accordingly redefine the derivative of a vector by

$$B_{i;k} \equiv \frac{\partial B_i}{\partial x^k} - \Gamma_{ik}^l B_l \equiv B_{i,k} - \Gamma_{ik}^l B_l. \tag{2.34}$$

This derivative, by definition, must transform as a tensor. It is called the *covariant derivative* and will be denoted by a semicolon, as against the ordinary derivative, which is denoted by a comma.

If $B_{i;k}$ must transform as a tensor, the coefficients Γ_{kl}^i have to transform according to the following law:

$$\Gamma_{k\,l}^{\prime\,i} = \frac{\partial x^{\prime i}}{\partial x^m} \frac{\partial x^n}{\partial x^{\prime k}} \frac{\partial x^p}{\partial x^{\prime l}} \Gamma_{np}^m + \frac{\partial^2 x^p}{\partial x^{\prime k} \partial x^{\prime l}} \frac{\partial x^{\prime i}}{\partial x^p}. \tag{2.35}$$

This result can be verified after some straightforward but tedious calculation.

A scalar, of course does not change under parallel transport, which is why $\partial\phi/\partial x^k$ transform as a vector. If we use this result we see that for a vector A_i, $(A_i A^i)_{,k}$ is a vector. This property enables us to construct the covariant derivative of a contravariant vector A^i:

$$A^i_{\,;k} \equiv \frac{\partial A^i}{\partial x^k} + \Gamma_{lk}^i A^l \equiv A^i_{\,,k} + \Gamma_{lk}^i A^l. \tag{2.36}$$

The rule of covariant differentiation of a tensor of arbitrary rank is easily obtained: we introduce a $(+\Gamma)$ term for each contravariant index and a $(-\Gamma)$ term for each covariant index. Thus for the metric tensor we have

$$g_{ik;l} = \frac{\partial g_{ik}}{\partial x^l} - \Gamma^p_{il} g_{pk} - \Gamma^p_{kl} g_{ip}. \qquad (2.37)$$

2.4 Riemannian geometry

Einstein used the non-Euclidean geometry developed by Riemann to describe his theory of gravitation. The Riemannian geometry introduces the additional simplification that

$$\Gamma^i_{kl} = \Gamma^i_{lk}; \qquad g_{ik;l} \equiv 0. \qquad (2.38)$$

Going back to (2.37), we see that $g_{ik;l} = 0$ gives us 40 linear equations for the 40 unknowns Γ^i_{kl}. These equations have a unique solution. For, from (2.37) and (2.38) we get

$$\Gamma_{k|il} + \Gamma_{i|kl} = g_{ik;l},$$

where

$$\Gamma_{k|il} = g_{pk} \Gamma^p_{il}.$$

Rotate the indices cyclically to obtain two more relations:

$$\Gamma_{l|ki} + \Gamma_{k|li} = g_{kl,i}, \qquad \Gamma_{i|lk} + \Gamma_{l|ik} = g_{li,k}.$$

Use the symmetry condition (2.38) to eliminate $\Gamma_{l|ki} = \Gamma_{l|ik}$ and $\Gamma_{k|il} = \Gamma_{k|li}$ from the above three relations to get

$$2\Gamma_{i|kl} = g_{ik,l} + g_{li,k} - g_{kl,i}.$$

Raising the index i, we get the required solution:

$$\Gamma^i_{kl} = \tfrac{1}{2} g^{im} \left(\frac{\partial g_{mk}}{\partial x^l} + \frac{\partial g_{ml}}{\partial x^k} - \frac{\partial g_{kl}}{\partial x^m} \right). \qquad (2.39)$$

We next consider some particular properties of these symbols that are useful in various manipulations. If we differentiate the determinant of the metric tensor we get

$$\mathrm{d}g = g g^{ik} \, \mathrm{d}g_{ik}. \qquad (2.40)$$

This relation is useful in expressing some Γ^i_{kl} and covariant derivatives in relatively simple forms. Thus, using (2.39) and (2.40), it is possible to prove the following relations:

$$\Gamma^l_{il} = \frac{1}{(-g)^{1/2}} \frac{\partial}{\partial x^i} (-g)^{1/2};$$

$$\Gamma^l_{ik} g^{ik} = -\frac{1}{(-g)^{1/2}} \frac{\partial}{\partial x^m} [(-g)^{1/2} g^{ml}];$$

(2.41)

$$A^i_{;i} = \frac{1}{(-g)^{1/2}} \frac{\partial}{\partial x^i} [(-g)^{1/2} A^i];$$

$$F^{ik}_{;k} = -\frac{1}{(-g)^{1/2}} \frac{\partial}{\partial x^k} [(-g)^{1/2} F^{ik}] \text{ for } F^{ik} = -F^{ki}.$$

(Here A^i and F^{ik} are respectively vector and tensor fields.) For example, to prove the first relation note that (2.39) gives

$$\Gamma^i_{il} = \tfrac{1}{2} g^{im}(g_{mi,l} + g_{ml,i} - g_{il,m}).$$

Since $(g_{ml,i} - g_{il,m})$ is antisymmetric in (i,m), its product with the symmetric g^{im} vanishes. The result then follows when we recall (2.40).

The symmetry condition (2.38) enables us to choose special coordinates in which the Christoffel symbols all vanish at a given point. Suppose we start with $\Gamma^m_{np} \neq 0$ in the coordinate system (x^i) at point P. Let the coordinates of P be given by x^i_P. Now define new coordinates in the neighbourhood of P by

$$x'^k = -\tfrac{1}{2}\Gamma^k_{nm}(x^n - x^n_P)(x^m - x^m_P).$$

(2.42)

Then we have, at P,

$$x'^i_P = 0, \qquad \frac{\partial x'^i}{\partial x^m} = 0, \qquad \frac{\partial^2 x'^i}{\partial x^n \partial x^m} = -\Gamma^i_{nm},$$

with the result that, from (2.35),

$$\Gamma'^i_{mn}\big|_P = 0.$$

Further, by a linear transformation we can arrange to have a coordinate system with

$$g_{ik} = \eta_{ik} = \text{diag}(+1, -1, -1, -1), \qquad \Gamma^i_{kl} = 0$$

(2.43)

at any chosen point P. Such a coordinate system is called a *locally inertial* coordinate system, for reasons that will become clear later. Apart from its physical implications in general relativity, the locally inertial coordinate system is often useful as a mathematical device for simplifying calculations. We also warn the reader that the operative word is 'local': the simplifications implied in (2.43) cannot be achieved globally. What prevents us from achieving a globally inertial coordinate system? In seeking an answer to this question we encounter the most crucial aspect in which a non-Euclidean geometry differs from its Euclidean counterpart.

2.5 Spacetime curvature

Figure 2.4 repeats the previous example of non-Euclidean geometry on the surface of a sphere. We have the triangle ABC of Figure 2.2(a) whose three angles are each 90°. Consider what happens to a vector (shown by a dotted arrow) as it is parallelly transported along the three sides of this triangle. As shown in the figure, this vector is originally perpendicular to AB when it starts its journey at A. When it reaches B it lies along CB. So it keeps pointing along this line as it moves from B to C. At C it is again perpendicular to AC. So, as it moves along CA from C to A, it maintains this perpendicularity with the result that when it arrives at A it is pointing along AB. In other words, one circuit around this triangle has resulted in a change of direction of the vector by 90°, although at each stage it was being moved parallel to itself!

A similar experiment with a triangle drawn on a flat piece of paper will tell us that there is no resulting change in the direction of the vector when it moves parallel to itself around the triangle. So our physical triangle behaves differently from the flat Euclidean triangle.

The phenomenon illustrated in Figure 2.4 can also be described as follows. If we had moved our vector from A to C along two different routes – along AC and along AB followed by BC – we would have found it pointing in two different directions. In fact, if we had taken any arbitrary curves from A to C we would have found that the parallel transport of a vector from A to C varies from curve to curve; that is, the outcome depends on the path of transport from A to C.

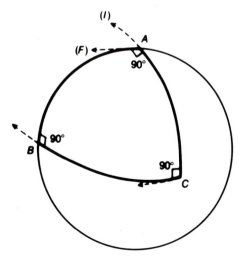

Fig. 2.4 Parallel transport on a spherical surface.

This is one of the properties that distinguishes a curved space from a flat space. Let us consider it in more general terms for our four-dimensional spacetime. Let a vector B_i at P be transported parallelly to Q and let us ask for the condition that the answer should be *independent* of the curve joining P to Q. We have seen that under parallel transport from a point $\{x^i\}$ to a neighbouring point $\{x^i + \delta x^i\}$, the components of the vector change according to (2.32). If it were possible to transport B_i from P to Q without the result depending on which path is taken, then we would be able to generate a vector field $B^i(x^k)$, satisfying the differential equation

$$\frac{\partial B_i}{\partial x^k} = \Gamma^l_{ik} B_l. \tag{2.44}$$

So the answer to our question depends on whether we can find a nontrivial solution to (2.44).

The necessary condition for the existence of a solution is easily derived. We differentiate (2.44) with respect to x^n to get

$$\frac{\partial^2 B_i}{\partial x^n \partial x^k} = \frac{\partial}{\partial x^n}(\Gamma^l_{ik} B_l) = \frac{\partial \Gamma^l_{ik}}{\partial x^n} B_l + \Gamma^l_{ik} \frac{\partial B_l}{\partial x^n}$$

$$= \left(\frac{\partial \Gamma^m_{ik}}{\partial x^n} + \Gamma^l_{ik} \Gamma^m_{ln} \right) B_m.$$

We now interchange the order of differentiation with respect to x^n and x^k and use the result $B_{i,nk} = B_{i,kn}$. We then get the required necessary condition as

$$R_i{}^m{}_{kn} \equiv \frac{\partial \Gamma^m_{ik}}{\partial x^n} - \frac{\partial \Gamma^m_{in}}{\partial x^k} + \Gamma^l_{ik} \Gamma^m_{ln} - \Gamma^l_{in} \Gamma^m_{lk} = 0. \tag{2.45}$$

It is not obvious simply from the above expression that $R_i{}^m{}_{kn}$ should be a tensor. Yet our result, in order to be significant, must clearly hold whatever coordinates we employ to derive it. So we do expect $R_i{}^m{}_{kn}$ to be a tensor. A simple calculation shows that, for any twice differentiable vector field B_i,

$$B_{i;nk} - B_{i;kn} \equiv R_i{}^m{}_{kn} B_m. \tag{2.46}$$

Since the left-hand side is a tensor, so is the right-hand side, and, B_m being an arbitrary vector, we have by the *quotient law* (see Exercise 10) the result that $R_i{}^m{}_{kn}$ are the components of a tensor.

This tensor, known as the *Riemann Christoffel tensor* (or, more commonly, the *Riemann tensor*), plays an important role in specifying the geometrical properties of spacetime. Although we have derived (2.45) as a necessary condition, a slightly more sophisticated technique shows that

(2.45) is also the sufficient condition that a vector field $B_i(x^k)$ can be defined over the spacetime by parallel transport.

Spacetime is said to be *flat* if its Riemann tensor vanishes everywhere. Otherwise, it is said to be *curved*. Exercises 26 and 27 illustrate two other ways in which this tensor distinguishes the properties of a curved spacetime from those of a flat spacetime.

2.5.1 Symmetries of R_{iklm}

It is more convenient to lower the second index of the Riemann tensor to study its symmetry properties. Since the symmetry or antisymmetry of a tensor does not depend on what coordinates are used, it is more convenient to write (2.45) in the locally inertial coordinates (2.43). We then get

$$R_{iklm} = \tfrac{1}{2}(g_{kl,im} + g_{im,kl} - g_{km,il} - g_{il,km}). \tag{2.47}$$

From this expression the following symmetries are immediately obvious:

$$R_{iklm} = -R_{kilm} = -R_{ikml} = -R_{lmik}. \tag{2.48}$$

We also get relations of the following type:

$$R_{iklm} + R_{imkl} + R_{ilmk} \equiv 0. \tag{2.49}$$

If we take all these symmetries into account, we find that of the $4^4 = 256$ components of the Riemann tensor, only 20 at most are independent! Moreover, we will soon see that there are identities linking their derivatives.

2.5.2 The Ricci and Einstein tensors

By the process of contraction we can construct lower rank tensors from R_{iklm}. The tensor

$$R_{kl} = g^{im} R_{iklm} \equiv R^m{}_{klm} \tag{2.50}$$

is called the *Ricci tensor*. If we use the locally inertial coordinate system, we see immediately that

$$R_{kl} = R_{lk}. \tag{2.51}$$

Owing to the symmetries of (2.48), there are no other independent second-rank tensors that can be constructed out of R_{iklm}.

By further contraction we get a scalar:

$$R = R_{ik} \equiv R^k{}_k. \tag{2.52}$$

R is called the *scalar curvature*. The tensor

$$G_{ik} \equiv g^{kl}R_{kl} - \tfrac{1}{2}g_{ik}R \tag{2.53}$$

will turn out to have a special role to play in Einstein's general relativity. This tensor is called the *Einstein tensor*.

2.5.3 Bianchi identities

The expression (2.47) suggests another symmetry for the components of R_{iklm}. This symmetry is not algebraic, but involves calculus. In covariant language we may express it as follows:

$$R_{iklm;n} + R_{iknl;m} + R_{ikmn;l} \equiv 0. \tag{2.54}$$

These relations are known as the *Bianchi identities*. Their proof is most easily given in the locally inertial system, as in (2.47).

But multiplying (2.54) by $g^{im}g^{kn}$, and using (2.50)–(2.52), we can deduce from these identities another that is of importance to relativity:

$$(R^{ik} - \tfrac{1}{2}g^{ik}R)_{;k} \equiv 0. \tag{2.55}$$

In other words, *the Einstein tensor G^{ik} has zero divergence*.

2.6 Geodesics

So far we have talked about non-Euclidean geometries without mentioning whether they have the equivalents of straight lines in Euclidean geometry. We now show how equivalent concepts do exist in the Riemannian geometry under consideration here.

There are two properties of a straight line that can be generalized: the property of 'straightness' and the property of 'shortest distance'. Straightness means that as we move along the line, its direction does not change. Let us see how we can generalize this concept first.

Let $x^i(\lambda)$ be the parametric representation of a curve in spacetime. Its tangent vector is given by

$$u^i = \frac{\mathrm{d}x^i}{\mathrm{d}\lambda}. \tag{2.56}$$

Our straightness criterion demands that u^i should not change along the curve. In going from λ to $\lambda + \delta\lambda$, the change in u^i is given by

$$\Delta u^i = \frac{\mathrm{d}u^i}{\mathrm{d}\lambda}\,\delta\lambda + \Gamma^i_{kl}u^k\,\delta x^l.$$

The second expression on the right-hand side arises from the change produced by parallel transport through a coordinate displacement δx^l.

However, $\delta x^l = u^l \, \delta\lambda$. Therefore the condition of no change of direction u^i implies $\Delta u^i = 0$; that is,

$$\frac{\mathrm{d}u^i}{\mathrm{d}\lambda} + \Gamma^i_{kl} u^k u^l = 0. \tag{2.57}$$

This is the condition that our curve must satisfy in order to be straight.

The second property of a straight line in Euclidean geometry is that it is the curve of shortest distance between two points. Let us generalize this property in the following way. Let the curve, parametrized by λ, connect two points P_1 and P_2 of spacetime, with parameters λ_1 and λ_2 respectively. Then the 'distance' of P_2 from P_1 is defined as

$$s(P_2, P_1) = \int_{\lambda_1}^{\lambda_2} \left(g_{ik} \frac{\mathrm{d}x^i}{\mathrm{d}\lambda} \frac{\mathrm{d}x^k}{\mathrm{d}\lambda} \right)^{1/2} \mathrm{d}\lambda \equiv \int_{\lambda_1}^{\lambda_2} L \, \mathrm{d}\lambda. \tag{2.58}$$

We now demand that $s(P_2, P_1)$ be 'stationary' for small displacements of the curve connecting P_1 and P_2, these displacements vanishing at P_1 and P_2.

This is a standard problem in the calculus of variations, and its solution leads to the familiar Euler–Lagrange equations

$$\frac{\mathrm{d}}{\mathrm{d}\lambda} \left(\frac{\partial L}{\partial \dot{x}^i} \right) - \frac{\partial L}{\partial x^i} = 0, \tag{2.59}$$

where $\dot{x}^i \equiv \mathrm{d}x^i/\mathrm{d}\lambda$ and $L \equiv [g_{ik}(\mathrm{d}x^i/\mathrm{d}\lambda)(\mathrm{d}x^k/\mathrm{d}\lambda)]^{1/2}$ is a function of x^i and $\dot{x}^i$. It is easy to see that (2.59) leads to

$$\frac{\mathrm{d}}{\mathrm{d}\lambda} \left(g_{ik} \frac{1}{L} \frac{\mathrm{d}x^k}{\mathrm{d}\lambda} \right) - \tfrac{1}{2} g_{mn,i} \frac{1}{L} \frac{\mathrm{d}x^m}{\mathrm{d}\lambda} \frac{\mathrm{d}x^n}{\mathrm{d}\lambda} = 0.$$

If we substitute

$$\mathrm{d}s = L \, \mathrm{d}\lambda \tag{2.60}$$

and use (2.39), we get the above equation in the form

$$\frac{\mathrm{d}^2 x^i}{\mathrm{d}s^2} + \Gamma^i_{kl} \frac{\mathrm{d}x^k}{\mathrm{d}s} \frac{\mathrm{d}x^l}{\mathrm{d}s} = 0. \tag{2.61}$$

There are a few loose ends to be sorted out in the above derivation. First, L would be real only for timelike curves. Thus if we want to use a real parameter along the curve, then for spacelike curves we must replace $\mathrm{d}s$ by

$$\mathrm{d}\sigma = i \, \mathrm{d}s, \qquad i = (-1)^{1/2}. \tag{2.62}$$

For null curves, $L = 0$. The above treatment therefore breaks down. It is then more convenient to replace the integral (2.58) by another:

$$I = \int_{\lambda_1}^{\lambda_2} L^2 \, \mathrm{d}\lambda, \tag{2.63}$$

and consider $\delta I = 0$. We can always choose a new parameter $\lambda' = \lambda'(\lambda)$ such that the equation of the curve takes the same form as (2.61), with λ' replacing s.

It is easy to see that (2.61) is the same as (2.57). Although s in (2.61) has the special meaning 'length along the curve', while λ in (2.57) appears to be general, it is not difficult to see that if (2.57) is satisfied then λ must be a constant multiple of s. This is because (2.57) has the first integral

$$g_{ik} \frac{dx^i}{d\lambda} \frac{dx^k}{d\lambda} = C, \qquad C = \text{constant}. \qquad (2.64)$$

These curves of 'stationary distance' are called *geodesics*. For timelike curves $C > 0$, for spacelike curves $C < 0$, while for null curves $C = 0$. λ is called an *affine parameter*.

Example Let us calculate the null geodesics from $t = 0$, $r = 0$ to the point $t = T$, $r = R$, $\theta = \theta_1$, $\phi = \phi_1$ in the de Sitter spacetime

$$ds^2 = c^2 \, dt^2 - e^{2Ht}[dr^2 + r^2(d\theta^2 + \sin^2 \theta \, d\phi^2)],$$

where $H = $ constant. It is not difficult to verify that the θ and ϕ equations of (2.61) are satisfied by $\theta = \theta_1$, $\phi = \phi_1$. That is, our straight line moves in the fixed (θ, ϕ) direction. The t equation simplifies to

$$\frac{d^2 t}{d\lambda^2} + \frac{H}{c^2} e^{2Ht} \left(\frac{dr}{d\lambda}\right)^2 = 0.$$

The first integral (2.64) gives, on the other hand,

$$c^2 \left(\frac{dt}{d\lambda}\right)^2 = e^{2Ht} \left(\frac{dr}{d\lambda}\right)^2.$$

The two equations can be easily solved to give

$$t = \frac{1}{H} \ln \left(1 + \frac{\lambda}{\lambda_0}\right), \qquad r = \frac{c}{H} \frac{\lambda}{\lambda + \lambda_0},$$

where λ_0 is determined from the boundary condition that when $r = R$, $t = T$. Note that a solution is possible only if R and T are related by the condition

$$R = \frac{c}{H} (1 - e^{-HT}).$$

We next consider the special role played by geodesics in general relativity.

2.7 The principle of equivalence

Having described the machinery of vectors and tensors, and having outlined the salient features of Riemannian geometry, we now make our first contact with physics and introduce the so-called *principle of equivalence*, which has played the key role in general relativity.

Let us go back to the purely mathematical result embodied in the relations shown in (2.43) and attempt to describe their physical meaning. These relations tell us that special (locally inertial) coordinates exist in the neighbourhood of any point P in spacetime that behave like the coordinates (t, x, y, z) of special relativity. Physically, these coordinates imply a special frame of reference in which a momentary illusion is created at P and in a small neighbourhood of P that the geometry is of special relativity. The illusion is momentary and local to P because we have seen that the relations of (2.43) cannot be made to hold everywhere and at all times.

In view of the assertion made in section 2.1 that gravitation manifests itself as non-Euclidean geometry, we would have to argue that in the above locally inertial frame gravitation has been transformed away momentarily and in a small neighbourhood of P. How does this happen in practice? Consider Einstein's celebrated example of the freely falling lift. A person inside such a lift feels weightless. The accelerated frame of reference of the lift provides the locally inertial frame in the small neighbourhood of the falling person. Similarly, a spacecraft circling around the Earth is in fact freely falling in the Earth's gravity, and the astronauts inside it feel weightless.

It should be emphasized that this feeling of weightlessness in a falling lift or a spacecraft is limited to local regions: there is no universal frame that transforms away Earth's gravity everywhere, at all times. If we demand that the relations of (2.43) hold at all points of spacetime, we would need to have $\partial \Gamma^i_{kl}/\partial x^m = 0$ everywhere, leading to $R^i_{klm} = 0$ – that is, to a flat spacetime. Thus a curved spacetime with a non-vanishing Riemann tensor is necessary to describe the genuine effects of gravitation. (See Exercise 27.)

The *weak principle of equivalence* states that effects of gravitation can be transformed away locally and over small intervals of time by using suitably accelerated frames of reference. Thus it is the physical statement of the mathematical relations given by (2.43). It is possible, however, to go from here to a much stronger statement, the so-called *strong principle of equivalence*, which states that any physical interaction (other than gravitation, which has now been identified with geometry) behaves in a locally inertial frame as if gravitation were absent. For example, Maxwell's equations will have their familiar form (of special relativity) in a locally inertial frame. Thus an observer performing a local experiment in a freely falling lift would measure the speed of light to be c.

The strong principle of equivalence enables us to extend any physical law that is expressed in the covariant language of special relativity to the more general form it would have in the presence of gravitation. The law is usually expressed in vectors, tensors, or spinors in the Minkowski spacetime of special relativity. All we have to do is to write it in terms of the corresponding entities in curved spacetime. Thus in the flat spacetime of special relativity, the Maxwell electromagnetic field F^{ik} is related to the current vector j^k by

$$F^{ik}_{,i} = 4\pi j^k. \tag{2.65}$$

In curved spacetime the ordinary tensor derivative is replaced by the covariant derivative:

$$F^{ik}_{;i} = 4\pi j^k. \tag{2.66}$$

Notice that the effect of gravitation enters through the Γ^i_{kl} terms that are present in (2.66). This generalization of (2.65) to (2.66) is called the *minimal coupling* of the field with gravitation, since it is the simplest one possible.

So in order to describe how other interactions behave in the presence of gravitation, we use the covariance under the general coordinate transformation as the criterion to be satisfied by their equations. It is immediately clear from the example of the electromagnetic field that a light ray describes a null geodesic.

In the same vein we can now describe a moving object that is acted on by no other interaction except gravitation – for example, a probe moving in the gravitational field of the Earth. *In the absence of gravity*, this object would move in a straight line with uniform velocity; that is, with the equation of motion,

$$\frac{\mathrm{d}u^i}{\mathrm{d}s} = 0, \qquad u^i = \text{4-velocity}. \tag{2.67}$$

In the presence of gravity, (2.67) is modified to our geodesic equation (2.61).

We end this section with another example that provides a clue about how gravitational effects show up in spacetime geometry according to general relativity. Consider the Minkowski spacetime with the standard line element

$$\mathrm{d}s^2 = c^2\,\mathrm{d}t^2 - \mathrm{d}x^2 - \mathrm{d}y^2 - \mathrm{d}z^2. \tag{2.68}$$

If we make the coordinate transformation for a constant g,

$$x = \frac{c^2}{g}\left(\cosh\frac{gt'}{c} - 1\right) + x'\cosh\frac{gt'}{c}, \tag{2.69}$$

$$y = y', \qquad z = z', \qquad t = \frac{c}{g}\sinh\frac{gt'}{c} + \frac{x'}{c},$$

this leads to the line element

$$ds^2 = \left(1 + \frac{gx'}{c^2}\right)^2 dt'^2 - dx'^2 - dy'^2 - dz'^2. \tag{2.70}$$

What interpretation can we give to (2.70)? The origin of the (x', y', z') system has a world line whose parametric form in the old coordinates is given by

$$x = \frac{c^2}{g}\left(\cosh\frac{gt'}{c} - 1\right), \qquad y = 0, \qquad z = 0, \qquad t = \frac{c}{g}\sinh\frac{gt'}{c}. \tag{2.71}$$

Using the kinematics of special relativity, we can easily see that (2.71) describes the motion of a point that has a uniform acceleration g in the x-direction, a point that is momentarily at rest at the origin of (x, y, z) at $t = 0$. We may interpret the line element (2.70) and the new coordinate system as describing the spacetime in the rest frame of the uniformly accelerated observer.

Direct calculation shows that not all Γ^i_{kl} are zero in (2.70) at $x' = 0$, $y' = 0$, $z' = 0$. The frame is therefore non-inertial. For the neighbourhood of the origin, the metric component

$$g_{00} \approx 1 + \frac{2gx'}{c^2} = 1 + \frac{2\phi}{c^2}, \tag{2.72}$$

where ϕ is the Newtonian gravitational potential for a uniform gravitational field that induces an acceleration due to gravity $= -g$. We have here the reverse situation to that of the falling lift: we seem to have generated a pseudogravitational field by choosing a suitably accelerated observer. The prefix 'pseudo-' is used because the gravitational field is not real – it is an illusory effect arising from the choice of coordinates. The Riemann tensor is zero. Nevertheless, the relation (2.72) is also suggestive of the real gravitational field, as we will see in section 2.9.

2.8 Action principle and the energy tensors

Before examining relativity proper, let us see how we can write the laws of physics in the covariant language in Riemannian spacetime using the strong principle of equivalence. We take the familiar example of charged particles interacting with the electromagnetic field. The physical laws can be derived from an action principle. First we write the action in Minkowski spacetime:

$$\mathscr{A} = -\sum_a cm_a \int ds_a - \frac{1}{16\pi c} \int F_{ik} F^{ik} \, d^4 x - \sum_a \frac{e_a}{c} \int A_i \, da^i; \qquad (2.73)$$

here A_i are the components of the 4-potential, which are related to the field tensor F_{ik} by

$$A_{k,i} - A_{i,k} = F_{ik}, \qquad (2.74)$$

while e_a and m_a are the charge and rest mass of particle a, whose coordinates are given by a^i and the proper time by s_a with

$$ds_a^2 = \eta_{ik} \, da^i \, da^k. \qquad (2.75)$$

How do we generalize (2.73) to Riemannian spacetime? First, we note that η_{ik} in (2.75) are replaced by g_{ik}. Next, starting from the covariant vector A_i, we generate F_{ik} by the covariant generalization of (2.74):

$$A_{k;i} - A_{i;k} = F_{ik}. \qquad (2.76)$$

However, since the expression (2.76) is antisymmetric in (i, k), the extra terms involving the Christoffel symbols vanish and we are back to (2.74)! The volume integral in (2.73) is modified to

$$\int F_{ik} F^{ik} (-g)^{1/2} \, d^4 x. \qquad (2.77)$$

The extra factor $(-g)^{1/2}$ has crept in because the combination

$$(-g)^{1/2} \, dx^1 \, dx^2 \, dx^3 \, dx^0 = \tfrac{1}{24} e_{ijkl} \, dx^i \, dx^j \, dx^k \, dx^l$$

acts as a scalar. We therefore have the generalized form of (2.73):

$$\mathscr{A} = -\sum_a cm_a \int ds_a - \frac{1}{16\pi c} \int F_{ik} F^{ik} (-g)^{1/2} \, d^4 x - \sum \frac{e_a}{c} \int A_i \, da^i. \qquad (2.78)$$

The variation of the world line of particle a gives its equation of motion,

$$\frac{d^2 a^i}{ds_a^2} + \Gamma^i_{kl} \frac{da^k}{ds_a} \frac{da^l}{ds_a} = \frac{e_a}{m_a} F^i_l \frac{da^l}{ds_a}, \qquad (2.79)$$

while the variation of A_i gives the field equations (2.66).

The transition from (2.73) to (2.78) has, however, introduced an additional independent feature into the action, besides the particle world lines and the potential vector. The new feature is the spacetime geometry typified by the metric tensor g_{ik}. What will happen if we demand that the

g_{ik} are also dynamical variables and that the action $\mathcal{A}$ remains stationary for small variations of the type

$$g_{ik} \to g_{ik} + \delta g_{ik}? \tag{2.80}$$

From the generalized action principle, should we expect to get the equations that determine the spacetime geometry? Let us investigate.

A glance at the action (2.78) shows that the last term does not contribute anything under (2.80) if we keep the worldlines and A_i fixed in spacetime. The first two terms, however, do make contributions. Let us consider them in that order. First note that

$$\delta(\mathrm{d}s_a^2) = \delta g_{ik} \, \mathrm{d}a^i \, \mathrm{d}a^k.$$

That is,

$$\delta(\mathrm{d}s_a) = \tfrac{1}{2}\delta g_{ik} \frac{\mathrm{d}a^i}{\mathrm{d}s_a} \frac{\mathrm{d}a^k}{\mathrm{d}s_a} \, \mathrm{d}s_a.$$

Therefore,

$$\delta \sum_a cm_a \int \mathrm{d}s_a = \tfrac{1}{2} \sum_a c \int m_a \frac{\mathrm{d}a^i}{\mathrm{d}s_a} \frac{\mathrm{d}a^k}{\mathrm{d}s_a} \, \mathrm{d}s_a \, \delta g_{ik}. \tag{2.81}$$

Let us consider this variation in a small 4-volume $\mathcal{V}$ near a point P. If we consider a locally inertial coordinate system near P we can identify the above expression in a more familiar form. Let us first identify

$$p^i_{(a)} = cm_a \frac{\mathrm{d}a^i}{\mathrm{d}s_a}$$

as the 4-momentum of particle a. Then $cp^0_{(a)} = E_a = $ energy of the particle, and we get

$$\tfrac{1}{2}cm_a \frac{\mathrm{d}a^i}{\mathrm{d}s_a} \frac{\mathrm{d}a^k}{\mathrm{d}s_a} \, \mathrm{d}s_a = \frac{c^2}{2E_a} p^i_{(a)} p^k_{(a)} \, \mathrm{d}t_a = \frac{c}{2E_a} p^i_{(a)} p^k_{(a)} \, \mathrm{d}x^0_a.$$

Figure 2.5 shows the volume $\mathcal{V}$ as a shaded region in the neighbourhood of P, t being the local time coordinate and x^μ ($\mu = 1, 2, 3$) the local rectangular coordinates. The expression (2.81) can then be looked upon as a volume integral over $\mathcal{V}$ of the form

$$\delta \sum_a cm_a \int \mathrm{d}s_a = \frac{1}{2c} \int_{\mathcal{V}} \delta g_{ik} \, T^{ik}_{(m)} \, \mathrm{d}^4 x, \tag{2.82}$$

where $T^{ik}_{(m)}$ is the sum of expressions

$$\frac{c^2}{E_a} p^i_{(a)} p^k_{(a)}$$

for each particle a that crosses a unit volume of the shaded region near P.

General relativity

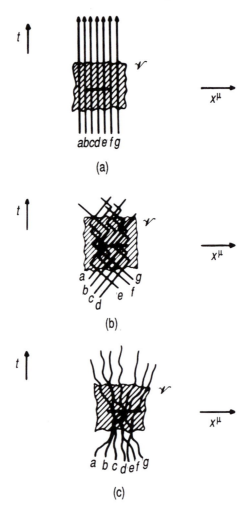

abcdefg

(a)

(b)

a b c d e f g

(c)

Fig. 2.5 Three cases of particle motion in the locally inertial region T near a typical point P of spacetime. The thick line on the x^μ-axes in each case represents a unit 3-volume. All particles $a, b, c, d, \ldots$ crossing this volume are counted for computing T^{ik}. (a) Particle world lines $a, b, c, \ldots$ are nearly parallel. This is the dust approximation. (b) The particles move at random with speeds near the speed of light, frequently changing directions through collisions. This is the relativistic case. (c) The intermediate situation, in which the particles collide, change directions, and generate pressures, but their motions are nonrelativistic. This is the case of a fluid.

2.8.1 Energy tensor of matter

This expression for T_{ik} is none other than the usual expression for the energy tensor of matter. Since we will need this tensor frequently, it is derived below for three different types of matter.

Dust: This is the simplest situation, in which all the world lines going through the shaded region in Figure 2.5(a) are more or less parallel, indicating that the particles of matter are moving without any relative motion in the neighbourhood of P. If we write the typical 4-velocity as u^i and using a Lorentz transformation to make $u^i = (1, 0, 0, 0)$ (that is, transforming to the rest frame of the dust) then the only non-zero component of the energy tensor is

$$T^{00} = \sum_a m_a c^2 = \rho_0 c^2,$$

where the summation is over a unit volume in the neighbourhood of P. Here ρ_0 is the rest mass density of dust. In any other Lorentz frame we get

$$\underset{(m)}{T^{ik}} = \rho_0 c^2 u^i u^k, \tag{2.83}$$

an expression that is easily generalized to any (non-Lorentzian) coordinate system.

Relativistic particles: This situation represents the opposite extreme. Here we have highly relativistic particles moving at random through $\mathcal{V}$ (see Figure 2.5(b)). The 4-momentum of a typical particle is then approximated to the form

$$p^i = \left(\frac{E}{c}, P\right), \qquad E^2 = c^2 P^2 + m^2 c^4 \approx c^2 P^2, \qquad P = |P|.$$

Using the fact that the particles are moving randomly, we find that the energy tensor has pressure components also:

$$T^{00} = \sum E = \varepsilon,$$

$$T^{11} = T^{22} = T^{33} = \sum \frac{P^2 c^2}{3E}. \tag{2.84}$$

The factor $\frac{1}{3}$ comes from randomizing in all directions. These are the only nonzero pressure components. Here ε is the energy density. Thus for extreme relativistic particles we get

$$\underset{(m)}{T^{ik}} = \mathrm{diag}\left(\varepsilon, \tfrac{1}{3}\varepsilon, \tfrac{1}{3}\varepsilon, \tfrac{1}{3}\varepsilon\right). \tag{2.85}$$

This form is also applicable to randomly moving neutrinos or photons.

Fluid: This situation is illustrated in Figure 2.5(c) and consists of a collection of particles with small (nonrelativistic) random motions. If we choose the frame in which the fluid as a whole is at rest as the frame of reference, we can evaluate the components of $T^{ik}_{(m)}$ as follows. Let a typical particle have the momentum vector given by

$$p^0 = \frac{mc^2}{\left(1 - \dfrac{v^2}{c^2}\right)^{1/2}}, \quad p^\mu = \frac{m\mathbf{v}}{\left(1 - \dfrac{v^2}{c^2}\right)^{1/2}} \qquad (\mu = 1, 2, 3). \qquad (2.86)$$

Then

$$T^{00} = \sum mc^2 \left(1 - \frac{v^2}{c^2}\right)^{-1/2} \approx \sum mc^2 \left(1 + \frac{v^2}{2c^2}\right) = \rho c^2,$$

$$(2.87)$$

$$T^{11} = T^{22} = T^{33} = \tfrac{1}{3} \sum mv^2 \left(1 - \frac{v^2}{c^2}\right)^{-1/2} \approx p.$$

Here ρ and p are the density and pressure of the fluid. In a frame of reference in which the fluid as a whole has a 4-velocity u^i, the energy tensor becomes

$$T^{ik}_{(m)} = (p + \rho c^2)u^i u^k - p\eta^{ik}. \qquad (2.88)$$

The generally covariant form of (2.88) is obviously

$$T^{ik}_{(m)} = (p + \rho c^2)u^i u^k - pg^{ik}. \qquad (2.89)$$

Note that ρ is not just the rest-mass density, but also includes energy density of internal motion, as seen in (2.87).

We may now relax our restriction to the locally inertial coordinate system at P. The generalized form of (2.82) is then

$$\delta \sum_a cm_a \int ds_a = \frac{1}{2c} \int T^{ik}_{(m)} (-g)^{1/2} \, \delta g_{ik} \, d^4x. \qquad (2.90)$$

2.8.2 Energy tensor of the electromagnetic field

We next consider the variation of the second term of (2.73). If we keep A_i fixed, the F_{ik}, as given by (2.76) or (2.74), remain unchanged under the variation of g_{ik}. Hence

$$\delta(F_{ik}F^{ik}(-g)^{1/2}) = F_{ik}F_{lm}\delta(g^{il}g^{km}(-g)^{1/2}).$$

From (2.24) we get

$$\delta g^{ik}g_{kl} = -g^{ik}\delta g_{kl},$$

that is,

$$\delta g^{ik} = -g^{lm}g^{kn}\delta g_{mn}. \qquad (2.91)$$

Also, from (2.40) we have

$$\delta (-g)^{1/2} = \tfrac{1}{2} g^{ik} (-g)^{1/2} \delta g_{ik}. \tag{2.92}$$

Substituting these expressions into the variation of the second term of the action gives

$$\delta \frac{1}{16\pi c} \int_{\mathcal{V}} F_{ik} F^{ik} (-g)^{1/2} \, \mathrm{d}^4 x = \frac{1}{2c} \int_{\mathcal{V}} \underset{(em)}{T^{ik}} (-g)^{1/2} \delta g_{ik} \, \mathrm{d}^4 x, \tag{2.93}$$

with the electromagnetic energy tensor given by

$$\underset{(m)}{T^{ik}} = \frac{1}{4\pi} \left(\tfrac{1}{4} F_{mn} F^{mn} g^{ik} - F^i{}_l F^{lk} \right) \tag{2.94}$$

It is obvious from our two examples that the energy tensor of any term in the action of the form Λ is related to the variation of Λ by

$$\delta \Lambda = \frac{1}{2c} \int \underset{(\Lambda)}{T^{ik}} (-g)^{1/2} \delta g_{ik} \, \mathrm{d}^4 x. \tag{2.95}$$

In theories defined only in Minkowski space the appearance of energy tensors is somewhat *ad hoc*. They do not enter explicitly into any dynamic or field equations. They appear only through their divergences, the typical conservation of energy and momentum being given by

$$T^{ik}{}_{,k} = 0. \tag{2.96}$$

In our curved spacetime framework the T^{ik} find a natural expression through the variation of g_{ik}. It was this variation of the metric tensor that led Hilbert to derive the field equations of general relativity shortly after Einstein had proposed them from heuristic considerations. We now turn our attention to this topic.

2.9 Gravitational equations

The preceding section showed that the variation of the action $\mathcal{A}$ with respect to g_{ik} leads us to the energy tensor of various interactions. We still do not have dynamic equations that tell us how to determine the g_{ik} in terms of the distribution of matter and energy. It was Einstein's conjecture that the energy tensors should act as the 'sources' of gravity. Following the general trend of nineteenth-century physics, especially the Maxwell equations, Einstein looked for an expression that would act like a wave equation for g_{ik}, with T_{ik} as the source. It is immediately clear that the standard wave equation in the covariant form

$$g^{mn} g_{ik;mn} = \kappa T_{ik}, \tag{2.97}$$

where κ is a constant, will not do, for the left-hand side vanishes identically. Is there a second-rank tensor symmetric in its indices (like the T_{ik}) that involves second derivatives of g_{ik}? Clearly, if the tensor is to bring out the special feature of curvature of spacetime, it must be related to the Riemann tensor. After trial and error, Einstein finally arrived at the tensor G_{ik} of (2.53). His field equations of general relativity, published in 1915, took the form

$$R_{ik} - \tfrac{1}{2}g_{ik}R \equiv G_{ik} = -\kappa T_{ik}. \tag{2.98}$$

These equations have the added advantage that in view of the Bianchi identities in (2.55) we must have

$$T^{ik}{}_{;k} \equiv 0. \tag{2.99}$$

That is, the law of conservation of energy and momentum follows naturally from (2.98).

Although there are 10 Einstein equations for 10 unknown g_{ik}, the divergence condition of (2.99) reduces the number of independent equations to 6. This underdeterminacy of the problem is due to the general covariance of the theory: if g_{ik} is a solution, then so is any tensor transform of g_{ik} obtained through a change of coordinates.

The expression (2.99) follows for any T^{ik} obtained from an action principle by the variation of g_{ik} (see Exercise 33). It is therefore pertinent to ask whether the Einstein tensor can also be derived from an action principle. This problem was solved by Hilbert soon after Einstein proposed his equations of gravitation. Hilbert's problem can be posed as follows. Consider the variation of the term

$$\int_{\mathcal{V}} R\,(-g)^{1/2}\,\mathrm{d}^4 x$$

for $g^{ik} \to g^{ik} + \delta g^{ik}$ with the restriction that δg^{ik} and $\delta g^{ik}{}_{,l}$ vanish on the boundary of $\mathcal{V}$. It can be shown (see Exercise 34 and 35) that

$$\delta \int_{\mathcal{V}} R(-g)^{1/2}\,\mathrm{d}^4 x = \int_{\mathcal{V}} \delta g^{ik}(R_{ik} - \tfrac{1}{2}g_{ik}R)(-g)^{1/2}\,\mathrm{d}^4 x$$

$$= -\int_{\mathcal{V}} \delta g_{ik}(R^{ik} - \tfrac{1}{2}g^{ik}R)(-g)^{1/2}\,\mathrm{d}^4 x. \tag{2.100}$$

Thus it follows that Einstein's equations can be derived from an action principle if we add to $\mathcal{A}$ the term

$$\frac{1}{2\kappa c} \int_{\mathcal{V}} R(-g)^{1/2}\,\mathrm{d}^4 x. \tag{2.101}$$

If to the scalar R we add a constant (2λ, say) that is trivially a scalar, we get a modified set of field equations:

$$R_{ik} - \tfrac{1}{2}g_{ik}R + \lambda g_{ik} = -\kappa T_{ik}. \tag{2.102}$$

We will consider these equations only when we discuss cosmology, since the extra term (the λ-term) has cosmological significance. For the time being we return to (2.98) and relate κ to known physical constants.

2.9.1 Newtonian approximation

We now come to the important question of the magnitude of κ and the relationship between general relativity and Newtonian gravitation. The first hint of the connection between Newtonian gravitation and the present theory was provided by (2.72), where we saw that provided g_{00} did not differ significantly from unity then the difference $(g_{00} - 1)$ is proportional to the Newtonian gravitation potential. We now seek to formalize this relationship and thereby determine κ. We will show that in the so-called slow motion + weak field approximation, general relativity reduces to Newtonian gravitation.

This approximation is specified by the following assumptions:

1. The motions of particles are nonrelativistic: $v \ll c$. In this case we are back to Newtonian mechanics.
2. The gravitational fields are weak in the sense that
$$g_{ik} = \eta_{ik} + h_{ik}, \qquad |h_{ik}| \ll 1. \tag{2.103}$$
The inequality suggests that we ignore powers of h_{ik} higher than 2 in the action principle and higher than 1 in the field equations.
3. The fields change slowly with time. This means we ignore time derivatives in comparison with space derivatives.

Let us now see how the action is simplified under these approximations. First note that with $x^0 = ct$,
$$\begin{aligned} ds^2 &= (\eta_{ik} + h_{ik}) \, dx^i \, dx^k \\ &\approx (1 + h_{00}) c^2 \, dt^2 - v^2 \, dt^2, \end{aligned}$$
that is,
$$ds \approx \left(1 + h_{00} - \frac{v^2}{c^2}\right)^{1/2} c \, dt \approx \left(1 + \tfrac{1}{2} h_{00} - \frac{v^2}{2c^2}\right) c \, dt. \tag{2.104}$$

We next look at the term involving the scalar curvature. The linearized expression for the Riemann tensor [see (2.47)] is
$$R_{iklm} \approx \tfrac{1}{2}(h_{kl,im} + h_{im,kl} - h_{km,il} - h_{il,km}). \tag{2.105}$$
The corresponding values of R can also be calculated. However, care is needed if we are to look at the action principle rather than the field equations in this approximation, for we anticipate quadratic expressions in the h_{ik} to appear in the geometrical term (2.101).

Item 3 above eliminates time derivatives altogether. Further, the ratios of typical space and time displacements are $\delta x^\mu/\delta x^0 = v^\mu/c$, where v^μ are typical Newtonian velocities. Thus h_{00} is more important than any other h_{ik}, at least by the factor (c/v). We will henceforth ignore all other h_{ik} in comparison with h_{00}. We then get

$$g^{00} \approx 1 - h_{00}, \tag{2.106}$$
$$(-g)^{1/2} \approx 1 + \tfrac{1}{2}h_{00} \tag{2.107}$$

and

$$R(-g)^{1/2} \approx -(1 - \tfrac{1}{2}h_{00})\nabla^2 h_{00}. \tag{2.108}$$

Using these relations we finally get the approximate action as

$$\mathscr{A} \approx -\frac{1}{2\kappa} \iint (1 - \tfrac{1}{2}h_{00})\nabla^2 h_{00}\, d^3\mathbf{x}\, dt - \sum \tfrac{1}{2}mc^2 \int h_{00}\, dt$$
$$+ \sum \tfrac{1}{2}m \int v^2\, dt + \text{constant}. \tag{2.109}$$

The constant represents path-independent terms that can be ignored in a variational problem. Here we have dropped particle labels a, b, ... and used the 3-vector $\mathbf{x}$ to denote x^μ $(\mu = 1, 2, 3)$. We can use Green's theorem and ignore surface terms. Thus in

$$\int_{\text{3-volume}} (1 - \tfrac{1}{2}h_{00})\nabla^2 h_{00}\, d^3\mathbf{x} = \int_{\text{2-surface}} (1 - \tfrac{1}{2}h_{00})\nabla h_{00}\, d\mathbf{S}$$
$$+ \tfrac{1}{2}\int_{\text{3-volume}} (\nabla h_{00})^2\, d^3\mathbf{x}$$

we can ignore the surface term. Hence

$$\mathscr{A} \approx -\frac{1}{4\kappa} \iint (\nabla h_{00})^2\, d^3\mathbf{x}\, dt - \sum \tfrac{1}{2}mc^2 \int h_{00}\, dt + \sum \tfrac{1}{2}m \int v^2\, dt. \tag{2.110}$$

Now compare this with the Newtonian action

$$\mathscr{A}_{\text{N}} \approx -\frac{1}{8\pi G} \iint (\nabla \phi)^2\, d^3\mathbf{x}\, dt - \sum m \int \phi\, dt + \sum \tfrac{1}{2}m \int v^2\, dt, \tag{2.111}$$

with ϕ as the gravitational potential. Clearly, (2.110) becomes the same as (2.111) if we put

$$\phi = \tfrac{1}{2}c^2 h_{00}, \qquad \kappa = \frac{8\pi G}{c^4}. \tag{2.112}$$

Thus we have completed our project of evaluating κ and relating the relativistic framework to Newtonian gravitation. Assumptions 1 to 3 above are known as the *Newtonian approximation*. It leads to the linear gravitation theory of Newton, which has wide applications ranging from the tidal phenomenon of the Earth's oceans to motions of planets of the Solar System to motions of stars and galaxies in clusters. Provided these three assumptions hold, general relativity does not add anything new. If assumptions 1 and 3 are dropped but 2 is retained then we are in the domain of the weak field theory of *gravitational radiation*. For, in the weak field limit it is seen that spacetime curvature effects propagate as waves with the speed of light. If this text were devoted primarily to general relativity, we would have discussed this intriguing phenomenon in detail. A few properties of gravitational radiation are outlined in Exercises 37 to 40. To get the full effects of general relativity, however, we must drop all three assumptions and face the nonlinear equations of (2.98) in their most general form. Naturally this is a complicated task, and after more than six decades of this theory there are only a handful of exact solutions of direct physical relevance. We will end this chapter with a discussion of the earliest, simplest and the most important of these solutions.

2.10 The Schwarzschild solution

Shortly after Einstein published his equations of general relativity, Karl Schwarzschild solved them to find the spacetime geometry outside a spherical distribution of matter of mass M. The corresponding problem in Newtonian gravitation yields the solution for the gravitational potential as

$$\phi = -\frac{GM}{r}, \tag{2.113}$$

r being the distance from the centre of the spherical distribution.

At a large distance from the centre, we expect the gravitational field to be weak. So under the Newtonian approximation we expect

$$g_{00} \sim 1 - \frac{2GM}{c^2 r} \tag{2.114}$$

We will now show how the Schwarzschild solution is obtained and how this exact solution takes the above form.

The problem can be solved by making use of symmetry arguments. If the spacetime outside such a spherical distribution is empty, then its geometry should be spherically symmetric about the centre O of the distribution. So we start with the most general form of the line element that fulfils this requirement of spherical symmetry.

It can be shown that the most general form of such a line element is

$$ds^2 = e^{\nu} c^2 \, dt^2 - e^{\lambda} \, dr^2 - r^2 (d\theta^2 + \sin^2 \theta \, d\phi^2), \tag{2.115}$$

where v and λ are functions of r and t. If $v = \lambda = 0$, we get the Minkowski line element in spherical polar space coordinates. The non-Euclidean effects are therefore contained in the functions λ and v. Although in this case r ceases to measure the radial distance from O, it still has the meaning that the spherical surface $r = \text{constant} = r_0$ (for example) has the surface area $4\pi r_0^2$. The arguments leading to (2.115) are group-theoretic, involving the invariance of spacetime under rotations about the point O. The techniques describing these arguments are beyond the scope of this text: see the classic book by Eisenhart listed in the bibliography for these details.

Given the line element (2.115), the next step is to calculate g^{ik}, $(-g)^{1/2}$, and Γ^i_{kl}. We then calculate R_{kl}, which are given by (2.50) and are expressible in the following form:

$$R_{kl} = -\frac{\partial \Gamma^i_{kl}}{\partial x^i} + \frac{\partial^2 (\ln (-g)^{1/2})}{\partial x^k \partial x^l} + \Gamma^m_{kn}\Gamma^n_{lm} - \frac{\partial}{\partial x^n}(\ln(-g)^{1/2}) \cdot \Gamma^n_{kl}.$$

$$(2.116)$$

Since the space outside the distribution is empty, it has $T_{kl} = 0$. Therefore the contraction of the field equations in (2.98) gives $R = 0$, and these equations reduce to

$$R_{ik} = 0. \qquad (2.117)$$

The (00) and (11) components ($r = x^1$) give, after some manipulation, the following equations:

$$e^{-\lambda}\left(\frac{\lambda'}{r} - \frac{1}{r^2}\right) + \frac{1}{r^2} = 0. \qquad (2.118)$$

$$-e^{-\lambda}\left(\frac{v'}{r} + \frac{1}{r^2}\right) + \frac{1}{r^2} = 0. \qquad (2.119)$$

From these we get

$$v' + \lambda' = 0,$$

that is,

$$v + \lambda = f(t).$$

(Here a prime denotes differentiation with respect to r, and an overdot denotes differentiation with respect to t.) The arbitrary function $f(t)$ can, however, be set to equal zero since we still have an arbitrary time transformation

$$t = g(\bar{t})$$

at our disposal, which changes v to

$$\bar{v} = v + 2 \ln \frac{dg}{d\bar{t}}$$

and preserves the form of the line element (2.115). Therefore we can take without loss of generality

$$v + \lambda = 0. \tag{2.120}$$

However, we also have, from $R_{01} = 0$,

$$\dot{\lambda} = 0. \tag{2.121}$$

Thus both λ and $v\ (= -\lambda)$ are functions of r only. The equations (2.118) and (2.119) then yield the solution

$$e^v = e^{-\lambda} = 1 - \frac{A}{r}, \qquad A = \text{constant}. \tag{2.122}$$

However, if we are given the mass of the object to be M, we may use the boundary condition (2.114) to set $A = 2GM/c^2$.

Thus we get out required solution as the line element

$$ds^2 = \left(1 - \frac{2GM}{c^2 r}\right) c^2 \, dt^2 - \left(1 - \frac{2GM}{c^2 r}\right)^{-1} dr^2 - r^2 (d\theta^2 + \sin^2 \theta \, d\phi^2).$$

$$\tag{2.123}$$

This is known as the *Schwarzschild line element*. It turns out that because of the symmetries of the problem the other field equations are automatically satisfied: we only need the (11), (00) and (01) components to arrive at the solution. Also, the solution (2.123) is manifestly static. Thus there is no scope for a dynamic solution such as one involving gravitational radiation, even if our spherical source is expanding, contracting, oscillating. This remarkable result is known as *Birkhoff's theorem*.

We now consider a few implications of this solution.

2.10.1 Experimental tests of general relativity

Most of the present tests of general relativity are based on the Schwarzschild solution, and they seek to measure the fine differences between the predictions of Newtonian gravitation and those of general relativity. These are discussed briefly below.

Before confronting the experimental situation, however, it is necessary to clarify how to attach meanings to measurements in a spacetime that is non-Euclidean. We have already seen that coordinates have no absolute status and hence to rely on them blindly might lead to incorrect results. The Schwarzschild metric (2.123) can be used to illustrate the concept of measurement.

Suppose, for example, an observer is located at a point with $r = $ constant, $\theta = $ constant, $\phi = $ constant. How does he relate the time kept by his watch to the coordinate t? From the principle of equivalence we know that since $d\tau = ds/c$ measures the observer's proper time in a locally inertial frame, being a scalar, it does so in all frames. For our observer, $dr = 0$, $d\theta = 0$, $d\phi = 0$, so from (2.123) we get

$$d\tau = \left(1 - \frac{2GM}{c^2 r}\right)^{1/2} dt.$$

This gives the required answer.

If instead of being stationary the observer is moving radially and his radial coordinate at time t is given by $r = f(t)$, his proper time interval corresponding to dt is

$$d\tau = \left(1 - \frac{2GM}{c^2 r}\right)^{1/2} dt \left\{1 - \left[\frac{(df(t)/dt)c^{-1}}{1 - 2GM/c^2 f(t)}\right]^2\right\}^{1/2}.$$

The gravitational redshift

Consider any static line element – that is, one in which g_{ik} do not depend on $x^0 \equiv ct$. Suppose we have two observers A and B with world lines

$$x^\mu = \text{constant} = a^\mu, b^\mu \qquad (2.124)$$

respectively. Let Γ be a null geodesic from A to B, with parametric equations given by

$$x^i = x^i(\lambda), \qquad (2.125)$$

with $x^\mu(0) = a^\mu$, $x^\mu(1) = b^\mu$, $x^0(0) = ct_A$, $x^0(1) = ct_B$. What does our geodesic correspond to in physical terms?

It describes a light ray leaving observer A at time t_A and reaching observer B at time t_B. Because of the static nature of the line element, we also have another null geodesic solution given by

$$x^\mu = x^\mu(\lambda), \qquad \mu = 1, 2, 3,$$
$$x^0 = x^0(\lambda) + \Delta, \qquad \Delta = \text{constant.} \qquad (2.126)$$

This describes a light ray leaving A at $t_A + \Delta/c$ and reaching B at $t_B + \Delta/c$.

Now in the rest frame of A, the time interval Δ/c corresponds to a proper time interval (measured by A) of

$$\frac{\Delta}{c}[g_{00}(a^\mu)]^{1/2}.$$

If n light waves have left A in this time interval, then the frequency of these waves as measured by A is

$$v_A = \frac{cn}{\Delta} [g_{00}(a^\mu)]^{-1/2}.$$

Since the same *number* of waves are received by B in the corresponding time interval, we get the ratio of frequencies measured by B and A as

$$\frac{v_B}{v_A} = \left[\frac{g_{00}(a^\mu)}{g_{00}(b^\mu)} \right]^{1/2}. \tag{2.127}$$

This is also the ratio of the wavelengths $\lambda_A : \lambda_B$ measured by A and B respectively.

If in the Schwarzschild solution A is an observer located on the surface of a star, at $r = r_s$, say, and B is a distant observer with $r \gg 2GM/c^2$, we get

$$\frac{\lambda_B}{\lambda_A} \approx \left(1 - \frac{2GM}{c^2 r_s} \right)^{-1/2}. \tag{2.128}$$

Thus spectral lines from a massive compact star should be redshifted. For $2GM/c^2 r_s$ small compared to unity, the redshift

$$z = \frac{\lambda_B - \lambda_A}{\lambda_A} \approx \frac{GM}{c^2 r_s}. \tag{2.129}$$

White dwarf stars like Sirius B and 40 Eridani B do show redshifts in the range of 10^{-4} to 10^{-5}, which are of the right order of magnitude. More reliable and quantitatively accurate measurements, however, are possible only in a terrestrial experiment. For example, in 1960 Pound and Rebka measured the change in the frequency of a γ-ray photon emitted by an excited iron nucleus as it fell from a height of 60 to 70 feet. As such a photon falls through a height h, the Newtonian potential increases by gh, where g is the acceleration due to gravity on the Earth's surface. From (2.129) we see that the photon should undergo a blueshift; that is, its frequency increases by a fraction gh/c^2. Although this fraction is as small as 10^{-15}, it can be measured by modern laboratory techniques. The Pound–Rebka experiment and later work have confirmed the gravitational redshift effect.

The perihelion precession of Mercury

If we treat the Sun as the mass M in the Schwarzschild solution and the planets as probes moving in the curved spacetime around the Sun, then in the first approximation each planet will move along a timelike geodesic. The equations of motion of a planet are therefore easily obtained (see Exercise 45). In the Newtonian approximation, the planet describes an ellipse given by its polar equation

$$\frac{l}{r} = 1 + e \cos (\theta - \theta_0). \tag{2.130}$$

Here l is the semi latus retum, e the eccentricity, and θ_0 the direction in which its perihelion (point of closest approach to the Sun) lies.

A more careful solution of the equations (see Exercise 47) shows, however, that θ_0 is not a constant, but changes its magnitude at a steady rate illustrated in Figure 2.6. This precession of perihelion is at a rate

$$n = \frac{6\pi G M_\odot}{lTc^2}, \tag{2.131}$$

where $M_\odot$ = mass of the Sun and T = period of the planet. The value of n is largest for Mercury, which of all the planets has the most eccentric and closest orbit to the Sun. The rate for Mercury, $n \approx 43$ arc second per century, explained exactly the rate of precession, which had long remained unaccounted for in the Newtonian theory.

Recently a more dramatic example of such a precession was observed for the binary star system that houses the pulsar PSR 1913 + 16. Here the gravitational effects are stronger than in the Sun–Mercury system, and the precession rate is as high as 4.23 degrees per year – about 3.6×10^4 times the value of Mercury.

The bending of light

Just as timelike geodesics determine the tracks of planets, we can calculate the track of a light ray by determining the equations of null geodesics. These equations are straightforward to write down (see Exercise 48) and integrate (see Exercise 49). The most dramatic effects arise when a null geodesic goes very close to the mass distribution.

For a light ray grazing the solar limb, the spatial direction of the ray changes by an angle

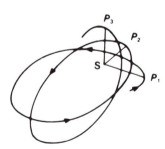

Fig. 2.6 P is the point of closest approach to the Sun (S) in the orbit of Mercury. As successive orbits are completed, the point P advances steadily from P_1 to P_2 and so on. (The advance rate per orbit is actually much smaller than shown here.)

$$\alpha = \frac{4GM_\odot}{c^2 R_\odot} \simeq 1.75 \text{ arc second,} \qquad (2.132)$$

where $R_\odot$ = radius of the Sun. The bending angle is indeed very small, and its measurement was first attempted by Eddington and his colleagues in 1919 at the time of a solar eclipse. (The experiment involves measuring the apparent change in the direction of a star as its line of sight grazes the solar limb. For obvious reasons, optical astronomers have to wait for a total solar eclipse to perform this experiment.) That measurement and subsequent attempts by optical astronomers have yielded somewhat inconclusive results, largely because the refractory effects of the gas in the outer parts of the Sun also contribute to the bending. In the 1970s measurements with microwaves (for which the effects of refraction are very small), confirmed the above bending angle much more precisely with only about 5 per cent experimental error.

It is worth pointing out with regard to the gravitational redshift and the bending of light that since the strict Newtonian theory did not predict any effects of gravity on light, their observation implies a disproof of Newtonian gravitation. However, it is possible to enlarge the scope of the Newtonian framework and argue that light is made of particles (photons) that are also subject to the inverse-square law. (Indeed, Newton himself had speculated on this possibility.) We can then show that such an enlarged theory gives (2.129) for gravitational redshift (see Exercise 52), the same as in relativity, but half the relativistic value for the bending of light (see Exercise 51). The observed bending of microwaves therefore rules out such a theory.

Radar echo delay

Just as the direction of a light ray is altered by the Sun's gravity, so is its apparent travel time. This effect can also be calculated in a straightforward manner. In the 1970s measurements were made by bouncing radar signals emitted from the spacecrafts Mariner 6 and 7 off the surface of the Earth as the signals grazed the solar limb. The expected delays on the order of 200 microseconds were observed within 3 per cent error bars.

The equality of inertial and gravitational mass

An important consequence of the principle of equivalence is the equality of inertial and gravitational mass. A little thought will convince us that Galileo's experiment from the Leaning Tower of Pisa, which demonstrated that all bodies fall freely with equal rapidity, is an essential part of

Einstein's thought experiment involving the freely falling lift. Both experiments are possible because the same quantity enters the law of motion as inertial mass and the law of gravitation as gravitational mass.

Recent experiments with lunar laser ranging have been successful in measuring the distance of the Moon from the Earth within a few centimeters. Such experiments also demonstrate that the Moon moves around the Earth as predicted by the equations of general relativity. In particular, these experiments ruled out certain alternative theories of gravitation, like the Brans–Dicke theory, that allow for the variation of inertial mass with distance from another mass.

Laboratory experiments of the torsion balance type have been conducted very accurately with different materials to establish this equality with high sensitivity. Such experiments place stringent upper limits on the possible presence of a 'fifth force' operating at a range of a few metres.

Precession of a gyroscope

Although the Schwarzschild solution describes the gravitational effects of the Sun or the Earth with great accuracy, there is scope for further improvement. For instance, a rotating mass would introduce a $d\phi \, dt$ term in the metric. Although the effects of such terms are very small for the Earth or the Sun, modern technology can measure them.

One experiment that can measure the effect of a rotating mass makes use of gyroscopes. The axis of a gyroscope sent on an equatorial orbit around the Earth will slowly precess. An estimated rate of precession of ~ 7 arc seconds per year can be detected with present technology, and such an experiment has been planned.

Gravitational radiation

Calculations based on the weak field theory show that the magnitude of gravitational radiation from terrestrial apparatus is extremely small and beyond the scope of present technology. Celestial objects, however, can and do emit appreciable quantities of gravitational radiation, and attempts are being made to devise detectors to measure them. Typical sources of gravity waves are supernova explosions, binary stars, and, possibly, primordial developments in the very early universe (see Chapter 6).

2.10.2 Black holes

All the effects discussed above are those of *weak* gravity. For the Sun the ratio $2GM_\odot/c^2 R_\odot$ is as low as $4 \cdot 10^{-6}$, and for the Earth it is even smaller. Can we visualize an object that is so compact that its mass M and radius R are related by

$$\frac{2GM}{c^2 R} \sim 1?$$

A glance at the Schwarzschild solution will show that for such an object the spacetime geometry near the surface will be markedly non-Euclidean. Several unexpected results arise if such objects exist (see, for example, Exercises 53 and 54).

An object whose Schwarzschild radius R satisfies the condition

$$R \leqslant \frac{2GM}{c^2} \qquad (2.133)$$

is called a *black hole*. As its name implies, such an object is dark because its strong gravity traps light and prevents it from getting away. A glance at (2.128) shows that the gravitational redshift of a black hole is infinite. Since redshift, z, implies a decrease in the energy of a light photon by the factor $(1 + z)^{-1}$, no photon with finite energy can come out of a black hole.

Considerable work was done between 1965 and 1975 in attempts to study black holes and their weird properties. Since the emphasis of this text is on cosmology, we must be content with the superficial description given here. It is to a discussion of cosmology that we must now proceed.

Exercises

1 Verify that a piece of string stretched across the spherical globe of the Earth lies along the arc of a great circle, which is a 'straight line' on the spherical surface. Check whether lines of latitude and longitude are straight.

2 One of the 'self-evident truths' on which Euclid's geometry is based is the so-called parallel postulate. This states that given a straight line l and a point P not on it, one and only one line parallel to l can be drawn through P (that is, a line that does not meet l if both lines are extended indefinitely.) What happens to this postulate in the geometry on the surface of a sphere and on a saddle-shaped surface?

3 Calculate tangent vectors at typical points on the curves in spacetime given by the following relations ($x^1 = x$, $x^2 = y$, $x^3 = z$, and $x^0 = ct$):

(a) $x = ct_0 \cos(t/t_0)$, $y = ct_0 \sin(t/t_0)$,
     $z = ct$, $t_0 = $ constant.
(b) $x = 0$, $y = 0$, $z^2 - c^2 t^2 = 0$.
(c) $x = ct \cos(t/t_0)$, $y = ct \sin(t/t_0)$, $z = $ constant, $t_0 = $ constant.
Determine whether these tangent vectors are spacelike, timelike, or null in Minkowski spacetime.

4   Calculate the components of the normals of the following surfaces and determine whether they are spacelike, timelike, or null in Minkowski spacetime ($x^1 = x$, $x^2 = y$, $x^3 = z$, $x^0 = ct$):
      (a) $x^2 + y^2 + z^2 - \lambda^2 t^2 = $ constant, $\lambda = $ constant.
      (b) $x^2 + y^2 - \lambda^2 t^2 = $ constant, $\lambda = $ constant.
      (c) $x^2 - \lambda^2 t^2 = $ constant, $\lambda = $ constant.

5   Which of the following expressions are invalid with respect to the summation convention? Simplify those expressions that are valid.
      (a) $A_{ij} B^{jk} A_{jl}$. (b) $g_{ik} g^{ik}$. (c) $R_{ik} g_{ik}$. (d) $e_{iklm} e^{iklm}$.
      (e) $T^{ik} g_l^k$.

6   A_{ik} is a tensor such that the matrix $\|A_{ik}\|$ is nonsingular. Show that the components of the inverse matrix transform as a tensor. (An example of this result is the tensor g^{ik}.)

7   Show that the property of symmetry or antisymmetry with respect to indices of a tensor is invariant under coordinate transformations.

8   Construct, with the help of g_{ik} only, a fourth-rank tensor that is symmetric with respect to the interchange of any two of its indices.

9   Verify that if F_{ik} is an antisymmetric tensor field, then $F_{ik,l} + F_{kl,i} + F_{li,k}$ is a third-rank tensor.

10   Prove the *quotient law* in the following form: if $A_{ik} B^k$ is a vector for any arbitrary vector B^k, then A_{ik} must transform as a tensor.

11   Use two-dimensional polar coordinates (r, θ) on a Euclidean plane. Let A_r and A_θ be the radial and transverse components of a vector A at a typical point P chosen with respect to locally Cartesian axes, with directions coinciding with $\theta = $ constant and $r = $ constant, respectively. Show that a parallel transport of the vector at P to a neighbouring point $Q(r + \delta r, \theta + \delta \theta)$ gives the two components of the vector at Q as

$$A_r + \delta\theta A_\theta, \qquad A_\theta - \delta\theta A_r.$$

12   By using the requirement that $B_{i;k}$ transforms as a tensor, deduce the transformation relation (2.35) for Γ_{kl}^i.

13   Deduce the form (2.36) for $A^i{}_{;k}$, using (2.34) for $B_{l;k}$ and assuming

that the covariant derivative of a scalar equals its ordinary derivative.

14 Suppose two metrics are defined on the same spacetime, and let Γ^i_{kl} and $\bar{\Gamma}^i_{kl}$ be the two corresponding Riemannian affine connections. Deduce that the quantities

$$Q^i_{kl} = \Gamma^i_{kl} - \bar{\Gamma}^i_{kl}$$

transform as a tensor. (The coordinates are the same in the two cases.)

15 Show that to arrive at a locally inertial system it is necessary to have $\Gamma^i_{kl} = \Gamma^i_{lk}$.

16 Deduce the relations shown in (2.41) from first principles.

17 Show that for a scalar field ϕ, the wave operator takes the form

$$\Box \phi = g^{ik}\phi_{;ik} = \frac{1}{(-g)^{1/2}} \frac{\partial}{\partial x^k}\left((-g)^{1/2} g^{ik} \frac{\partial\phi}{\partial x^i}\right).$$

18 The line element on the surface of a sphere of radius a in Euclidean space is given by

$$ds^2 = a^2(d\theta^2 + \sin^2\theta\, d\phi^2).$$

For this space calculate Γ^i_{kl}; $i, k, l = 1, 2$ (with $\theta = x^1$ and $\phi = x^2$) and verify the result discussed in the text regarding the change of direction of a vector under parallel displacement around the 3-right-angled triangle ABC.

19 Prove from first principles that $B_{i;nk} - B_{i;kn} = R_i{}^m{}_{kn}B_m$.

20 For an antisymmetric tensor field F_{ik}, show that $F^{ik}_{;ik} = 0$.

21 A_i is a vector field satisfying $A^i{}_{;i} = 0$. If $F_{ik} = A_{k;i} - A_{i;k}$, show that

$$F^{ik}_{;i} = g^{mn}A^k{}_{;mn} + R^k_m A^m.$$

22 Deduce the form (2.47) taken by R_{iklm} in the locally inertial coordinate system. Use the same coordinates to deduce the symmetric nature of R_{ik}.

23 Show by direct enumeration that the number of algebraically independent components of R_{iklm} is 20.

24 Using a locally inertial coordinate system, deduce the Bianchi identities. From these identities in their covariant form show that

$$R^l_{k;l} = \tfrac{1}{2} R_{,k}.$$

25 Show that the first integral of (2.57) is

$$g_{ik}\frac{dx^i}{d\lambda}\frac{dx^k}{d\lambda} = \text{constant}.$$

26   Show that the change in the direction of a vector under a parallel
      displacement around a closed infinitesimal curve can be expressed
      in terms of the Riemann tensor and the area spanned by the curve.

27   Let a bundle of geodesics be specified by a parameter μ, so that a
      typical point on the μ-geodesic has the coordinates $x^k(\lambda, \mu)$, λ
      being the affine parameter. The vector $v^k = \partial x^k / \partial \mu$ denotes the
      rate of deviation from one geodesic to another across the bundle.
      Deduce the following relations:
      (a) $v^k_{;l} u^l = u^k_{;l} v^l$, where $u^k = \partial x^k / \partial \lambda$.
      (b) $d^2 v^k / d\lambda^2 + R^k_{lmn} u^l v^m u^n = 0$.
      The latter is the equation of *geodetic deviation*.

28   Construct a Newtonian analogue of geodetic deviation by compar-
      ing the deviation of two test particles falling on the (spherical)
      Earth along two neighbouring radial trajectories.

29   Verify the existence of the $\frac{1}{3}$ factor in (2.84).

30   Show that the results of (2.87) are based on standard kinetic
      theory.

31   Calculate the form of the energy tensor for a plane electromagnetic
      wave.

32   Show that if the Lagrangian density of a physical interaction in
      curved spacetime is L, so that its contribution to action is

$$\int L \, (-g)^{1/2} \, d^4x,$$

      then, provided L depends on the geometry only through g_{ik}, $g_{ik,l}$
      the energy tensor of the interaction is given by

$$T^{ik} = - \frac{2c}{(-g)^{1/2}} \left[\frac{\partial L(-g)^{1/2}}{\partial g_{ik}} - \left(\frac{\partial L(-g)^{1/2}}{\partial g_{ik,l}} \right)_{,l} \right].$$

33   Show that from the scalar nature of L in Exercise 32 above it is
      possible to deduce that

$$T^{ik}_{;k} = 0.$$

      Hint: Use (2.95) and the fact that an infinitesimal change in the
      coordinates $x^i \to x^i + \xi^i$ gives $\delta g_{ik} = -(\xi_{i;k} + \xi_{k;i})$.

34   Show that under $g_{ik} \to g_{ik} + \delta g_{ik}$, $\delta \Gamma^i_{kl}$ transforms as a tensor.

35   Show that

$$\delta \int_V R(-g)^{1/2} \, d^4x = \int_V (R_{ik} - \tfrac{1}{2} g_{ik} R) \delta g^{ik} (-g)^{1/2} \, d^4x$$

      for variations of the metric that vanish along with their derivatives on the boundary of V. *Hint*: Write $R = R_{ik}g^{ik}$ so that $\delta R = \delta R_{ik} + R_{ik}\delta g^{ik}$. Use a locally inertial coordinate system to deduce that

$$(-g)^{1/2}g^{ik}\,\delta R_{ik} = -(-g)^{1/2}[(g^{ik}\,\delta\Gamma^l_{ik})_{;l} - (g^{il}\,\delta\Gamma^k_{ik})_{;l}]$$
$$= (-g)^{1/2}w^k{}_{;k},$$

where w^k (from Exercise 34) is a vector. Then use Green's theorem.

36　From the Newtonian approximation of Einstein's field equations and the geodesic equations, deduce Poisson's equation and the Newtonian equations of motion in a gravitational field.

37　Show that in the weak field approximation for gravitational radiation it is possible to make a coordinate transformation to ensure a gauge condition,

$$\psi^k_{i,k} = 0,$$

where

$$\psi^k_i = h^k_i - \tfrac{1}{2}h^l_l\,\delta^k_i.$$

Further, show that the ψ^k_i satisfy the wave equation in flat space

$$\Box\psi^k_i = -\frac{16\pi G}{c^4}\,T^k_i.$$

38　Compare the above linearized theory of gravitational waves with the electromagnetic theory of Maxwell. Construct plane wave solutions in the case $T^k_i = 0$.

39　In a plane wave solution of the gravitational wave equation, estimate the components of the Riemann tensor. Show that in principle a gravitational wave can be detected by the measurement of the components of the Riemann tensor with the help of the equation of geodetic deviation.

40　Just as the second time derivative of a changing electric dipole moment acts as the elementary source of electromagnetic radiation, the third time derivative of a changing quadrupole moment (of mass) acts as the simplest source of gravitational radiation. Use this result and dimensional arguments to show why the emission of gravitational waves is energetically very weak in laboratory conditions.

41　Show that if we apply the line element (2.115) to the interior $(r \leqslant r_s)$ of the spherically symmetric matter distribution, we get from the (00) component of the field equations

$$e^{-\lambda} = 1 - \frac{2GM(r)}{c^2},$$

where $M(r) = \int_0^r 4\pi\rho^2 T_0^0 \,d\rho$. The quantity $M(r_s)$ may be identified with the gravitational mass M that appears in the exterior solution $(r > r_s)$.

42   Discuss gravitational redshift and blueshift.

43   Calculate the ratio $GM/c^2 r_s$ in order that the entire visible spectrum (4000 Å to 8000 Å) in the light from the surface of the spherical object is just about redshifted out.

44   Calculate the proportionate increase in the frequency of a γ-ray photon as it descends a vertical height of 100 metres to the surface of the Earth.

45   Show that the equations of a timelike geodesic in the Schwarzschild spacetime are given by

$$\frac{d^2t}{ds^2} + \frac{dv}{dr}\frac{dr}{ds}\frac{dt}{ds} = 0,$$

$$\frac{d^2\theta}{ds^2} + \frac{2}{r}\frac{dr}{ds}\frac{d\theta}{ds} - \sin\theta\cos\theta\left(\frac{d\phi}{ds}\right)^2 = 0.$$

$$\frac{d^2\phi}{ds^2} + \frac{2}{r}\frac{dr}{ds}\frac{d\phi}{ds} + 2\cot\theta\,\frac{d\phi}{ds}\frac{d\theta}{ds} = 0,$$

$$\frac{d^2r}{ds^2} - \frac{1}{2}\frac{dv}{dr}\left(\frac{dr}{ds}\right)^2 - re^v\left(\frac{d\theta}{ds}\right)^2 - r^2\sin^2\theta e^v\left(\frac{d\phi}{ds}\right)^2$$
$$+ \frac{1}{2}e^{2v}\frac{dv}{dr}\left(\frac{dt}{ds}\right)^2 c^2 = 0,$$

with $e^v = 1 - \dfrac{2GM}{c^2 r}$.

46   Show that the equations in Exercise 45 may be integrated as follows without loss of generality:

$$\theta = \frac{\pi}{2}, \qquad \frac{dt}{ds} = \left(1 - \frac{2GM}{c^2 r}\right)^{-1} E, \qquad r^2\frac{d\phi}{ds} = h,$$

where E and h are constants of motion. What other integral of these equations is known?

47   Show that for the Sun–Mercury system an approximate solution of the equations of Exercise 46 is provided by

$$r = l[1 + e\cos(\phi - \phi_0)]^{-1}$$

where ϕ_0 is a slowly increasing function of time. Evaluate $d\phi_0/dt$ and relate the result to the observed precession of the perihelion of Mercury.

48   Show that null geodesics (describing light rays, for example) in the Schwarzschild spacetime are given by the following equations:

$$\frac{d^2t}{d\lambda^2} + \frac{dv}{dr}\frac{dr}{d\lambda}\frac{dt}{d\lambda} = 0,$$

$$\frac{d^2\theta}{d\lambda^2} + \frac{2}{r}\frac{dr}{d\lambda}\frac{d\theta}{d\lambda} - \sin\theta\cos\theta\left(\frac{d\phi}{d\lambda}\right)^2 = 0,$$

$$\frac{d^2\phi}{d\lambda^2} + \frac{2}{r}\frac{dr}{d\lambda}\frac{d\phi}{d\lambda} + 2\cot\theta\frac{d\phi}{d\lambda}\frac{d\theta}{d\lambda} = 0,$$

$$c^2\left(\frac{dt}{d\lambda}\right)^2 e^v = e^{-v}\left(\frac{dr}{d\lambda}\right)^2 + r^2\left(\frac{d\theta}{d\lambda}\right)^2 + r^2\sin^2\theta\left(\frac{d\phi}{d\lambda}\right)^2.$$

49   Show how to obtain first integrals of the equations in Exercise 48, analogous to those of Exercise 46.

50   Show that the following is an approximate solution of the null geodesic equations:

$$r\cos\phi = r_s - \frac{GM}{c^2 r_s}(r\cos^2\phi + 2r\sin^2\phi).$$

Interpret this solution as describing bending of light by a massive object.

51   Considering the light photon as a projectile moving under the Newtonian inverse-square law, calculate the bending of light produced by a massive object. Show that the net bending is half that given by general relativity.

52   Show how Newtonian gravitation can be adapted suitably to describe the phenomenon of gravitational redshift.

53   A star of solar mass slowly contracts from initial radius $R_\odot$. Show how its gravitational redshift increases as a function of its radial coordinate r_s. What happens when $r_s < 2GM_\odot/c^2$?

54   A light ray describes a circular trajectory around a black hole. Show how this is possible and calculate the size of the orbit.

3

From relativity to cosmology

3.1 Historical background

In 1915 Einstein put the finishing touches to the general theory of relativity. The Schwarzschild solution described in Chapter 2 was the first physically significant solution of the field equations of general relativity. It showed how spacetime is curved around a spherically symmetric distribution of matter. The problem solved by Schwarzschild is basically a local problem, in the sense that the distortions of spacetime geometry from the Minkowski geometry of special relativity gradually diminish to zero as we move further and further away from the gravitating sphere. This result can be easily verified from the line element (2.123) by letting the radial coordinate r go to infinity. In technical jargon, a spacetime satisfying this property is called *asymptotically flat*. In general any spacetime geometry generated by a local distribution of matter is expected to have this property. Even from Newtonian gravity we expect an analogous result: that the gravitational field of a local distribution of matter will die away at a large distance from the distribution. Can the universe be approximated by a local distribution of matter?

Einstein felt that the answer to the above question would be in the negative. Rather, he expected the universe to be filled with matter, however far we are able to probe it. A Schwarzschild-type solution cannot therefore provide the correct spacetime geometry of such a distribution of matter. Since we can never get away from gravitating matter, the concept of asymptotic flatness must break down. A new type of solution is therefore needed to describe a universe filled everywhere with matter. Einstein published such a solution in 1917.

Before we consider Einstein's solution, it is worth noting that more than two centuries earlier Newton also had attempted a solution describing a matter-filled universe of infinite extent. A highly symmetric distribution of matter does lead to a solution in Newtonian gravity. Imagine, for example, a uniform distribution of matter filling the infinite Euclidean space. An observer viewing the universe from any vantage point will find that it looks the same in all directions and that it presents the same aspect from all vantage points. These two properties are known as *isotropy* and *homogeneity*, and they will turn out to play simplifying roles in relativistic cosmology as well. Newton found that such a universe would be static, for, any particle of matter is being attracted equally in all directions, so it should stay put where it is.

On the other hand, homogeneity precludes any pressure gradients in the universe. And we know that any finite distribution of pressure-free matter would tend to shrink under its own gravity. Stars are able to maintain a stationary shape only because they have large enough pressure gradients inside to withstand their own gravity. Clearly, in going from a finite to an infinite universe something new has entered the argument: the boundary conditions at infinity. Considerable ambiguity arises in Newtonian theory when we try to interpret these boundary conditions.

Newton also found his solution to be unstable: any local inhomogeneity would precipitate gravitational contraction that would tend to augment the local inhomogeneity. Newton compared the instability of the solution to that of a set of needles finely balanced on their points.

Nevertheless, in 1934 E. A. Milne and W. H. McCrea showed how some of the problems of Newtonian cosmology can be resolved. The reader interested in this approach may find some properties of Newtonian cosmology outlined in Exercises 1 to 3 at the end of this chapter and also in Chapter 4.

We will now return to Einstein's solution of 1917.

3.2 The Einstein universe

It is evident from the field equations of general relativity derived in Chapter 2 that their solution in the most general form – the solution of an interlinked set of nonlinear partial differential equations – is beyond the present range of techniques available to applied mathematics. It is necessary to impose simplifying symmetry assumptions in order to make any progress towards a solution. Just as Schwarzschild assumed spherical symmetry in his local solution, Einstein assumed homogeneity and isotropy in his cosmological problem. He further assumed, like Schwarzschild, that spacetime is static. This enabled him to choose a time coordinate t such that the line element of spacetime could be described by

$$ds^2 = c^2\, dt^2 - \alpha_{\mu\nu}\, dx^\mu\, dx^\nu, \tag{3.1}$$

where $\alpha_{\mu\nu}$ are functions of space coordinates $x^\mu (\mu, \nu = 1, 2, 3)$ only.

Note that constraint of homogeneity implies that the coefficient of dt^2 can only be a constant, which we have normalized to c^2. Similarly, the condition of isotropy tells us that there should be no terms of the form $dt\, dx^\mu$ in the line element. This can be seen easily in the following way. If we had terms like $g_{0\mu}\, dt\, dx^\mu$ in the line element, then spatial displacements dx^μ and $-dx^\mu$ would contribute oppositely to ds^2 over a small time interval dt, and such directional variation is forbidden by isotropy. Can we say anything more about $\alpha_{\mu\nu}$?

Einstein believed that the universe has so much matter as to 'close' the space. And this assumption led him to a specific form for $\alpha_{\mu\nu}$. We will now elaborate a little on the notion of closed space and on how to arrive at $\alpha_{\mu\nu}$. Let us begin with examples from lower-dimensional spaces.

As the simplest example of an open space is the Euclidean straight line extending indefinitely in both directions, we can use a real variable r to denote a typical point on the line with $-\infty < r < \infty$. Figure 3.1(a) shows such a straight line. Figure 3.1(b) shows an example of a closed curve Σ_1. It has no boundary, but if we use a real variable r to denote points on the curve then we will find that a finite range of r will suffice. If we go beyond this range we will begin to go over the curve again and again. A familiar simple example of this is the circle S_1 of radius S shown in Figure 3.1(c). If we use the Euclidean measure of distance to locate a point and denote by r the distance of this point from a fixed point N, we find that the range $0 \leqslant r < 2\pi S$ describes all the points on the circle.

While both the curves in Figure 3.1(b) and 3.1(c) are closed, the circle evidently has more symmetries than the curve Σ_1. This can be demonstrated as follows. If we take a small section (an arc) of the circle and slide it along the circle, it will always lie flush on it. We cannot do the same for the curve Σ_1. We can express this by saying that the circle S_1 describes homogeneous space, while the curve Σ_1 does not.

Figure 3.2 illustrates the corresponding situation in two dimensions. Two coordinates r and ϕ $(0 \leqslant r < \infty, 0 \leqslant \phi < 2\pi)$ are needed to locate a point on the Euclidean plane of Figure 3.2(a). The surface Σ_2 shown in Figure 3.2(b) and the sphere S_2 of radius S shown in Figure 3.2(c) are closed surfaces, of which S_2 is homogenous but Σ_2 is not. This latter property can be easily verified by our technique of sliding a small section of each surface along itself.

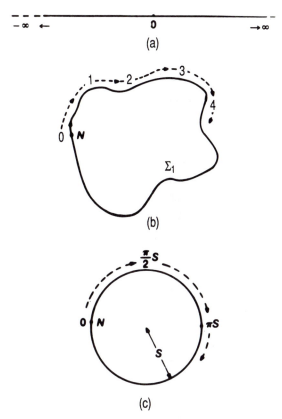

Fig. 3.1 Curves in one-dimensional space. (a) A straight line extending from $-\infty$ to ∞. This is an example of open space. (b) A closed curve Σ_1. Starting from a point N on it as the origin, we can use the length r along the curve to label points on it. If the length of the curve is L, when $r = L$ we come back to the starting point. This is a closed space. (c) A closed space S_1 that is homogeneous: it is a circle. If it has radius S, $L = 2\pi S$.

There is another symmetry inherent in the spherical surface, which can be demonstrated as follows. At any point O on it draw a small arc lying on the surface and then rotate this arc around the point O, trying all the while to keep the arc lying flush on the surface. Again the spherical surface S_2 allows you to do this, but Σ_2 does not. This means that the surface S_2 shows isotropy about O.

We can now see how to construct the homogenous and isotropic closed space of three dimensions that Einstein wanted for his model of the universe. It is S_3, the 3-surface of a four-dimensional hypersphere of radius S. The equation of such a 3-surface is given in Cartesian coordinates x_1, x_2, x_3, and x_4 by

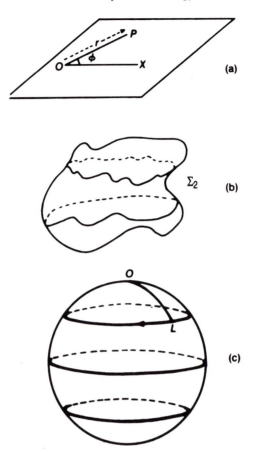

Fig. 3.2 (a) The plane is an open two-dimensional space. From any point O on it draw the straight line OX in any direction in the plane. The coordinates (r, ϕ) in the illustration show how to specify any point P on the plane. (b) An arbitrary closed surface Σ_2. (c) A closed surface S_2 that is homogeneous and isotropic. It is a sphere. Take any point O on S_2 and draw a small arc of a great circle OL lying on S_2. As OL is rotated around O, the point L moves along a small circle on S_2 and the arc always stays on S_2. This is an example of isotropy: as seen from O, the surface S_2 shows no preferential direction.

$$(x_1)^2 + (x_2)^2 + (x_3)^2 + (x_4)^2 = S^2. \tag{3.2}$$

To use coordinates *intrinsic* to the surface we define

$$x_4 = S \cos \chi, \qquad x_1 = \sin \chi \cos \theta, \qquad x_2 = S \sin \chi \sin \theta \cos \phi,$$
$$x_3 = S \sin \chi \sin \theta \sin \phi. \tag{3.3}$$

The spatial line element *on* the surface S_3 is therefore given by

$$\begin{aligned} \mathrm{d}\sigma^2 &= (\mathrm{d}x_1)^2 + (\mathrm{d}x_2)^2 + (\mathrm{d}x_3)^2 + (\mathrm{d}x_4)^2 \\ &= S^2[\mathrm{d}\chi^2 + \sin^2 \chi(\mathrm{d}\theta^2 + \sin^2 \theta \,\mathrm{d}\phi^2)]. \end{aligned} \tag{3.4}$$

The ranges of θ, ϕ, and χ are given by

$$0 \leqslant \chi \leqslant \pi, \qquad 0 \leqslant \theta \leqslant \pi, \qquad 0 \leqslant \phi \leqslant 2\pi. \tag{3.5}$$

At this stage it is worth pointing that there are two alternatives open to us. The first is what we have tacitly taken for granted, that χ takes the entire range $0 \leqslant \chi \leqslant \pi$, and this gives us what is commonly known as *spherical space*. If, however, we identify the antipodal points, the space is called *elliptical space*.

Another way to express $d\sigma^2$ is through coordinates r, θ, ϕ, with $r = \sin\chi(0 \leqslant r \leqslant 1)$. In elliptical space r runs through this range once: in spherical space it does so twice:

$$d\sigma^2 = S^2\left[\frac{dr^2}{1 - r^2} + r^2(d\theta^2 + \sin^2\theta\, d\phi^2)\right]. \tag{3.6}$$

The constant S is called the 'radius' of the universe. The line element for the Einstein universe is therefore given by

$$\begin{aligned}
ds^2 &= c^2\, dt^2 - d\sigma^2 \\
&= c^2\, dt^2 - S^2[d\chi^2 + \sin^2\chi(d\theta^2 + \sin^2\theta\, d\phi^2)] \tag{3.7} \\
&= c^2\, dt^2 - S^2\left[\frac{dr^2}{1 - r^2} + r^2(d\theta^2 + \sin^2\theta\, d\phi^2)\right].
\end{aligned}$$

Note that we have derived the line element (3.7) entirely from the various assumptions of symmetry. The field equations have not yet been used. We will now see what happens when we use the above line element to compute the left-hand side of Einstein's equations.

This is easily done with the machinery developed in Chapter 2. We write $x^0 = ct$, $x^1 = r$, $x^2 = \theta$, $x^3 = \phi$, so that

$$g_{00} = 1, \qquad g_{11} = -\frac{S^2}{1 - r^2}, \qquad g_{22} = -S^2 r^2, \qquad g_{33} = -S^2 r^2 \sin^2\theta.$$

$$g^{00} = 1, \qquad g^{11} = -\frac{1 - r^2}{S^2}, \qquad g^{22} = -\frac{1}{S^2 r^2}, \qquad g^{33} = -\frac{1}{S^2 r^2 \sin^2\theta}.$$

Elementary calculus then tells us that the only nonzero components of Γ^i_{kl} are the following:

$$\Gamma^1_{11} = \frac{r}{1 - r^2}, \qquad \Gamma^2_{12} = \Gamma^3_{13} = \frac{1}{r}, \qquad \Gamma^1_{22} = -r(1 - r^2),$$

$$\Gamma^1_{33} = -r(1 - r^2)\sin^2\theta, \qquad \Gamma^2_{33} = -\sin\theta\cos\theta, \qquad \Gamma^3_{23} = \cot\theta.$$

Next, using the formulae given in the last chapter, we find the following nonzero components of the Einstein tensor:

$$R^0_0 - \tfrac{1}{2}R = -\frac{3}{S^2}, \tag{3.8}$$

$$R_1^1 - \tfrac{1}{2}R = R_2^2 - \tfrac{1}{2}R = R_3^3 - \tfrac{1}{2}R = -\frac{1}{S^2}. \tag{3.9}$$

To complete the field equations, Einstein used the energy tensor for dust derived in (2.83). For dust at rest in the above frame of reference, u^i has only one component, the time component, nonzero. We therefore get

$$T_0^0 = \rho_0 c^2,$$
$$T_1^1 = T_2^2 = T_3^3 = 0. \tag{3.10}$$

Thus the two equations (3.8) and (3.9) lead to two independent equations:

$$-\frac{3}{S^2} = -\frac{8\pi G}{c^2}\rho_0, \qquad -\frac{1}{S^2} = 0. \tag{3.11}$$

Clearly, no sensible solution is possible from these equations, thus suggesting that no static homogeneous isotropic and dense model of the universe is possible under the Einstein equations.

It was his inability to generate such a model that led Einstein to modify his equations (2.98) to (2.102), thus introducing the now famous (or infamous) λ-term. If we introduce this additional constant into the picture, our equations in (3.11) are modified to

$$\lambda - \frac{3}{S^2} = -\frac{8\pi G}{c^2}\rho_0 \tag{3.12}$$

and

$$\lambda - \frac{1}{S^2} = 0. \tag{3.13}$$

We now do have a sensible solution. We get

$$S = \left(\frac{1}{\lambda}\right)^{1/2} = \frac{c}{2(\pi G\rho_0)^{1/2}}. \tag{3.14}$$

Einstein considered this solution as justifying his conjecture that with sufficiently high density it should be possible to 'close' the universe. In (3.14) we have the radius S of the universe as given by the matter density ρ_0, with the result that the larger the value of ρ_0, the smaller is the value of S. However, if λ is a given universal constant like G, both ρ_0 and S are determined in terms of λ (as well as G and c). How big is λ?

In 1917 very little information was available about ρ_0, from which λ could be determined. The value of

$$S \approx 10^{26} - 10^{27} \text{ cm}$$

quoted in those days is therefore only of historical interest. If we take ρ_0

as $\sim 10^{-31}$ g cm^{-3} as the rough estimate of mass density in the form of galaxies (see Chapter 9), we get $S \sim 10^{29}$ cm and $\lambda \approx 10^{-58}$ cm^{-2}.

The λ-term introduces a force of repulsion between two bodies that increases in proportion to the distance between them. The above value of λ is too small to make any detectable difference from the prediction of standard general relativity (that is, with $\lambda = 0$) in any of the Solar System tests mentioned in Chapter 2. Thus the Einstein universe faced no threat from the local tests of gravity. The model, however, did not survive much longer than a decade, for reasons discussed below.

3.3 The expanding universe

In the late nineteenth century the philosopher and scientist Ernst Mach raised certain conceptual objections to Newton's laws of motion. Mach critically examined the role of a background against which motion is to be measured and argued that unless there is a material background it is not possible to attach any meaning to the concepts of rest or motion. Einstein was greatly influenced by Mach's discussion. The Einstein universe described above includes matter-filled space and thus a background of distant matter against which a local observer can measure motion and formulate laws of mechanics. In fact, as we have just seen, the density of matter determines the precise geometrical nature of spacetime in the Einstein model.

Einstein believed this to be a unique feature of general relativity. He felt that the presence of matter was essential to have a meaningful spacetime geometry. However, his expectation that general relativity can yield only such matter-filled spacetimes as solutions of the field equations was proved wrong shortly after the publication of his paper in 1917. For in 1917 W. de Sitter published another solution of the field equations in (2.102) with the line element given by

$$ds^2 = c^2 \left(1 - \frac{H^2 R^2}{c^2}\right) dt^2 - \frac{dR^2}{1 - \left(\dfrac{H^2 R^2}{c^2}\right)} - R^2(d\theta^2 + \sin^2 \theta \, d\phi^2),$$

$$(3.15)$$

where H is a constant related to λ by

$$\lambda = \frac{3H^2}{c^2}.$$

$$(3.16)$$

The remarkable feature of the de Sitter universe is that *it is empty*.

Moreover, although the above coordinates give the impression that the universe is static, it is possible to find a new set of coordinates (t, r, θ, ϕ) in terms of which the line element (3.15) takes the form

$$\mathrm{d}s^2 = c^2\,\mathrm{d}t^2 - \mathrm{e}^{2Ht}[\mathrm{d}r^2 + r^2(\mathrm{d}\theta^2 + \sin^2\theta\,\mathrm{d}\phi^2)]. \qquad (3.17)$$

It is easy to verify that test particles with constant values of (r, θ, ϕ) follow timelike geodesics in this model. Thus the proper separation between any two particles measured at a given time t increases with time as e^{Ht}. That is, these particles are all moving apart from one another.

However, these particles have no material status. They have no masses and they do not influence the geometry of spacetime. In the dynamic sense the universe is empty, although in the kinematic sense it is expanding. As Eddington once put it, the de Sitter universe has motion without matter, in contrast to the Einstein universe, which has matter without motion.

The de Sitter universe showed, however, that empty spacetimes could be obtained as solutions of general relativity. For reasons discussed above, a universe of this type fails to meet Mach's criterion that there should be a background of distant matter against which local motion can be measured. Although the property of emptiness of the de Sitter universe was embarrassing, its property of expansion turned out to contain the germ of the truth. For by the end of the third decade of this century, the observations of Hubble and Humason indicated that the universe is not static but is indeed expanding.

Chapter 1 summarized these observations. The phenomenon of nebular redshift observed by Hubble and Humason in the 1920s has now been observed in practically all extragalactic objects. As mentioned in section 1.8, a Newtonian interpretation of such redshifts involves the Doppler effect. How can we express this phenomenon in the language of general relativity? Can we generate models of the universe that combine de Sitter's notion of expansion with Einstein's notion of nonemptiness? The Friedmann models to be discussed in Chapter 4 do just that, and were in fact obtained by Friedmann between 1922 and 1924, five years *before* Hubble's data became well known.

The rest of this chapter outlines the kinematic features of the expanding models of the universe. We will first describe how to generalize the arguments that led Einstein to the static line element (3.7). This generalization will lead us to a nonstatic line element that preserves the properties of homogeneity and isotropy assumed by Einstein, but is potentially capable of explaining Hubble's data.

3.4 Simplifying assumptions of cosmology

Once we decide to generalize from a static to a non-static model of the universe, our task becomes more complicated. Figure 3.3(a) shows a spacetime diagram with a swarm of world lines representing particles moving in arbitrary ways. There is no order in this picture, and where two world lines intersect we have colliding particles. It would indeed be very

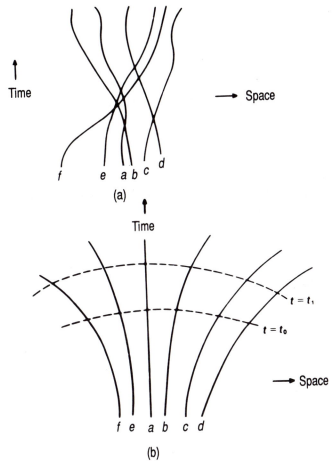

Fig. 3.3 (a) An arbitrary bundle of world lines $a, b, c, \ldots$ describes particles moving haphazardly. Intersecting world lines denote particle collisions. (b) Particles move along nonintersecting world lines $a, b, c, \ldots$ which have no wobbles or irregularities. This is the regularity expressed formally by the Weyl postulate. Note that this regularity enables us to construct a sequence of spacelike hypersurfaces orthogonal to the world lines of the bundle. These are hypersurfaces of constant cosmic time t. Thus the cosmologist can talk of cosmic epochs $t = t_0$, $t = t_1$, and so on in an unambiguous fashion.

difficult to solve the Einstein field equations for such a mess of gravitating matter. Fortunately, the real universe does not appear to be so messy.

Hubble's observations indicate that the universe is (or at least seems to be) an orderly structure in which the galaxies, considered as basic units, are moving apart from one another. Thus Figure 3.3(b) represents a typical spacetime section of the universe in which the world lines represent the histories of galaxies. These world lines, unlike those of Figure 3.3(a), are nonintersecting and form a funnel-like structure in which the separation between any two world lines is steadily increasing.

This intuitive picture of regularity is often expressed formally as the *Weyl postulate*, after the early work of the mathematician Hermann Weyl. The postulate states that the world lines of galaxies designated as *fundamental observers* form a 3-bundle of nonintersecting geodesics orthogonal to a series of spacelike hypersurfaces.

To appreciate the full significance of Weyl's postulate, let us try to express it in terms of coordinates and metric of spacetime. Accordingly we use three spacelike coordinates x^μ ($\mu = 1, 2, 3$) to label a typical world line in the 3-bundle of galaxy world lines. Further, let the coordinate x^0 label a typical member of the series of spacelike hypersurfaces mentioned above. Thus

$$x^0 = \text{constant}$$

is a typical spacelike hypersurface orthogonal to the typical world line given by

$$x^\mu = \text{constant}.$$

Although in practice the galaxies form a discrete set, we can extend the discrete set (x^μ) to a continuum by the *smooth fluid approximation*. This approximation is none other than the widely used device of going over from a discrete distribution of particles to a continuum density distribution. In this case we can treat the quantities x^μ as forming a continuum along with x^0 and use them as the four coordinates x^i to describe space and time.

It is worth emphasizing the importance of the nonintersecting world lines. If two galaxy world lines did intersect, our coordinate system above would break down, for we would than have two different values of x^μ specifying the same spacetime point (the point of intersection). In the next chapter we will, however, encounter an exceptional situation in which all world lines intersect at one singular point!

Let the metric in terms of these coordinates be given by the tensor g_{ik}.

What can we assert about this metric tensor on the basis of the Weyl postulate? The orthogonality condition tells us that

$$g_{0\mu} = 0. \tag{3.18}$$

Further, the fact that the line $x^\mu =$ constant is a *geodesic* tells us that the geodesic equations

$$\frac{d^2 x^i}{dx^2} + \Gamma^i_{kl} \frac{dx^k}{ds} \frac{dx^l}{ds} = 0 \tag{3.19}$$

are satisfied for $x^i =$ constant, $i = 1, 2, 3$. Therefore

$$\Gamma^\mu_{00} = 0, \qquad \mu = 1, 2, 3. \tag{3.20}$$

From (3.18) and (3.20) we therefore get

$$\frac{\partial g_{00}}{\partial x^\mu} = 0, \qquad \mu = 1, 2, 3. \tag{3.21}$$

Thus g_{00} depends on x^0 only. We can therefore replace x^0 by a suitable function of x^0 to make g_{00} constant. Hence we take, without loss of generality,

$$g_{00} = 1. \tag{3.22}$$

The line element therefore becomes

$$\begin{aligned} ds^2 &= (dx^0)^2 + g_{\mu\nu} dx^\mu dx^\nu \\ &= c^2 dt^2 + g_{\mu\nu} dx^\mu dx^\nu, \end{aligned} \tag{3.23}$$

where we have put $ct = x^0$. This time coordinate is called the *cosmic time*. It is easily seen that the spacelike hypersurfaces in Weyl's postulate are the surfaces of simultaneity with respect to the cosmic time. Moreover, t is the proper time kept by any galaxy.

The second important assumption of cosmology is embodied in the *cosmological principle*. This principle states that at any given cosmic time, the universe is homogeneous and isotropic. That is, the surfaces $t =$ constant exhibit the properties discussed earlier in connection with the Einstein universe. There we saw that the three-dimensional surface S_3 of a hypersphere has the requisite properties of homogeneity and isotropy. But is this the only alternative available?

Einstein, as we saw earlier, selected this alternative because he believed space to be closed. However, if we do not insist on closed space, two more alternatives are available to us, which can be seen in the following way. First let us consider an analogy in lower dimensions.

Figure 3.4 shows three surfaces. Figure 3.4(a) shows a section of the Euclidean plane, Figure 3.4(b) a spherical surface, Figure 3.4(c) a saddle-shaped surface. Suppose we try to cover these surfaces with a plain

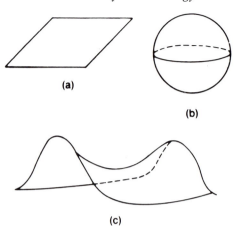

(a)

(b)

(c)

Fig. 3.4 Examples of surfaces of (a) zero curvature (b) positive curvature, and (c) negative curvature.

sheet of paper. We will find that our sheet fits exactly and smoothly on the plane surface. If we try to cover the spherical surface, the sheet of paper develops wrinkles, indicating that the sheet of paper has area in excess of that needed to cover the surface. Similarly, in trying to cover the saddle our paper will be torn, being short of the necessary covering area. These differences can be expressed in differential geometry by the notion of curvature. The plain surface has zero curvature, the spherical surface has positive curvature, and the saddle has negative curvature. Our paper-covering experiment tells us in general whether a given surface has a zero, positive, or negative curvature. These ideas can be extended to higher dimensions as well.

In the Einstein universe the space sections were the 3-surfaces of hyperspheres, and hence they had a constant *positive curvature*. The constancy of curvature is necessary to ensure the properties of homogeneity and isotropy; for if the curvature of space differs from place to place, physical measurements could be devised to detect the differences. We can similarly get other homogeneous and isotropic spaces by considering them as 3-surfaces of *constant negative curvature* or of *zero curvature*.

In terms of the Cartesian coordinates x_1, x_2, x_3, x_4 used earlier, a 3-surface of constant negative curvature is given by an equation of the form

$$x_1^2 + x_2^2 + x_3^2 - x_4^2 = -S^2. \tag{3.24}$$

where S is a constant. The substitution

$$x_1 = S \sinh \chi \cos \theta, \qquad x_2 = S \sinh \chi \sin \theta \cos \phi,$$
$$x_3 = S \sinh \chi \sin \theta \sin \phi, \qquad x_4 = S \cosh \chi \qquad (3.25)$$

gives

$$dx_1^2 + dx_2^2 + dx_3^2 - dx_4^2 = S^2[d\chi^2 + \sinh^2 \chi(d\theta^2 + \sin^2 \theta \, d\phi^2)]. \quad (3.26)$$

Notice the minus sign in front of dx_4^2. It means that we are embedding our 3-surface not in a Euclidean space but in a pseudo-Euclidean space. (In Euclidean space the Pythagoras theorem holds with the line-element given by $dx^2 = dx_1^2 + dx_2^2 + dx_3^2 + \cdots$. If some of the $+$ signs on the right-hand side are changed to $-$ signs, the result is a pseudo-Euclidean space. Thus Minkowski space is pseudo-Euclidean.) If we further substitute

$$r = \sinh \chi, \qquad (3.27)$$

(3.26) becomes

$$d\sigma^2 = S^2 \left[\frac{dr^2}{1 + r^2} + r^2(d\theta^2 + \sin^2 \theta \, d\phi^2) \right]. \qquad (3.28)$$

Compare this with the expression (3.6) for the space of positive curvature:

$$d\sigma^2 = S^2 \left[\frac{dr^2}{1 - r^2} + r^2(d\theta^2 + \sin^2 \theta \, d\phi^2) \right]. \qquad (3.29)$$

Both the expressions can be combined into a single expression by introducing a parameter k that takes values ± 1:

$$d\sigma^2 = S^2 \left[\frac{dr^2}{1 - kr^2} + r^2(d\theta^2 + \sin^2 \theta \, d\phi^2) \right]. \qquad (3.30)$$

Notice that if we set $k = 0$ we get the third alternative – the 3-surface of zero curvature:

$$d\sigma^2 = S^2[dr^2 + r^2(d\theta^2 + \sin^2 \theta \, d\phi^2)]. \qquad (3.31)$$

The right-hand side of (3.31) is simply the Euclidean line element scaled by the constant factor S.

The constant S can, however, depend on cosmic time, since we were considering a typical $t = \text{constant}$ hypersurface in the above argument. Thus the most general line element satisfying the Weyl postulate and the cosmological principle is given by

$$ds^2 = c^2 \, dt^2 - S^2(t) \left[\frac{dr^2}{1 - kr^2} + r^2(d\theta^2 + \sin^2 \theta \, d\phi^2) \right], \qquad (3.32)$$

where the 3-spaces $t = \text{constant}$ are Euclidean for $k = 0$, closed with positive curvature for $k = \pm 1$, and open with negative curvature for $k = -1$. For reasons that will become clearer later, the scale factor $S(t)$ is often called the *expansion factor*.

The line element (3.32) that we have obtained using partly intuitive and partly heuristic arguments was rigorously derived in the 1930s by H. P. Robertson and A. G. Walker (independently). If is often referred to as the *Robertson–Walker line element*.

The Robertson–Walker line element is sometimes expressed in a slightly different form with the help of the following radial coordinate transformation:

$$\bar{r} = \frac{2r}{1 + (1 - kr^2)^{1/2}}. \tag{3.33}$$

We then get the line element as

$$ds^2 = c^2 dt^2 - \frac{S^2(t)}{\left(1 + \dfrac{k\bar{r}^2}{4}\right)} [d\bar{r}^2 + \bar{r}^2(d\theta^2 + \sin^2\theta\, d\phi^2)] \tag{3.34}$$

This line element is manifestly isotropic in $\bar{r}$, θ, ϕ. We will, however continue to use (3.32).

Notice how the simplifying postulates of cosmology have reduced the number of unknowns in the metric tensor from 10 to the single function $S(t)$ and the discrete parameter k that characterize the Robertson–Walker metric. The task of the relativist is now simplified to solving an ordinary differential equation in the independent variable t. We will defer the solution of this problem to the next chapter.

We end this chapter with a discussion of some of the important observational features of a typical Robertson–Walker spacetime. These features show how a non-Euclidean geometry can substantially alter conclusions based on naive Euclidean concepts.

3.5 The redshift

Let us first try to understand how the nebular redshift found by Hubble and Humason is accounted for by the Robertson–Walker model. We begin by recalling that the basic units of Weyl's postulate are galaxies with constant coordinates x^μ. We can easily identify the x^μ with the (r, θ, ϕ) of Robertson–Walker spacetime. Thus each galaxy has a constant set of coordinates (r, θ, ϕ). This coordinate frame is often referred to as the *cosmological rest frame*. As observers we are located in our Galaxy, which also has constant (r, θ, ϕ) coordinates. Later on, in Chapter 9, we will show that this remark is only approximately correct, because our Galaxy has a small motion relative to this cosmological frame. Without loss of generality we can take $r = 0$ for our vantage point.

Although this assumption suggests that we are placing ourselves at the centre of the universe, this does not confer any special status on us. Because of the assumption of homogeneity, *any* galaxy could be chosen to have $r = 0$. Our particular choice is simply dictated by convenience.

Consider a galaxy G_1 at (r_1, θ_1, ϕ_1) emitting light waves towards us. Let us denote by t_0 the present epoch of observation. At what time should a light wave leave G_1 in order to arrive at $r = 0$ at time $t = t_0$? To find the answer to this question we need to know the path of the wave from G_1 to us. Since light travels along null geodesics, as described in Chapter 2, we need to calculate the null geodesic from G_1 to us.

From the symmetry of a spacetime we can guess that a null geodesic from $r = 0$ to $r = r_1$ will maintain a constant spatial direction. That is, we expect to have $\theta = \theta_1$, $\phi = \phi_1$ all along the null geodesic. This guess proves to be correct when we substitute these values into the geodesic equations. Accordingly we will assume that only r and t change along the null geodesic. Next we recall that a first integral of the null geodesic equation is simply $ds = 0$. For the Robertson–Walker line element this gives us

$$c\, dt = \pm \frac{S\, dr}{(1 - kr^2)^{1/2}}. \tag{3.35}$$

Since r decreases as t increases along this null geodesic, we should take the minus sign in the above relation. Suppose the null geodesic left G_1 at time t_1. Then we get from the above relation

$$\int_{t_1}^{t_0} \frac{c\, dt}{S(t)} = \int_0^{r_1} \frac{dr}{(1 - kr^2)^{1/2}}. \tag{3.36}$$

Thus if we know $S(t)$ and k, we know the answer to our question.

However, consider what happens to successive wave crests emitted by G_1. Suppose the wave crests were emitted at t_1 and $t_1 + \Delta t_1$ and received by us at t_0 and $t_0 + \Delta t_0$, respectively. Then, comparably to (3.36), we have

$$\int_{t_1 + \Delta t_1}^{t_0 + \Delta t_1} \frac{c\, dt}{S(t)} = \int_0^{r_1} \frac{dr}{(1 - kr^2)^{1/2}}. \tag{3.37}$$

If $S(t)$ is a slowly varying function so that it effectively remains unchanged over the small intervals Δt_0 and Δt_1, we get by subtraction of (3.36) from (3.37)

$$\frac{c\, \Delta t_0}{S(t_0)} - \frac{c\, \Delta_1}{S(t_1)} = 0,$$

that is,

$$\frac{c\,\Delta t_0}{c\,\Delta t_1} = \frac{S(t_0)}{S(t_1)} \equiv 1 + z. \qquad (3.38)$$

It is not difficult to see that the quantity z defined above is the redshift. The term $c\,\Delta t_1$ is the wavelength λ_1 measured by an observer at rest in the galaxy G_1, while $c\,\Delta t_0$ is the wavelength λ_0 measured by an observer at rest in our Galaxy, since in the Robertson–Walker spacetime the cosmic time measures the proper time kept by any galaxy. Thus the wavelength of the light wave increases by a fraction z in the transmission from G_1 to us, provided $S(t_0) > S(t_1)$. In other words, Hubble's observations of redshift are explained if we assume $S(t)$ to be an increasing function of time.

It is worth commenting on the way this redshift has been arrived at. Our derivation above shows that the effect arises from the passage of light through a non-Euclidean spacetime. It does *not* arise from the Doppler effect, since in our coordinate frame all galaxies have constant (r, θ, ϕ) coordinates. In a non-Euclidean spacetime it is not possible to attach an unambiguous meaning to the relative velocity of two objects separated by a great distance. People are often tempted to relate z to velocity by the special relativistic relation

$$1 + z = \left(\frac{1 + \dfrac{v}{c}}{1 - \dfrac{v}{c}}\right)^{1/2} \qquad (3.39)$$

Such an interpretation is not valid in our present framework because, as we saw in Chapter 2, special relativity applies only in a local region of spacetime.

It is also necessary to contrast (3.38) with the gravitational redshift described in Chapter 2. The gravitational redshift is characterized by the fact that if light from object B to object A is redshifted then the light from A to B is blueshifted. In the present case, if light from galaxy A to galaxy B is redshifted then that from B to A will also be redshifted provided $S(t)$ is increasing during the transmission of light.

We will refer to the present redshift as *cosmological redshift*.

3.6 Apparent magnitude

The redshift discussed above shows up in the spectrum of a galaxy. The astronomer measures another quantity associated with the galaxy – its apparent magnitude. Let us now see how the apparent magnitude is related to the luminosity of the galaxy and its distance from us in the expanding universe described by Robertson–Walker spacetime.

Let L be the total energy emitted by the galaxy G_1 in unit time at the epoch t_1 when light left it in order to reach us at the present epoch t_0. The redshift z of the galaxy is therefore given by (3.38). It is now necessary to specify the wavelength range of observation. To fix ideas: suppose the intensity distribution of G_1 over wavelengths λ is given by the normalized function $I(\lambda)$. Thus

$$dL = LI(\lambda)\,d\lambda \qquad (3.40)$$

is the energy emitted by G_1 per unit time over the bandwidth $(\lambda, \lambda + d\lambda)$. If instead of wavelengths we wanted to use frequencies, the corresponding intensity function $J(v)$ is related to $I(\lambda)$ by

$$cJ(v) = \lambda^2 I(\lambda). \qquad (3.41)$$

Both $J(v)$ and $I(\lambda)$ are used by the astronomer, depending on convenience.

In the case of isotropic light emission by G_1, by the time its light reaches us it is distributed uniformly across a sphere of coordinate radius r_1 centred on G_1 (see Figure 3.5). What is the proper surface area of this sphere?

In the Robertson–Walker line element, put $t = \text{constant}$ and $r = \text{constant}$ to get

$$ds^2 = -r^2 S^2 (d\theta^2 + \sin^2\theta\,d\phi^2).$$

This is the line element on the surface of a Euclidean sphere of radius rS. Hence the answer to the above question is that light from G_1 is distributed over a total surface area $4\pi r_1^2 S^2(t_0)$ at time t_0. We now need to know how much light is received per unit time by us across unit proper area held perpendicular to the line of sight to G_1, and over a bandwidth $(\lambda_0, \lambda_0 + \Delta\lambda_0)$. Denote this quantity by $\mathscr{F}(\lambda_0)\,\Delta\lambda_0$.

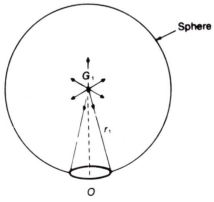

Fig. 3.5 The radiation emitted by galaxy G_1 is distributed uniformly across a sphere of coordinate radius r_1 with G_1 as the centre. The observer O (that is, ourselves) located on this sphere would expect to receive a proportionate quantity of this radiation across a unit area held normal to the direction G_1O.

Note first that because of redshift the arriving light with wavelengths in the range $(\lambda_0, \lambda_0 + \Delta\lambda_0)$ left G_1 in the wavelength range

$$\left(\frac{\lambda_0}{1 + z}, \frac{\lambda_0 + \Delta\lambda_0}{1 + z}\right).$$

Now the total amount of energy that leaves G_1 between the epochs t_1 and $t_1 + \Delta t_1$ in the above frequency range is

$$LI\left(\frac{\lambda_0}{1 + z}\right) \cdot \frac{\Delta\lambda_0}{1 + z} \cdot \Delta t_1.$$

How many photons carry the above quantity of energy? For a small enough bandwidth, we may assume that a typical photon had, at emission, the wavelength $\lambda_0/(1 + z)$, a frequency $(1 + z)c/\lambda_0$, and hence an energy equal to $(1 + z)ch/\lambda_0$, where h is Planck's constant. Therefore the required number of photons is

$$\delta\mathcal{N} = LI\left(\frac{\lambda_0}{1 + z}\right) \frac{\Delta\lambda_0}{1 + z} \frac{\Delta t_1}{(1 + z)ch/\lambda_0}$$

$$= \frac{L\lambda_0}{ch} \cdot \frac{1}{(1 + z)^2} \cdot I\left(\frac{\lambda_0}{1 + z}\right) \Delta\lambda_0 \, \Delta t_1.$$

At the epoch of reception, these photons are distributed across a surface area of $4\pi r_1^2 S^2(t_0)$ and are received over a time interval $(t_0, t_0 + \Delta t_0)$. Thus the number of photons received by us per unit area held normal to the line of sight and per unit time is given by

$$\frac{L\lambda_0}{ch} \cdot \frac{1}{(1 + z)^2} \, I\left(\frac{\lambda_0}{1 + z}\right) \Delta\lambda_0 \cdot \frac{\Delta t_1}{\Delta t_0} \cdot \frac{1}{4\pi r_1^2 S^2(t_0)}.$$

At this epoch, each photon has been degraded in energy by the factor $(1 + z)^{-1}$. Thus each photon now has the energy ch/λ_0. If we multiply the above expression by this factor, we get the quantity we were after:

$$\mathcal{F}(\lambda_0)\Delta\lambda_0 = L \frac{1}{(1 + z)^2} \cdot \frac{\Delta t_1}{\Delta t_0} \cdot I\left(\frac{\lambda_0}{1 + z}\right) \cdot \frac{1}{4\pi r_1^2 S^2(t_0)} \cdot \Delta\lambda_0.$$

However, we note from (3.38) that $\Delta t_1/\Delta t_0$ gives us another factor $(1 + z)^{-1}$ in the denominator. Thus finally we get

$$\mathcal{F}(\lambda_0) = \frac{LI(\lambda_0/1 + z)}{(1 + z)^3 4\pi r_1^2 S^2(t_0)}. \tag{3.42}$$

In terms of frequencies the result is quoted as *flux density*

$$\widetilde{\mathcal{F}}(\nu_0) = \frac{LJ(\nu_0 \cdot 1 + z)}{(1 + z)4\pi r_1^2 S^2(t_0)}. \tag{3.43}$$

Here $\widetilde{\mathscr{F}}(v_0)\,\Delta v_0$ is the amount of radiation perpendicular to unit area in unit time across a frequency range $(v_0, v_0 + \Delta v_0)$.

The optical astronomer uses this result in the form (3.42), while the radio astronomer uses it in the form (3.43). The X-ray astronomer uses energies instead of frequencies, so that (3.43) is scaled by h. We will have occasion to use these expressions frequently, since they occur in the various observational tests of cosmology. We will end this section by deriving a few results of interest to optical astronomy.

The expression (3.42) integrated over all wavelengths gives

$$\mathscr{F}_{\text{bol}} = \frac{L_{\text{bol}}}{4\pi r_1^2 S^2(t_0)(1 + z)^2}, \qquad (3.44)$$

where L_{bol} $(=L)$ is the absolute *bolometric* luminosity of G_1. $\mathscr{F}_{\text{bol}}$ is correspondingly the apparent bolometric luminosity of G_1. On the logarithmic scale of magnitudes familiar to the optical astronomer, (3.44) becomes

$$m_{\text{bol}} = -2.5 \log\left(\frac{\mathscr{F}_{\text{bol}}}{\mathscr{F}_0}\right),$$

$$M_{\text{bol}} = -2.5 \log\left(\frac{L_{\text{bol}}}{L_\odot}\right) + 4.75, \qquad (3.45)$$

$$m_{\text{bol}} - M_{\text{bol}} = 5 \log D_1 - 5,$$

where

$$\mathscr{F}_0 = 2.48 \times 10^{-5} \text{ erg cm}^{-2}\,\text{s}^{-1},$$

$$L_\odot = \text{solar luminosity} = 2 \times 10^{33} \text{ erg s}^{-1}, \qquad (3.46)$$

and

$$D_1 = r_1 S(t_0)(1 + z) \text{ measured in parsecs.}$$

D_1 is called the luminosity distance of G_1. If we are interested in a magnitude, and so on, we may similarly use (3.42) in the logarithmic form, with the apparent magnitude defined by

$$m(\lambda_0) = -2.5 \log \mathscr{F}(\lambda_0) + \text{constant},$$

the constant depending on the filter used. It is customary to indicate the filter by a suffix attached to m. Thus m_{pg} stands for photographic magnitude, m_{v} for visual magnitude, m_{b} for blue magnitude, and so on.

We will use this relation in Chapters 9 and 10. Note, however, that because of redshift the astronomer has to apply a correction to include the effect of the term $I(\lambda_0/1 + z)$. Thus an astronomer using a red filter may be actually receiving the photons that originated in the blue part of the spectrum of G_1 if $z \approx 1$. This correction, which is crucial to many cosmological observations, is called the *K-correction*.

3.7 Hubble's law

Hubble's law was derived for galaxies of low redshifts. The largest redshift in Hubble's 1929 paper was $z \approx 0.003$. At these small redshifts we can use the Taylor expansion to derive a simple linear relation between D_1 and z, the relation arrived at by Hubble from his early observations for $z \ll 1$.

$$D_1 \approx r_1 S(t_0). \tag{3.47}$$

We also get by the Taylor expansion (3.36)

$$\int_0^{r_1} \frac{dr}{(1 - kr^2)^{1/2}} \approx r_1, \tag{3.48a}$$

$$\int_{t_1}^{t_0} \frac{c \, dt}{S(t)} \approx \frac{c(t_0 - t_1)}{S(t_0)}, \tag{3.48b}$$

$$S(t_1) \approx S(t_0) - (t_0 - t_1) \cdot \left(\frac{\dot{S}}{S}\right)_{t_0} S(t_0), \tag{3.48c}$$

$$S(t_1) = \frac{S(t_0)}{1 + z} \approx S(t_0)(1 - z). \tag{3.48d}$$

From these relations and from (3.47) we get

$$D_1 \approx r_1 S(t_0) \approx c(t_0 - t_1)$$

$$\approx \left[\left(\frac{\dot{S}}{S}\right)_{t_0}\right]^{-1} \cdot cz,$$

which can be expressed in the form

$$cz = H_0 D_1, \tag{3.49}$$

with H_0, the Hubble constant, given by

$$H_0 = \left(\frac{\dot{S}}{S}\right)_{t=t_0}. \tag{3.50}$$

From a Doppler shift point of view, cz may be identified with the velocity of recession at small z. In this form (3.49) is sometimes called the *velocity–distance relation*. Expressed as part of the velocity–distance relation, the Hubble constant has the unit of velocity per unit distance, the most common unit in usage being kilometres per second per megaparsec. In many calculations of observational and physical cosmology we will use

$$H_0 = h_0 \times 100 \, \text{km s}^{-1} \, \text{Mpc}^{-1}. \tag{3.51}$$

Although Hubble originally obtained $h_0 \sim 5.3$, the present estimate of h_0 lies in the range $0.5 \leqslant h_0 \leqslant 1$. We will discuss in Chapter 9 how modern techniques arrive at the above result.

Another useful way of expressing H_0 is in units of reciprocal time; that is, by expressing

$$T_0 = H_0^{-1} \tag{3.52}$$

in units of time. A good time unit for T_0 is a billion years. The present estimate of T_0 is in the range of approximately 9 to 18 billion years.

3.8 Angular size

Figure 3.6 illustrates a somewhat unusual effect of the non-Euclidean geometry of Robertson–Walker spacetime. We consider our galaxy G_1 to have a linear extent d, as shown in Figure 3.6. What angle does this length d subtend at our location?

To decide the answer to this question, consider two neighbouring null geodesics (representing light rays) from the two points A, B at the two extremities of G_1 directed towards our Solar System. Without loss of generality we can choose our angular coordinates such that A has the coordinates θ_1, ϕ_1, while B has the coordinates $(\theta_1 + \Delta\theta_1, \phi_1)$. (Although we have used homogeneity to take $r = 0$ at our location, we can also use isotropy to choose any particular direction as the polar axis $\theta = 0$, $\theta = \pi$.)

According to the Robertson–Walker line element, the proper distance between A and B is obtained by putting $t = t_1 = $ constant, $r = r_1 = $ constant, $\phi = \phi_1 = $ constant, and $d\theta = \Delta\theta_1$ in (3.32). We then get

$$ds^2 = -r_1^2 S^2(t_1)(\Delta\theta_1)^2 = -d^2,$$

since in the rest frame of G_1, the spacelike separation $AB = d$. Thus

$$\Delta\theta_1 = \frac{d}{r_1 S(t_1)} = \frac{d(1 + z)}{r_1 S(t_0)} \tag{3.53}$$

gives the answer to our question.

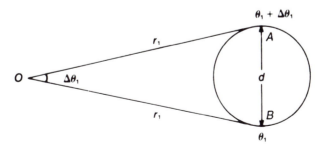

Fig. 3.6 The angle subtended by galaxy G_1 at the observer O.

Notice that as r_1 increases we are looking at more and more remote galaxies, which must therefore be seen at earlier and earlier epochs t_1. However, in an expanding universe $S(t_1)$ was smaller at earlier epochs t_1, so it is not obvious that $r_1 S(t_1)$ should get progressively larger as we look at more and more remote galaxies. Clearly, we need to know how fast $S(t_1)$ decreases as r_1 increases. Although (3.53) provides the answer in an implicit form, we still need to know $S(t)$ to be able to perform these integrations. All that we can say at present is that the observed angular size of galaxies need not be a monotonic decreasing function of their distance, as it would be in the Euclidean universe.

3.9 Source counts

The distribution of discrete luminous sources to great distances may give indications that spacetime geometry is non-Euclidean. How does the number of galaxies up to coordinate distance r_1 (that is, up to the distance of galaxy G_1) increase with r_1? Let us suppose that at any epoch t there are $n(t)$ galaxies in a *unit comoving coordinate volume* (using the r, θ, ϕ coordinates). The word 'comoving' indicates that although the galaxies individually retain the same coordinates (r, θ, ϕ), the proper separation between them at any epoch increases with epoch according to the scale factor $S(t_1)$. Thus the proper volume of any region bounded by such galaxies increases as S^3.

When we observe galaxies at radial coordinates between r and $r + dr$, we see them at times in the range t, $t + dt$, where from (3.36)

$$\int_t^{t_0} \frac{c\, dt'}{S(t')} = \int_0^r \frac{dr'}{(1 - kr'^2)^{1/2}}. \tag{3.54}$$

The number of galaxies seen in this shell is therefore

$$dN = \frac{4\pi r^2\, dr}{(1 - kr^2)^{1/2}} \cdot n(t), \tag{3.55}$$

where t is related to r through (3.54). Thus the required number of galaxies out of $r = r_1$ is given by

$$N(r_1) = \int_0^{r_1} \frac{4\pi r^2 n(t)\, dr}{(1 - kr^2)^{1/2}}. \tag{3.56}$$

If no galaxies are created or destroyed between $r = 0$ and $r = r_1$, we may take $n(t) = $ constant, and the integral can be explicitly evaluated. Clearly, the answer must depend on the parameter k. The function $N(r_1)$ increases faster than the Euclidean form ($\propto r_1^3$) in closed universes ($k = \pm 1$) and slower than this form in open universes of negative curvature. In Chapter 10 we will cast this idea in a somewhat different form to make it suitable for observations of galaxies, radio sources and quasars.

Having discussed some of the general properties of the Robertson–Walker universes, it is now appropriate for us to turn to specific models – the models first considered by Friedmann, which are described in Chapter 4.

Exercises

1   In Newtonian cosmology space is Euclidean and time has the meaning implicit in Newtonian dynamics. In such a universe let $r \equiv (x_1, x_2, x_3)$ denote the coordinates (Cartesian, of course!) of a typical galaxy G, and let $v = (v_1, v_2, v_3)$ denote its velocity relative to a galaxy G_0 located at the origin. An observer in G_0 observes a velocity–distance relation for galaxies like G of the following form:

$$\mathbf{v} = \mathbf{f}(\mathbf{r}).$$

If it is assumed that the same relation is observed by any other galaxy G_1 with coordinates $\mathbf{r}_1$ and velocity $\mathbf{v}_1$, then the function $\mathbf{f}$ must satisfy the condition

$$\mathbf{f}(\mathbf{r} - \mathbf{r}_1) = \mathbf{f}(\mathbf{r}) - \mathbf{f}(\mathbf{r}_1).$$

Deduce that $\mathbf{f}$ must be a linear function of $\mathbf{r}$.

2   Show that under the assumption of isotropy at any epoch t, the velocity–distance relation of Exercise 1 must take the form

$$\mathbf{v} = H(t)\mathbf{r}.$$

Show that the above equation of motion has a first integral

$$\mathbf{r} = S(t)\mathbf{r}_0,$$

where $\mathbf{r}_0$ is a constant for the galaxy. Relate $S(t)$ to $H(t)$.

3   Carrying our Newtonian discussion further, suppose every galaxy measures the speed of light to be c in its own rest frame. However, when the galaxy at the origin observes light coming towards it from another galaxy G_1 at $\mathbf{r}_1$, the speed of light is modified by the vector addition of the local velocity $\mathbf{v} = H(t)\mathbf{r}$ at the typical intermediate point $\mathbf{r}$. Thus for radial propagation of light from G_1 to G_0 the propagation equation is

$$\frac{\mathrm{d}r}{\mathrm{d}t} = -c + H(t)r.$$

Integrate this equation using the results of Exercise 2 and discuss how the phenomenon of redshift can be explained in this fashion. (*Caution*: The above velocity addition formula is Newtonian and is not consistent with special relativity.)

4   Verify by direct substitution that (3.4) follows from (3.3).

5   Show that the volume of the Einstein universe is $2\pi^2 S^3$. Comment on the statement that this universe is 'finite but unbounded'. Does the above volume refer to spherical or to elliptical space?

6   A ray of light is emitted in a given direction in the Einstein universe. How long will the ray take to make one circuit about the universe and return to its starting point?

7   Using the metric components and the Christoffel symbols given in the text, verify the relations (3.8) and (3.9).

8   Taking $\rho_0 = 10^{-31}$ g cm^{-3}, calculate the radius of the Einstein universe and its total mass in spherical space.

9   With the density given in Exercise 8, calculate the λ-term and estimate the fraction by which the Sun's attraction for the Earth is reduced because of the λ-repulsion. Comment on the effect of this force on the experimental tests of general relativity.

10   Given that (3.17) is the coordinate transform of (3.15), find the transformation law between (R, T) and (r, t).

11   Comment on why we cannot look upon the de Sitter universe as a static universe in spite of its apparently static line element (3.15).

12   The de Sitter universe has an *event horizon* in the following sense. If test particles with constant r, θ, ϕ emit light signals towards the origin $r = 0$, then at given time t there is a critical value r_0 such that signals from all particles with $r \geqslant r_0$ emitted at t will never reach their destination. Calculate r_0.

13   Suppose that in the Weyl postulate we drop the condition of orthogonality of the surfaces $t = $ constant with respect to the geodesics $x^\mu = $ constant, so that (3.18) does not hold. Show that $g_{0\mu}$ must be independent of t.

14   By calculating the 3-volume of space within the coordinate region $r = $ constant in the spaces with the spatial line element

$$d\sigma^2 = S^2 \left[\frac{dr^2}{1 - kr^2} + r^2 (d\theta^2 + \sin^2 \theta \, d\phi^2) \right], \qquad k = 0, 1, -1,$$

develop the three-dimensional analogue of the experiment for covering the surfaces of zero, positive, and negative curvature described in the text.

15   Derive from first principles the coordinate transformation $r = f(\bar{r})$

that takes the Robertson–Walker line element from the form (3.32) to the form (3.34).

16   Determine the affine parameter for the radial null geodesic from galaxy G_1 to the origin $r = 0$ in Robertson–Walker spacetime.

17   A particle of mass m is fired from a galaxy at $t = t_0$ with a linear momentum P_0. Show that the momentum of the particle when it reaches another galaxy at a later epoch t (as measured in the rest frame of that galaxy) is given by

$$P = P_0 \frac{S(t_0)}{S(t)}.$$

Compare this result with the cosmological redshift for photons.

18   A galaxy, instead of following a typical Weyl geodesic, has a small random velocity relative to it. Use the nonrelativistic version of Exercise 17 to find out how this velocity decreases with time.

19   In a universe with $S(t) \propto t^{2/3}$ and $k = 0$, a galaxy is observed to have a redshift $z = 1.25$. How long has light taken to travel from that galaxy to us? Express your answer in units of T_0.

20   How will the forms (3.42) and (3.43) look if the spectrum of the emitting source is given by $J(v) \propto v^{-\alpha}$, $\alpha = $ constant in the relevant range of observations?

21   For $S \propto \exp H_0 t$ and $k = 0$, $H_0 = $ constant, show that (3.44) takes the form

$$\mathscr{F}_{\text{bol}} = \frac{L_{\text{bol}}}{4\pi \left(\dfrac{c}{H_0}\right)^2 z^2 (1 + z)^2}.$$

22   Calculate the redshift magnitude relation for bolometric magnitudes in the universe of Exercise 21.

23   Work out the formula (3.44) for the universe with $S \propto t^{2/3}$ and $k = 0$, and compare with the result of Exercise 21. How much brighter is the galaxy in the universe of Exercise 21?

24   If the Hubble constant is given by h_0 in the units of $100 \text{ km s}^{-1} \text{ Mpc}^{-1}$, show that $T_0 \simeq 9.8 h_0^{-1}$ billion years.

25   Show that if $k = 0$, $S \propto t^{2/3}$, the apparent angular sizes of distant objects of the same linear size have a minimum at $z = 1.25$.

26   Repeat Exercise 25 for the universe with $k = 0$, $S \propto \exp H_0 t$. At what redshift does the minimum value of apparent angular size lie?

27   If in (3.56) $k = 1$ and $n(t)$ is constant and equal to n_0, show that the number of galaxies in the entire universe is given by

$2\pi^2(c/H_0)^3 n_0$. Clarify whether this answer refers to spherical space or to elliptical space.

28   In (3.56), put $k = -1$, $n(t) = n_0$ (constant) and show that·for $S(t) = ct$, the number of galaxies with redshifts less than z is given by

$$N(z) = \frac{2\pi}{3}\left(\frac{c}{H_0}\right)^3 n_0\left[\frac{(z^2 + 2z)(z^2 + 2z + 2)}{2(1 + z)^2} - \ln(1 + z)\right].$$

4

The Friedmann models

4.1 The Einstein field equations in cosmology

The work covered in Chapter 3 did not tell us two important items of information about the universe: (1) the rate at which it expands as given by the function $S(t)$, and (2) whether its spatial sections $t = \text{constant}$ are open or closed as indicated by the parameter k. To find answers to these questions it is necessary to go beyond the Weyl postulate and the cosmological principle. We need a dynamic theory to proceed any further, and Einstein's general relativity is one such theory. In Chapter 8 we will consider alternative approaches to cosmology but for the present we will rely on relativity.

Once we decide to use relativity, our procedure is cut and dried. We already have the line element to start with:

$$\mathrm{d}s^2 = c^2\,\mathrm{d}t^2 - S^2(t)\left[\frac{\mathrm{d}r^2}{1 - kr^2} + r^2(\mathrm{d}\theta^2 + \sin^2\theta\,\mathrm{d}\phi^2)\right]. \quad (4.1)$$

We use it to compute the Einstein tensor, and thereby formulate the Einstein field equations. To solve them we will require the energy tensor of the contents of the universe.

Accordingly, we set

$$x^0 = ct, \qquad x^1 = r, \qquad x^2 = \theta, \qquad x^3 = \phi, \quad (4.2)$$

so that the nonzero components of g_{ik} and g^{ik} are

$$g_{00} = 1, \qquad g_{11} = -\frac{S^2}{1 - kr^2}, \qquad g_{22} = -S^2 r^2, \qquad g_{33} = -S^2 r^2 \sin^2\theta,$$

$$g^{00} = 1, \qquad g^{11} = -\frac{1 - kr^2}{S^2}, \qquad g^{22} = -\frac{1}{S^2 r^2},$$

$$g^{33} = -\frac{1}{S^2 r^2 \sin^2\theta}, \quad (4.3)$$

$$(-g)^{1/2} = \frac{S^2 r^2 \sin \theta}{(1 - kr^2)^{1/2}}.$$

The nonzero components of Γ^i_{kl} are then as follows:

$$\Gamma^1_{01} = \Gamma^2_{02} = \Gamma^3_{03} = \frac{1}{c} \frac{\dot{S}}{S}.$$

$$\Gamma^0_{11} = \frac{S\dot{S}}{c(1 - kr^2)}, \qquad \Gamma^0_{22} = \frac{S\dot{S}r^2}{c}, \qquad \Gamma^0_{33} = \frac{S\dot{S}r^2 \sin^2 \theta}{c}.$$

$$\Gamma^1_{11} = \frac{kr}{1 - kr^2}, \qquad \Gamma^2_{12} = \Gamma^3_{13} = \frac{1}{r},$$

$$\Gamma^1_{22} = -r(1 - kr^2), \qquad \Gamma^1_{33} = -r(1 - kr^2) \sin^2 \theta,$$

$$\Gamma^2_{33} = -\sin \theta \cos \theta, \qquad \Gamma^3_{23} = \cot \theta. \tag{4.4}$$

Now we use the expression for the Ricci tensor (see Chapter 2), which may be put in the following form:

$$R_{ik} = \frac{\partial^2 \ln (-g)^{1/2}}{\partial x^i \partial x^k} - \frac{\partial \Gamma^l_{ik}}{\partial x^l} + \Gamma^m_{in} \Gamma^n_{km} - \Gamma^l_{ik} \frac{\partial \ln (-g)^{1/2}}{\partial x^l}. \tag{4.5}$$

Straightforward but tedious calculation then gives the following nonzero components of R^i_k:

$$R^0_0 = \frac{3}{c^2} \frac{\ddot{S}}{S}, \tag{4.6}$$

$$R^1_1 = R^2_2 = R^3_3 = \frac{1}{c^2} \left(\frac{\ddot{S}}{S} + \frac{2\dot{S}^2 + 2kc^2}{S^2} \right). \tag{4.7}$$

From these we get

$$R = \frac{6}{c^2} \left(\frac{\ddot{S}}{S} + \frac{\dot{S}^2 + kc^2}{S^2} \right), \tag{4.8}$$

and hence

$$G^1_1 \equiv R^1_1 - \tfrac{1}{2}R = -\frac{1}{c^2} \left(2\frac{\ddot{S}}{S} + \frac{\dot{S}^2 + kc^2}{S^2} \right) = G^2_2 = G^3_3, \tag{4.9}$$

$$G^0_0 \equiv R^0_0 - \tfrac{1}{2}R = -\frac{3}{c^2} \left(\frac{\dot{S}^2 + kc^2}{S^2} \right). \tag{4.10}$$

We have gone through the details of the calculation to illustrate how the techniques of general relativity developed in Chapter 2 can be applied to the problem of cosmology. The reader may check that putting $S = $ constant $= S_0$ and $k = +1$ gives us the formulae (3.8) and (3.9) obtained for the Einstein universe in Chapter 3.

Recalling now the Einstein equations, we get from (4.9) and (4.10) the only nontrivial equations of the set as

$$2\,\frac{\ddot{S}}{S} + \frac{\dot{S}^2 + kc^2}{S^2} = \frac{8\pi G}{c^2}\,T^1_1 = \frac{8\pi G}{c^2}\,T^2_2 = \frac{8\pi G}{c^2}\,T^3_3, \qquad (4.11)$$

$$\frac{\dot{S}^2 + kc^2}{S^2} = \frac{8\pi G}{3c^2}\,T^0_0. \qquad (4.12)$$

We next consider the energy tensor.

4.2 Energy tensors of the universe

Before we consider specific forms of T^i_k, it is worth noting that two properties must be satisfied by any energy tensor in the present framework of cosmology. The first is obvious from (4.11):

$$T^1_1 = T^2_2 = T^3_3 = -p. \qquad (4.13)$$

The fact that these three components of T^i_k are equal is hardly surprising when we recall that we have imposed the condition of isotropy on the universe. The second property is not quite so obvious, but is derivable from (4.11) and (4.12). It relates the pressure p to the energy density ε.

For this derivation it is convenient to write

$$T^0_0 = \varepsilon, \qquad (4.14)$$

and to note that if we differentiate (4.12) with respect to t we can express the resulting answer as a linear combination of (4.11) and (4.12). The result is in fact equivalent to the following identity:

$$\frac{d}{dt}[S(\dot{S}^2 + kc^2)] \equiv \dot{S}\,[2S\ddot{S} + \dot{S}^2 + kc^2],$$

that is,

$$\frac{d}{dS}(\varepsilon S^3) + 3pS^2 = 0. \qquad (4.15)$$

It is not necessary, however, to write down the full field equations (4.11) and (4.12) in order to arrive at (4.15). The above result is a direct consequence of the conservation law implicit in the Einstein equations:

$$T^i_{k;i} = 0. \qquad (4.16)$$

We now turn our attention to the specific forms of the energy tensor.

Present observations suggest that galaxies are the major constituents of the universe. If galaxies followed the Weyl postulate strictly, we would have the typical velocity vector of a galaxy as

$$u^i = (1, 0, 0, 0). \qquad (4.17)$$

In our smooth fluid approximation a velocity field like (4.17) represents an orderly motion with no pressure. Thus we have in this case the system of galaxies behaving like dust, with

$$p = 0, \qquad \varepsilon = \rho c^2, \tag{4.18}$$

ρ being the rest mass density of galaxies.

In practice galaxies do not follow the Weyl postulate strictly, and their velocity vectors depart from (4.17). Such velocity departures are measurable for galaxies in clusters and are of the order $\leqslant 1000\ \mathrm{km\,s^{-1}}$. If we take a typical velocity departure $v \approx 1000\ \mathrm{km\,s^{-1}}$ then we would have a nonzero value of p in (4.18) of the order

$$p \approx \frac{v^2}{c^2}\varepsilon \sim 10^{-5}\varepsilon. \tag{4.19}$$

Therefore we would be justified in ignoring the p-term at the present epoch, in comparison with the ε-term as in the idealized situation of (4.18).

What about the future and past epochs? To assess the importance of the pressure term we have to investigate how the random motions of galaxies change in an expanding universe. We may express the 4-velocity of a galaxy as

$$u^i \equiv [1, u^\mu], \qquad u^\mu \ll 1.$$

The $\ll$ sign implies that the squares of u^μ are to be neglected in comparison with 1. Therefore the requirement $u_i u^i = 1$ is satisfied, and we also have the built-in assumption that the random motions are small. In the absence of any external forces, therefore, the velocity u^i satisfies the geodesic equation

$$\frac{\mathrm{d}u^i}{\mathrm{d}s} + \Gamma^i_{kl}u^k u^l = 0.$$

Substitution of the Γ^i_{kl} for the Robertson–Walker line element then gives the result

$$u^\mu S^2 = \text{constant}.$$

However, u^μ measures the velocity in the comoving (r, θ, ϕ) coordinates. The proper distances are obtained from the coordinate distances by the multiplication of the scale factor S. Thus proper random velocities v change with S as S^{-1}.

Hence, so long as S goes on increasing beyond the present epoch the approximation $p \ll \varepsilon$ will continue to apply. If, however, we turn towards the past epoch, the galaxy motions become more and more turbulent, since v was larger in the past. Thus if we use $S \approx 10^{-2}S_0$ (S_0 being the value of S at the present epoch), (4.19) would give $p \sim 10^{-1}\varepsilon$. Clearly the p-term would no longer be negligible at this epoch and prior to it.

At such epochs we have to abandon our simplified picture of cosmology and ask whether galaxies existed as single units then. Obviously, galaxies were formed at some stage in the past and in a proper theory of cosmology we have to say how and when they were formed. We will return to this question in Chapter 7. At present we simply state that the present cosmological framework of galaxies receding from one another breaks down, as does the dust approximation (4.19) at epochs like these.

If, however, we simply extrapolate $v \propto S^{-1}$ to very low values of S then v becomes comparable to c and our nonrelativistic approximation that led to $v \propto S^{-1}$ breaks down. The correct formula (see Exercise 3.17) then tells us that the 3-momentum P goes as S^{-1}. In this domain we have to use the formula (2.85), and we set

$$p = \tfrac{1}{3}\,\varepsilon. \tag{4.20}$$

Thus if S continues to increase from very small values, then (4.20) would hold for the early epochs, just as (4.18) holds in the recent epochs. The transition between the two epochs is through a rather messy phase when neither (4.18) nor (4.20) holds.

If (4.18) holds, then from (4.15) we get

$$\frac{\mathrm{d}}{\mathrm{d}S}\,(\rho S^3) = 0, \tag{4.21}$$

which integrates to

$$\rho = \rho_0 \frac{S_0^3}{S^3}, \tag{4.22}$$

ρ_0 and S_0 being the values of ρ and S at the present epoch.

Similarly, substitution of (4.20) into (4.15) gives

$$\frac{\mathrm{d}}{\mathrm{d}S}\,(\varepsilon S^4) = 0, \tag{4.23}$$

giving

$$\varepsilon \propto S^{-4}. \tag{4.24}$$

We therefore have the following picture. For a distribution of matter (4.24) holds when S was very small compared with S_0, and (4.22) holds in the more recent epochs. If, however, on top of matter we also have electromagnetic radiation present in the universe, it will also contribute to T^i_k. For small S, (4.20) holds uniformly for matter (moving relativistically) and for radiation. However, as S increases we have to be more careful in distinguishing between the contributions of matter and radiation to T^i_k. We will go into these details more fully in Chapter 5.

For the present discussion let us assume that beyond a certain epoch given by $S = \bar{S}$, radiation and matter decouple from each other, each going its own way. Thus we can write

$$T^i_k = T^i_{k\,\mathrm{matter}} + T^i_{k\,\mathrm{radiation}} \qquad (4.25)$$

and assume that the divergence of each energy tensor separately vanishes. Thus (4.22) continues to hold up to the present epoch for matter density. Since for the radiation energy tensor we have (for $\mu = 1, 2, 3$), say,

$$-T^\mu_{\mu\,\mathrm{radiation}} = \tfrac{1}{3} T^0_{0\,\mathrm{radiation}} = \tfrac{1}{3}\varepsilon, \qquad (4.26)$$

we get, for $S > \bar{S}$,

$$\varepsilon = \varepsilon_0 \, \frac{S^4_0}{S^4}. \qquad (4.27)$$

What is $\bar{S}$? Why, if at all, should matter decouple from radiation? What happened prior to $S = \bar{S}$? We defer a discussion of these questions to Chapter 5.

The present estimate of $\varepsilon_0 \approx 10^{-13}\,\mathrm{erg\,cm}^{-3}$, and that of $\rho_0 c^2 \gtrsim 10^{-10}\,\mathrm{erg\,cm}^{-3}$ mean that the matter density is more than 10^3 times the radiation density. Thus $\varepsilon_0 \ll \rho_0 c^2$, and we may ignore the contribution of radiation (in comparison with the contribution of matter) to the field equations (4.11) and (4.12) at the present epoch, and for $S > S_0$. However, for the past epochs with $S < S_0$, we have, from (4.22) and (4.27),

$$\frac{\varepsilon}{\rho c^2} = \frac{\varepsilon_0}{\rho_0 c^2} \cdot \frac{S_0}{S}. \qquad (4.28)$$

and we cannot ignore the contribution of radiation for, say, $S_0/S \sim 10^3$. Indeed, for small enough S the relative importance of radiation and matter is inverted: the radiation becomes more significant in deciding how S should vary with t.

From the above discussion we see that at $S \approx 10^{-3} S_0$, we have a transition from a radiation-dominated universe to a matter-dominated one. In the present chapter we will consider the matter-dominated epochs. The equations (4.11) and (4.12) are therefore to be solved with

$$T^1_1 = 0, \qquad T^0_0 = \rho_0 c^2 \, \frac{S^3_0}{S^3}. \qquad (4.29)$$

This simplification leads us to the classical models first considered by A. Friedmann between 1922 and 1924. Basically, these models ignore any contributions of electromagnetic radiation to T^i_k and suppose that the matter in the universe can be approximated by dust.

4.3 The solution of Friedmann's equations

From the Friedmann models, (4.11) and (4.12) become

$$2\frac{\ddot{S}}{S} + \frac{\dot{S}^2 + kc^2}{S^2} = 0, \tag{4.30}$$

$$\frac{\dot{S}^2 + kc^2}{S^2} = \frac{8\pi G\rho_0}{3} \cdot \frac{S_0^3}{S^3}. \tag{4.31}$$

In view of the conservation law given in (4.21), the above two equations are not independent, and only one of them is sufficient to determine $S(t)$. Since it is of lower order, we will choose (4.31) for our solution, and consider the three cases $k = 0, 1, -1$ separately.

4.3.1 Euclidean sections ($k = 0$)

This is the simplest case, and is also known as the *Einstein–de Sitter model*, since it was given by Einstein and de Sitter in a joint paper in 1932. Equation (4.31) becomes

$$\dot{S}^2 = \frac{8\pi G\rho_0}{3}\frac{S_0^3}{S}. \tag{4.32}$$

We now recall from Chapter 3 that the present value of Hubble's constant is given by

$$\left.\frac{\dot{S}}{S}\right|_{t_0} = H_0. \tag{4.33}$$

Hence applying (4.32) to the present epoch, we get

$$\rho_0 = \frac{3H_0^2}{8\pi G} \equiv \rho_c. \tag{4.34}$$

For reasons that will become clear later, ρ_c is often called the *closure density*. With the range of values of H_0 quoted in Chapter 3, we have

$$\rho_c = 2 \times 10^{-29} h_0^2 \text{ g cm}^{-3}. \tag{4.35}$$

These values are considerably higher than the matter density actually observed at present, and we will take up this point in detail in Chapters 7 and 9.

Returning to (4.32), we see that it is easy to verify that it has the solution

$$S = S_0 \left(\frac{t}{t_0} \right)^{2/3}. \tag{4.36}$$

(An arbitrary constant that arises from the integration of the differential equation can be set equal to zero by assuming that $S = 0$ at $t = 0$.) We also get the age of the universe as

$$t_0 = \frac{2}{3H_0}. \tag{4.37}$$

The constant S_0 is not determined. It has the dimensions of length, and it can be absorbed in the unit of length chosen. Figure 4.1 illustrates this solution. If we drop the suffix 0, the results (4.33), (4.34), and (4.37) hold at any arbitrary epoch t.

4.3.2 Closed sections ($k = 1$)

Equations (4.30) and (4.31) now take the form

$$2 \frac{\ddot{S}}{S} + \frac{\dot{S}^2 + c^2}{S^2} = 0, \tag{4.38}$$

$$\frac{\dot{S}^2 + c^2}{S^2} - \frac{8\pi G \rho_0 S_0^3}{3 S^3} = 0. \tag{4.39}$$

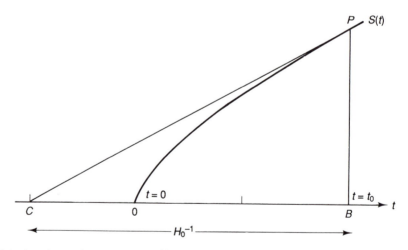

Fig. 4.1 A schematic graph of $S(t)$ as a function of t for the Einstein–de Sitter model. The present epoch t_0 is denoted by the point B on the t-axis. The ordinate at B, $PB = S_0$, the present value of the scale factor. The present value of the Hubble constant is given by the ratio $1/BC$, where C is the point common to the t-line and the tangent to the $S(t)$-curve at P. Thus $BC = H_0^{-1}$. As can be verified from the figure, the age of the universe, represented by OB, is two-thirds of time intercept BC.

It is convenient to introduce the quantities $q(t)$ and $H(t)$ through the relations:

$$\frac{\ddot{S}}{S} = -q(t)[H(t)]^2, \qquad H(t) = \frac{\dot{S}}{S}, \tag{4.40}$$

with their present values denoted by q_0 and H_0. We have already come across H_0, the Hubble constant. The second parameter, q_0, is called the *deceleration parameter*, and it is useful in expressing ρ_0 in terms of the closure density.

With the above definitions, (4.38) and (4.39) take the following forms when applied at the present epoch:

$$\frac{c^2}{S_0^2} = (2q_0 - 1)H_0^2, \tag{4.41}$$

$$\rho_0 = \frac{3}{8\pi G}\left(H_0^2 + \frac{c^2}{S_0^2}\right) = \frac{3H_0^2}{4\pi G}q_0. \tag{4.42}$$

The density ρ_0 is sometimes expressed in the following form:

$$\rho_0 = \rho_c\Omega_0 \tag{4.43}$$

so that from (4.42), (4.43) and (4.34) we get the *density parameter*

$$\Omega_0 = 2q_0. \tag{4.44}$$

Since the left-hand side of (4.41) is positive, we must have

$$q_0 > \tfrac{1}{2}, \qquad \Omega_0 > 1. \tag{4.45}$$

Thus our closed model has density exceeding the so-called closure density ρ_c. This explains the name 'closure density'. It is the value of the universal density that must be exceeded if the model is to describe a closed universe. We mention at this stage the result (to be proved shortly) that for the open models ($k = -1$) the inequalities of (4.45) are reversed.

Using (4.41) and (4.42) to eliminate S_0 and ρ_0 from (4.39), we get the differential equation

$$\dot{S}^2 = c^2\left(\frac{\alpha}{S} - 1\right), \tag{4.46}$$

with α given by

$$\alpha = \frac{2q_0}{(2q_0 - 1)^{3/2}} \cdot \frac{c}{H_0}. \tag{4.47}$$

The parameter α has the dimensions of length. Thus the model is characterized by the parameters H_0 and q_0.

Equation (4.46) can be integrated as follows. We get

$$ct = \int \frac{S^{1/2}\,\mathrm{d}S}{(\alpha - S)^{1/2}}.$$

Make the substitution

$$S = \alpha \sin^2 \frac{\Theta}{2} = \tfrac{1}{2}\alpha(1 - \cos\Theta). \tag{4.48}$$

Then the integral becomes

$$ct = \int \alpha \sin^2 \frac{\Theta}{2} \, d\Theta = \tfrac{1}{2}\alpha(\Theta - \sin\Theta). \tag{4.49}$$

Again, as in the case $k = 0$ we have taken $S = 0$ at $t = 0$ ($\Theta = 0$). We therefore get $t = t_0$ by requiring that $S = S_0$. From (4.41) and (4.47) we see that $S = S_0$ at $\Theta = \Theta_0$, where

$$\tfrac{1}{2}\alpha(1 - \cos\Theta_0) = \frac{c}{H_0}(2q_0 - 1)^{-1/2} = \frac{(2q_0 - 1)}{2q_0}\alpha,$$

that is,

$$\cos\Theta_0 = \frac{1 - q_0}{q_0}, \qquad \sin\Theta_0 = \frac{(2q_0 - 1)^{1/2}}{q_0}. \tag{4.50}$$

We therefore get from (4.49) the age of the universe as

$$\begin{aligned}
t_0 &= \frac{\alpha}{2c}(\Theta_0 - \sin\Theta_0) \\
&= \frac{q_0}{(2q_0 - 1)^{3/2}}\left[\cos^{-1}\left(\frac{1 - q_0}{q_0}\right) - \frac{(2q_0 - 1)^{1/2}}{q_0}\right] \cdot \frac{1}{H_0}.
\end{aligned} \tag{4.51}$$

For example, for $q_0 = 1$ we get

$$t_0 = \left(\frac{\pi}{2} - 1\right)H_0^{-1}. \tag{4.52}$$

Note that S reaches a maximum value at $\Theta = \pi$, when

$$S = S_{\max} = \alpha = \frac{2q_0}{(2q_0 - 1)^{3/2}} \cdot \frac{c}{H_0}. \tag{4.53}$$

Thus for $q_0 = 1$, the universe expands to twice its present size.

In closed models, therefore, expansion is followed by contraction and S decreases to zero. The value $S = 0$ is reached when $\Theta = 2\pi$; that is, when

$$t = t_L = \frac{\pi\alpha}{c} = \frac{2\pi q_0}{(2q_0 - 1)^{3/2}} \cdot \frac{1}{H_0}. \tag{4.54}$$

The quantity t_L may be termed the *lifespan* of this universe. For $q_0 = 1$, $t_L = 2\pi H_0^{-1} = 2\pi T_0$, where T_0 is defined by the relation (3.52).

Figure 4.2 illustrates the function $S(t)$ for the closed models for a number of parameter values q_0. All curves have been adjusted to have the same value of H_0 at point P. Notice that the value $S = 0$ is reached sooner in the past as q_0 is increased from just over $\tfrac{1}{2}$.

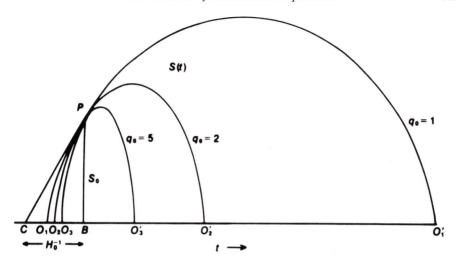

Fig. 4.2 The $S(t)$ curves for $q_0 = 1, 2, 5$. All curves have been scaled to touch at P, the present point, and they all have the common tangent PC. The intercept $BC = H_0^{-1}$. Notice that as q_0 increases the curves for $S(t)$ intersect the past section of the t-line at points $O_1, O_2, O_3, \ldots$ lying closer to B, implying that the age of the universe is reduced if q_0 is increased. The points $O_1', O_2', O_3', \ldots$ in the future section of the t-line show the singularities at which these models end their existence.

4.3.3 Open sections ($k = -1$)

Equations (4.30) and (4.31) become in this case

$$2\frac{\ddot{S}}{S} + \frac{\dot{S}^2 - c^2}{S^2} = 0,\tag{4.55}$$

$$\frac{\dot{S}^2 - c^2}{S^2} - \frac{8\pi G \rho_0 S_0^3}{3S^3} = 0.\tag{4.56}$$

We again use the definitions of (4.40) and apply them at the present epoch to get

$$\frac{c^2}{S_0^2} = (1 - 2q_0)H_0^2,\tag{4.57}$$

$$\rho_0 = \frac{3H_0^2}{4\pi G}q_0.\tag{4.58}$$

Thus instead of (4.45) we now have

$$0 \leqslant q_0 < \tfrac{1}{2}, \qquad 0 \leqslant \Omega_0 < 1.\tag{4.59}$$

And in place of (4.46) we get

$$\dot{S}^2 = c^2\left(\frac{\beta}{S} + 1\right),\tag{4.60}$$

with

$$\beta = \frac{2q_0}{(1 - 2q_0)^{3/2}} \frac{c}{H_0}. \tag{4.61}$$

The solution of (4.60) may be parametrized by an angle Ψ with

$$S = \tfrac{1}{2}\beta \, (\cosh \Psi - 1), \qquad ct = \tfrac{1}{2}\beta \, (\sinh \Psi - \Psi). \tag{4.62}$$

The present value of Ψ is given by

$$\cosh \Psi_0 = \frac{1 - q_0}{q_0}, \qquad \sinh \Psi_0 = \frac{(1 - 2q_0)^{1/2}}{q_0}. \tag{4.63}$$

We have set $t = 0$ at $S = 0$, as in the two preceding cases. The age of the universe is given by

$$\begin{aligned}
t_0 &= \frac{\beta}{2c} \, (\sinh \Psi_0 - \Psi_0) \\
&= \frac{q_0}{(1 - 2q_0)^{3/2}} \left\{ \frac{(1 - 2q_0)^{1/2}}{q_0} - \ln \left[\frac{1 - q_0 + (1 - 2q_0)^{1/2}}{q_0} \right] \right\} H_0^{-1}.
\end{aligned} \tag{4.64}$$

Like the Einstein–de Sitter model, these models continue to expand forever. The behaviour of $S(t)$ in these models is illustrated in Figure 4.3.

It is worth pointing out that the model with $q_0 = 0$, $S(t) = ct$ represents flat spacetime. In fact, by the following coordinate transformation we can change the line element to a manifestly Minkowski form:

$$R = ctr, \qquad T = t(1 + r^2)^{1/2},$$
$$ds^2 = c^2 \, dT^2 - dR^2 - R^2(d\theta^2 + \sin^2 \theta \, d\phi^2). \tag{4.65}$$

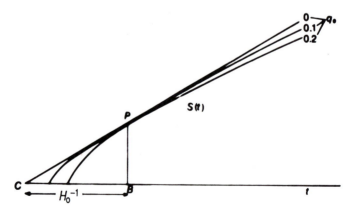

Fig. 4.3 The $S(t)$ curves for $q_0 = 0, 0.1$, and 0.2. As in Figure 4.2, all curves have the same value of H_0 at P. The age of the universe is seen to increase as q_0 decreases, being maximum ($= H_0^{-1}$) for $q_0 = 0$.

This model arose naturally in Milne's kinematic relativity, which was a cosmological theory with foundations different from those of general relativity. For this reason the above model is sometimes referred to as the *Milne model*.

4.3.4 Spacetime singularity

Figure 4.4 shows how t_0, the age of the universe, decreases as q_0 increases from 0 to ∞, the maximum value of t_0 being $T_0 = H_0^{-1}$. All the above Friedmann models have the common feature of having $S = 0$ at a certain epoch (which we have chosen to label by $t = 0$). As we approach $S = 0$, the Hubble constant increases rapidly, being infinite at $S = 0$, except in the special case of the Milne model $k = -1$, $q_0 = 0$. This epoch therefore indicates violent activity and is given the name *big bang*.

From a mathematical point of view, $S = 0$ describes a spacetime singularity. If we compute the components of R_{iklm} and construct invariants out of these, such as

$$R, \qquad R_{ik}R^{ik}, \qquad R_{iklm}R^{iklm}, \ldots,$$

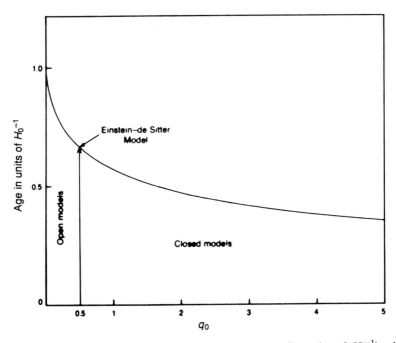

Fig. 4.4 The variation of the age of a Friedmann model (in units of H_0^{-1}) with the deceleration parameter q_0. The value $q_0 = \frac{1}{2}$ seperates the closed models from the open models.

these invariants diverge as S tends to 0. It is therefore meaningless to talk of a spacetime geometry at $S = 0$.

$S = 0$ also presents an insurmountable barrier to the physicist. If we use the strong principle of equivalence (see Chapter 2) to enable us to study how physical processes operate in strong gravitational fields, our procedure will break down at $S = 0$. Thus the singularity of the big bang is more significant (and perhaps more sinister) than the infinities that occur elsewhere in physics (such as in the radiative corrections of quantum electrodynamics).

Attempts to remove the singularity by modifying the energy tensor are not, however, likely to succeed if the modifications are of a conventional character. For the time being we will accept the existence of singularity as a fact of life under the regime of general relativity and learn to live with it.

4.4 The luminosity distance

Since the Friedmann models are frequently used to interpret cosmological observations, we will now derive some of the observable quantities in these models, starting with the luminosity distance defined in Chapter 3. Our aim is to express the final answer in terms of the two parameters that characterize a Friedmann model: H_0 and q_0.

4.4.1 The Einstein–de Sitter model

We use (3.36) to relate r_1, the radial coordinates of the galaxy G_1, to the time t_1 and to its redshift z.

$$r_1 = \int_{t_1}^{t_0} \frac{c \, dt}{S(t)} = \frac{c}{S_0} \int_{t_1}^{t_0} t_0^{2/3} t^{-2/3} \, dt$$

$$= \frac{c}{S_0} t_0^{2/3} 3(t_0^{1/3} - t_1^{1/3})$$

$$= \frac{3c}{S_0} t_0 \left[1 - \left(\frac{t_1}{t_0} \right)^{1/3} \right].$$

We now use (3.38) to note that

$$1 + z = \frac{S(t_0)}{S(t_1)} = \left(\frac{t_0}{t_1} \right)^{2/3},$$

so that with the help of (4.37)

$$r_1 = \frac{3ct_0}{S_0}\{1 - (1 + z)^{-1/2}\}$$

$$= \frac{2c}{S_0 H_0}\{1 - (1 + z)^{-1/2}\}. \tag{4.66}$$

The luminosity distance is therefore given by

$$D_1 = r_1 S_0(1 + z)$$

$$= \frac{2c}{H_0}[(1 + z) - (1 + z)^{1/2}]. \tag{4.67}$$

4.4.2 The closed model

This calculation is more involved. Equation (3.36) becomes

$$\int_0^{r_1} \frac{dr}{(1 - r^2)^{1/2}} = \int_{t_1}^{t_0} \frac{c\,dt}{S(t)}.$$

The left-hand side can be easily integrated. To integrate the right-hand side we use (4.46) to get

$$\int_{t_1}^{t_0} \frac{c\,dt}{S(t)} = \int_{S_1}^{S_0} \frac{dS}{[S(\alpha - S)]^{1/2}}.$$

Now use the parametric form (4.48): $S = \alpha \sin^2(\Theta/2)$. We then get

$$\int_{S_1}^{S_0} \frac{dS}{[S(\alpha - S)]^{1/2}} = \int_{\Theta_1}^{\Theta_0} d\Theta = \Theta_0 - \Theta_1.$$

Remembering that the integral on the left-hand side of (3.36) gives $\sin^{-1} r_1$, we have

$$r_1 = \sin(\Theta_0 - \Theta_1). \tag{4.68}$$

We must now relate this answer to z. We have

$$1 + z = \frac{S(t_0)}{S(t_1)} = \frac{\sin^2 \dfrac{\Theta_0}{2}}{\sin^2 \dfrac{\Theta_1}{2}},$$

giving

$$\sin \Theta_1 = \frac{2}{(1 + z)} \sin \frac{\Theta_0}{2}\left(z + \cos^2 \frac{\Theta_0}{2}\right)^{1/2},$$

$$\cos \Theta_1 = \frac{z + \cos \Theta_0}{1 + z}.$$

Also, from (4.50) we have

$$\sin\frac{\Theta_0}{2} = \left(\frac{2q_0 - 1}{2q_0}\right)^{1/2}, \qquad \cos\frac{\Theta_0}{2} = \left(\frac{1}{2q_0}\right)^{1/2}.$$

Putting all these together and performing algebraic simplification, we get

$$r_1 = \frac{[2q_0 - 1]^{1/2}}{q_0^2(1 + z)} \{q_0 z + (1 - q_0)[1 - (1 + 2zq_0)^{1/2}]\}. \qquad (4.69)$$

The luminosity distance is therefore given by

$$D_1 = r_1 S_0 (1 + z)$$

$$= \left(\frac{c}{H_0}\right)\frac{1}{q_0^2} \{q_0 z + (q_0 - 1)[(1 + 2zq_0)^{1/2} - 1]\}. \qquad (4.70)$$

4.4.3 The open model

The calculation is similar in this case to that for the closed model, with the difference that the trigonometric functions are replaced by hyperbolic ones. We will not go through the intermediate steps, but simply quote the final results:

$$r_1 = \frac{(1 - 2q_0)^{1/2}}{q_0^2(1 + z)} \{q_0 z + (1 - q_0)[1 - (1 + 2zq_0)^{1/2}]\}. \qquad (4.71)$$

$$D_1 = \left(\frac{c}{H_0}\right)\frac{1}{q_0^2} \{q_0 z + (q_0 - 1)[(1 + 2zq_0)^{1/2} - 1]\}. \qquad (4.72)$$

It is interesting to note that the final expressions for D_1 are the same for $k = \pm 1$, $|q_0 - \frac{1}{2}| > 0$. If we let $q_0 \to \frac{1}{2}$, it is easy to see that the Einstein–de Sitter model also has D_1 given by the same formula.

Figure 4.5 plots $D_1(q_0, z)$ as a function of z for various parametric values of q_0. Note that all curves start off with the linear Hubble law (3.49) for small z, but then branch out. As a rule we notice that for the same redshift the luminosity distance is larger for lower values of q_0. Thus for $q_0 = 1$ we have

$$D_1 = \left(\frac{c}{H_0}\right) z, \qquad (4.73a)$$

while for $q_0 = 0$ we get

$$D_1 = \left(\frac{c}{H_0}\right) z \left(1 + \frac{z}{2}\right). \qquad (4.73b)$$

So at $z = 1$, (4.73b) exceeds (4.73a) by as much as 50 per cent. In Chapter 10 we will discuss the feasibility of determining q_0 from Hubble-type observations of remote galaxies.

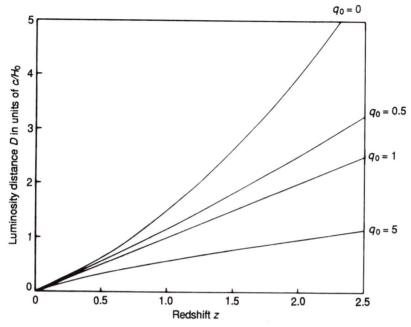

Fig. 4.5 The luminosity distance D_1 expressed (in units of c/H_0) as a function of the redshift z for $q_0 = 0, \frac{1}{2}, 1$ and 5. The relationship is linear, as predicted by Hubble's law for $q_0 = 1$. For $q_0 < 1$, D_1 increases with z faster than predicted by Hubble's linear law; while for $q_0 > 1$, D_1 increases more slowly with z. All curves merge for small z.

4.4.4 The particle horizon

It is pertinent to ask the following question: what is the limit on the proper distance up to which we can observe? This question is answered as follows. First calculate the limiting value of r_1 for $z \to \infty$, calling it r_L. The corresponding limiting proper distance is

$$R_L = S_0 \int_0^{r_L} \frac{dr}{(1 - kr^2)^{1/2}}.$$

It is then easy to verify that for the Friedmann models

$$R_L = \frac{c}{H_0} \times \begin{cases} 2 : (k = 0, q_0 = \frac{1}{2}) \\[2ex] \dfrac{2}{(2q_0 - 1)^{1/2}} \sin^{-1}\left(\dfrac{2q_0 - 1}{2q_0}\right)^{1/2} : (k = 1, q_0 > \frac{1}{2}) \\[2ex] \dfrac{2}{(1 - 2q_0)^{1/2}} \sinh^{-1}\left(\dfrac{1 - 2q_0}{2q_0}\right)^{1/2} : (k = -1, q_0 < \frac{1}{2}). \end{cases} \tag{4.74}$$

The existence of a finite value of R_L means that the universe has a *particle horizon*. Particles with $r_1 > r_L$ are not visible to us at present, no matter how good our techniques of observation are.

4.4.5 The event horizon

The particle horizon sets the limit to communications from the past. By contrast, the event horizon sets the limit on communications to the future. Let us ask the following question. An observer at $r = r_1$, $t = t_0$ sends a light signal to an observer at $r = 0$. Will the signal ever reach its destination? Suppose it does and let t_1 be the time of arrival. Then from (3.36) we get

$$\int_{t_0}^{t_1} \frac{c\,dt}{S(t)} = \int_0^{r_1} \frac{dr}{(1 - kr^2)^{1/2}}.$$

This relation determines t_1 for any given r_1 provided the integral on the left is large enough to match that on the right. Now it may happen that as $t_1 \rightarrow \infty$ the integral on the left converges to a finite value which corresponds to a value of the integral on the right for $r_1 = r_H$, say. In that case the above relation is not possible to satisfy for $r_1 > r_H$. In other words, the signal from the observer at $r_1 > r_H$ will *never* reach the observer at r_0. Thus no observer beyond a proper distance

$$R_H = S_0 \int_{t_0}^{\infty} \frac{c\,dt}{S(t)}. \qquad (4.75)$$

at $t = t_0$ can communicate with another observer.

This limit is called the *event horizon*. It does not exist for Friedmann models, but has the value c/H_0 for the de Sitter model.

4.5 Angular size

We now use the result derived in section 3.8 to study how apparent angular sizes vary with redshifts in different Friedmann models. We will assume that sources of a fixed linear size d are observed at different redshifts. Thus a source at (r_1, θ_1, ϕ_1) with redshift z will subtend at the observer at $r = 0$, the angle

$$\Delta\theta_1 = \frac{d}{r_1 S(t_1)} = \frac{d(1 + z)^2}{D_1}. \qquad (4.76)$$

Since we know D_1 from (4.72), $\Delta\theta_1$ is determined as a function of z and q_0. The interesting fact emerges that $\Delta\theta_1$ does not steadily decrease as z increases; but has a minimum at a certain value of z that depends on q_0.

It is easy to derive this result for $q_0 = \frac{1}{2}$. From (4.67) we get

$$\Delta\theta_1 = \frac{dH_0}{c} \frac{(1+z)^{3/2}}{(1+z)^{1/2} - 1}. \tag{4.77}$$

Straightforward differentiation gives us the result that the minimum value of $\Delta\theta_1$ ($= \theta_{min}$, say) and the redshift $z = z_m$ at which it occurs are given by

$$\theta_{min} = 6.75 \frac{dH_0}{c}$$

and

$$z_m = 1.25. \tag{4.78}$$

The cases $q_0 \gtrless \frac{1}{2}$ are more involved. We illustrate the case $q_0 > \frac{1}{2}$. Instead of using D_1 as given by (4.70), it is more convenient to use the parameter Θ introduced in (4.48) and (4.49) and to use the relation (4.68). We then get

$$\Delta\theta_1 = \frac{d}{r_1 S(t_1)} = \frac{2d}{\alpha} [(1 - \cos\Theta_1)\sin(\Theta_0 - \Theta_1)]^{-1}. \tag{4.79}$$

Differentiation with respect to Θ_1 tells us that the minimum occurs when

$$\sin\Theta_1 \sin(\Theta_0 - \Theta_1) - (1 - \cos\Theta_1)\cos(\Theta_0 - \Theta_1) = 0.$$

That is,

$$\sin\left(\Theta_0 - \frac{3\Theta_1}{2}\right) = 0,$$

thus giving

$$\Theta_1 = \frac{2\Theta_0}{3}, \qquad 1 + z_m = \frac{1 - \cos\Theta_0}{1 - \cos\dfrac{2\Theta_0}{3}}. \tag{4.80}$$

Using (4.47) we get

$$\Theta_{min} = \frac{(2q_0 - 1)^{3/2}}{q_0} \cdot \frac{1}{\left(1 - \cos\dfrac{2\Theta_0}{3}\right)\sin\dfrac{\Theta_0}{3}} \cdot \frac{dH_0}{c}. \tag{4.81}$$

The corresponding result for $q_0 < \frac{1}{2}$ is

$$\Theta_{min} = \frac{(1 - 2q_0)^{3/2}}{q_0} \frac{1}{\left(\cosh\dfrac{2\Psi_0}{3} - 1\right)\sinh\dfrac{\Psi_0}{3}} \cdot \frac{dH_0}{c}. \tag{4.82}$$

at the redshift z_m given by

$$1 + z_m = \frac{\cosh \Psi_0 - 1}{\cosh \dfrac{2\Psi_0}{3} - 1}. \tag{4.83}$$

Figure 4.6 plots $\Delta\theta_1$ as a function of z for different Friedmann models. Notice how the curves all start with the near-Euclidean result $\Delta\theta_1 \propto z^{-1}$ and then begin to differ from one another at larger z values. In principle this effect might be used to decide which particular Friedmann universe (if any!) comes closest to the actual universe.

4.6 Source counts

We now return to the general formula (3.55) and apply it to Friedmann models. It is more convenient to use redshift as the distance parameter

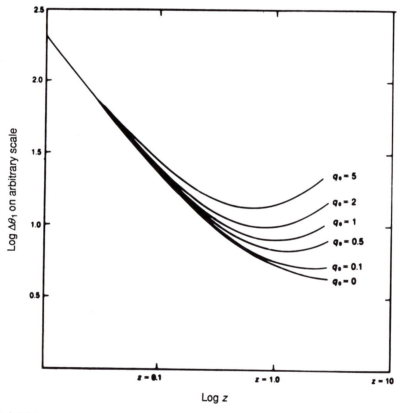

Fig. 4.6 This graph plots $\log \triangle\theta_1$ against $\log z$ for the Friedmann models with $q = 0$, 0.1, 0.5, 1, 2, and 5. All curves merge at small z into a straight line that describes the variation of $\triangle\theta_1$ with distance in a Euclidean universe.

instead of r or t. As before, we will work with the case $k = +1$. From (4.68) and the relations that follow it we have

$$r = \sin(\Theta_0 - \Theta_1),$$

$$\left| \frac{dr}{(1 - r^2)^{1/2}} \right| = |d\Theta_1|, \qquad 1 + z = \frac{\sin^2 \dfrac{\Theta_0}{2}}{\sin^2 \dfrac{\Theta_1}{2}},$$

$$\left| \frac{dz}{1 + z} \right| = \cot \frac{\Theta_1}{2} |d\Theta_1| = \left(\frac{1 + 2q_0 z}{2q_0 - 1} \right)^{1/2} |d\Theta_1|.$$

Therefore the number of astronomical sources with redshifts in the range $(z, z + dz)$ is given by

$$dN = 4\pi \sin^2(\Theta_0 - \Theta_1) \cdot n(t) \cdot \left| \frac{d\Theta_1}{dz} \right| dz.$$

Let us suppose that $n(t)$ is specified as a function $n(z)$ of z. Using (4.69) and some algebraic manipulation, we get

$$dN = 4\pi n(z) \frac{(2q_0 - 1)^{3/2}}{q_0^4} \frac{\{q_0 z + (q_0 - 1)[(1 + 2zq_0)^{1/2} - 1]\}^2 \, dz}{(1 + 2q_0 z)^{1/2}(1 + z)^3}.$$

(4.84)

Suppose $n(z)$ is expressed in a slightly different form. We recall that n was specified as the number of sources per unit *coordinate* volume, in terms of the comoving $(r, \theta, \phi,)$ coordinates. What is the relationship between n and the number of sources per unit *proper* volume? Denoting the latter by $\bar{n}$, we have

$$n = \bar{n} S^3 = \frac{\bar{n} S_0^3}{(1 + z)^3}.$$

(4.85)

From (4.41) we get

$$\frac{\bar{n}}{(1 + z)^3} = (2q_0 - 1)^{3/2} \left(\frac{H_0}{c} \right)^3 n.$$

(4.86)

Substitution into (4.84) gives

$$dN = 4\pi \left(\frac{c}{H_0} \right)^3 \frac{\{q_0 z + (q_0 - 1)[(1 + 2zq_0)^{1/2} - 1]\}^2 \bar{n} \, dz}{q_0^4 (1 + z)^6 (1 + 2q_0 z)^{1/2}}.$$

(4.87)

In this form (4.87) is applicable to all Friedmann models, even though our derivation assumed $q_0 > \frac{1}{2}$ and $k = 1$.

We will have occasion to use this result in connection with observations of galaxy counts and radio source counts.

4.7 Radiation background from sources

Let us use the above formulae to calculate the flux of radiation from sources distributed all over the universe. To fix ideas, let us suppose that there are $\bar{n}(z)\,dz$ sources per unit *proper* volume with redshifts in the range $(z, z + dz)$. Suppose a typical source at redshift z has a normalized intensity spectrum given by

$$J(v; z)$$

and total luminosity $L(z)$. Thus

$$\int_0^\infty J(v; z)\,dv = 1. \tag{4.88}$$

Consider now sources located in a thin solid angle $d\Omega$ in the direction $\theta = \theta_1$, $\phi = \phi_1$ from the origin of coordinates. Let

$$f(v_0)\,\Delta v_0\,d\Omega$$

denote the total flux of radiation received at $r = 0$ in the frequency range $(v_0, v_0 + \Delta v_0)$ from all the sources located in our solid angle.

Now the number of sources in a typical redshift range $(z, z + dz)$ is given by multiplying dN by $d\Omega/4\pi$, and the flux of radiation from a source in this range is given by the application of (3.43). Putting the two results together, we get

$$f(v_0) = \left(\frac{c}{H_0}\right)\frac{1}{4\pi}\int_0^\infty \frac{\bar{n}(z)L(z)J(v_0 \cdot \overline{1 + z}; z)\,dz}{(1 + z)^5(1 + 2q_0 z)^{1/2}}. \tag{4.89}$$

The formula will be found useful in estimating the contributions of sources to the cosmic radiation background. Note that the flux $\widetilde{\mathscr{F}}(v_0; z)$ from a typical source in the above calculation is related to the quantity dN/dz by the relation

$$\frac{dN}{dz} \cdot [\widetilde{\mathscr{F}}(v_0, z)] = \left(\frac{c}{H_0}\right) \cdot \frac{L(z)\bar{n}(z)J(v_0 \cdot \overline{1 + z}; z)}{(1 + z)^5(1 + 2q_0 z)^{1/2}}. \tag{4.90}$$

Now in a Euclidean space with a uniform distribution of sources, the number of sources up to a Euclidean distance R would be given by

$$N = \frac{4\pi}{3} R^3 \bar{n}_0,$$

$\bar{n}_0$ being the number density of sources, assumed constant.

Further, a typical source at a distance R and with a luminosity L would produce a flux at the origin given by

$$\widetilde{\mathscr{F}} = \frac{L}{4\pi R^2}.$$

We therefore get

$$\frac{\mathrm{d}N}{\mathrm{d}R} \cdot \widetilde{\mathscr{F}} = \bar{n}_0 L = \text{constant}. \tag{4.91}$$

To discover the analogue of this result in a Friedmann universe, we assume $\bar{n}(z) = \bar{n}_0(1 + z)^3$, corresponding to a constant number of sources in the unit coordinate volume. We also assume $L(z) = \text{constant}$ and integrate (4.90) over all ν_0. Then using (4.88) we get

$$\frac{\mathrm{d}N}{\mathrm{d}z} \cdot \widetilde{\mathscr{F}} = \left(\frac{c}{H_0}\right) \frac{L\bar{n}_0}{(1 + z)^3(1 + 2q_0z)^{1/2}}. \tag{4.92}$$

Thus the product on the left-hand side steadily decreases with increasing z in all Friedmann models. The redshift factors in the denominator see to it that the product of differential number count with flux is less for remote sources than for nearby ones.

We also see this effect in the contribution to the background in (4.89). The contribution of remote shells is steadily reduced by the redshift effect. This was therefore considered one way of resolving a long-standing paradox known as the *Olbers paradox*. In 1826 Heinrich Olbers, a German astronomer, computed the background from a uniform distribution of sources in a Euclidean universe of infinite extent in space and time. Using (4.91), Olbers concluded that the net flux is infinite! The Olbers paradox is often phrased as the question, 'Why is the sky dark at night?' By using (4.92) instead of (4.91), we see that attenuation at large redshifts results in $f(\nu_0)$ being finite. Various aspects of the Olbers paradox are discussed in Exercises 25 to 29, which demonstrate that the expanding universe is not the only way of arriving at a finite answer.

4.8 Cosmological models with the λ-term

Although our concern in this chapter is with the Friedmann models, we now discuss briefly another class of models that have a close relationship with the Friedmann models. These are the models given by the modified Einstein equations of (2.102) – the equations containing the cosmological constant λ. We have already discussed two special cases of this class of solutions in the last chapter, the static Einstein model and the empty de Sitter model. When Hubble's observations established the expanding universe picture, Einstein conceded that there was no special need for the λ-term in his equations. He even went so far as to say that introduction of this term was the 'biggest blunder' in his life. The Einstein–de Sitter model discussed in this chapter is the outcome of Einstein's collaboration with de Sitter after abandoning the λ-term.

Nevertheless, in the 1930s distinguished cosmologists like A. S. Eddington and the Abbé Lemaître felt that the λ-term introduced certain attractive features into cosmology and that models based on it should also be discussed at length. In modern cosmology the reception given to the λ-term varies from the hostile to the ecstatic. The inputs of particle physics in the early stages of the universe have provided a new interpretation for the λ-term, which we will discuss in Chapter 6.

Putting $\lambda \neq 0$ modifies (4.11) and (4.12) to the following:

$$2\frac{\ddot{S}}{S} + \frac{\dot{S}^2 + kc^2}{S^2} - \lambda c^2 = \frac{8\pi G}{c^2}\, T_1^1, \tag{4.93}$$

$$\frac{\dot{S}^2 + kc^2}{S^2} - \tfrac{1}{3}\lambda c^2 = \frac{8\pi G}{3c^2}\, T_0^0. \tag{4.94}$$

The conservation laws discussed in section 4.2 are not affected by the λ-term. If we restrict ourselves to dust only then (4.94) gives us the following differential equation in place of (4.31):

$$\frac{\dot{S}^2 + kc^2}{S^2} - \tfrac{1}{3}\lambda c^2 = \frac{8\pi G\rho_0}{3}\frac{S_0^3}{S^3}. \tag{4.95}$$

Similarly, (4.93) becomes

$$2\frac{\ddot{S}}{S} + \frac{\dot{S}^2 + kc^2}{S^2} - \lambda c^2 = 0. \tag{4.96}$$

Let us first recover the static model of Einstein. By setting $S = S_0$, $\dot{S} = 0$, $\ddot{S} = 0$ in (4.95) and (4.96), we get

$$\frac{kc^2}{S_0^2} - \tfrac{1}{3}\lambda c^2 = \frac{8\pi G\rho_0}{3}; \qquad \frac{kc^2}{S_0^2} = \lambda c^2.$$

From these relations it is not difficult to verify that $k = +1$, and we recover the relations obtained in section 3.2:

$$\lambda = \frac{1}{S_0^2} \equiv \lambda_c, \tag{4.97}$$

$$\rho_0 = \frac{\lambda_c c^2}{4\pi G}. \tag{4.98}$$

We shall denote by $\lambda = \lambda_c$ the critical value of λ for which a static solution is possible. It was pointed out by Eddington that the Einstein universe is unstable. A slight perturbation destroying the equilibrium conditions (4.97) and (4.98) leads to either a collapse to singularity

$(S \to 0)$ or an expansion to infinity $(S \to \infty)$. Eddington and Lemaître proposed instead a model in which λ exceeds λ_c by a small amount. In this case the universe erupts from $S = 0$ (the big bang) and slows down near $S = S_0$, staying thereabouts for a long time and then expanding away to infinity. It was argued that the quasistationary phase of the universe would be suitable for the formation of galaxies. This model is illustrated in Figure 4.7, which plots $S(t)$ for a range of values of λ for $k = +1$. Notice that for $\lambda < \lambda_c$ the universe contracts (as in the Friedmann case), while for $\lambda > \lambda_c$ it ultimately disperses to infinity, resembling the de Sitter universe.

Figure 4.7 also shows by dotted lines another series of models that contract from infinity to a minimum value of $S > 0$ and then expand back to $S \to \infty$. These models are sometimes called *oscillating models of the second kind*, to distinguish them from the models that shrink back to $S = 0$ and are called *oscillating models of the first kind*. This terminology is, however, not quite apt, since there is no repetition of phases in these models as implied by the word 'oscillating'.

The models with $k = 0$ or $k = -1$ do not show these different types of behaviour for $\lambda > 0$. We get from (4.95) a relation of the following type:

$$\dot{S}^2 = -kc^2 + \tfrac{1}{3}\lambda c^2 S^2 + \frac{8\pi G\rho_0 S_0^3}{3S}, \tag{4.99}$$

wherein each term on the right-hand side is nonnegative. Thus $\dot{S}$ does not change sign, and we get ever-expanding models. For $\lambda < 0$, however, we

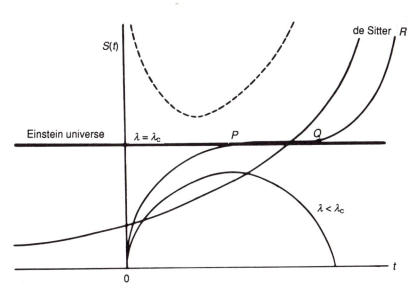

Fig. 4.7 The λ-cosmologies for $k = 1$. The quasistationary phase is from P to Q.

can get universes that expand and then recontract as in the $k = 1$ case for $\lambda < \lambda_c$.

This concludes our discussion of the general dynamic behaviour of the λ-cosmologies. We end this section by writing (4.95) and (4.96) at the present epoch in terms of H_0 and q_0. Thus in place of earlier relations we have

$$H_0^2 + \frac{kc^2}{S_0^2} - \tfrac{1}{3}\lambda c^2 = H_0^2 \Omega_0,$$

$$(1 - 2q_0) H_0^2 + \frac{kc^2}{S_0^2} - \lambda c^2 = 0.$$

From these we get

$$\Omega_0 = 2q_0 + \tfrac{2}{3}\lambda \frac{c^2}{H_0^2}. \tag{4.100}$$

Thus there is no unique relationship between q_0 and Ω_0: we have an additional parameter entering the relation. *Note also that it is possible to have negative q_0, that is, an accelerating expansion, if $\lambda > 0$.* This is because the λ-term introduces a force of cosmic *repulsion*.

4.9 Concluding remarks

Our discussion of the dynamic and geometric properties of the expanding universe is now complete. We started with general relativity – the theory that introduced the unique idea that gravitational effects are intimately connected with the non-Euclidean geometry of spacetime. Nowhere except in cosmology do we see examples of the large-scale effects of non-Euclidean geometry. The redshift, the dimming of light from distant sources, the peculiar behaviour of angular sizes, the existence of particle horizons, and the most dramatic of all, the spacetime singularity: these are all instances of such effects.

However, in the last analysis, cosmology is not an exercise in mathematical fancies, but a physical theory that must make predictions testable by observations. Hence we must now turn to physical cosmology and discuss the physical properties of the expanding universe. Do we have any relics of the early epochs just after the big bang? How did galaxies, which we have taken as the basic units of the universe, form in the first place? And how did matter itself come into existence in its elementary state? The next three chapters will deal with such issues.

Exercises

1   Verify the expressions for the Ricci tensor and the Einstein tensor for the Robertson–Walker line element.

2   Show how the assumptions of Weyl postulate and the cosmological principle reduce the number of independent Einstein equations from 10 to 2. What more can be deduced about these equations with the help of the conservation law?

3   Deduce (4.15) from (4.16).

4   Using the Einstein–de Sitter model, estimate the epoch at which the matter and radiation densities in the universe were equal. For this calculation take $\rho_0 = 10^{-29}$ g cm^{-3} and $\varepsilon_0 = 10^{-13}$ erg cm^{-3}, and express your answer as the fraction of the age of the universe.

5   What is the significance of the closure density? Show that there is a unique relationship between the deceleration parameter q_0 and the density parameter Ω_0 in a Friedmann universe. How is this relation modified by the λ-term?

6   A galaxy is observed with redshift 0.69. How long did light take to travel from the galaxy to us if we assume that we live in the Einstein–de Sitter universe with Hubble's constant $= 100$ km s^{-1} Mpc^{-1}?

7   Calculate from first principles the age of a Friedmann universe with $q_0 = 2$.

8   In the Friedmann universe with $q_0 = 1$, a galaxy is seen with redshift $z = 1$. How old was the universe at the time this galaxy emitted the light received today? ($H_0 = 100$ km s^{-1} Mpc^{-1}.)

9   A light ray is emitted at the present epoch in the closed Friedmann universe. Discuss the possibility of this ray making a round of the universe and coming back to its starting point.

10   Derive the formulae for r_1 and D_1 for the open Friedmann model with given q_0 and H_0.

11   Show that the expressions for D_1 in the cases $q_0 = \frac{1}{2}$ and $q_0 = 0$ can be obtained from (4.72) by suitable limiting processes.

12   Show that there is a unique value of q_0 for which the linear Hubble law holds exactly.

13   Invert the formula (4.72) to express z as a function of $D_1 H_0/c$.

14   Show by computing R_{iklm} that the Friedmann model with $q_0 = 0$ describes a flat spacetime.

15   Given that the Friedmann model with $q_0 = 0$ describes a flat spacetime, find coordinates in which its line element is manifestly that of Minkowski spacetime.

16   What is meant by a particle horizon? How does the size of the particle horizon depend on the epoch of observation in a given Friedmann model?

17   Show why the Friedmann models do not have event horizons.

18   In the Einstein–de Sitter model there are n sources in a unit comoving coordinate volume. Calculate the number of sources in principle visible to us at the present epoch, for $n = $ constant.

19   In a Friedmann model the minimum of angular size occurs at $z = 1$. Deduce from this the value of q_0.

20   The *surface brightness* of an astronomical object is defined by the flux received from the object divided by the angular area subtended by the object at the observation point. How does the surface brightness vary with redshift?

21   Show from first principles that the angular sizes of astronomical objects of fixed linear size will have a minimum at $z = 1.25$ in the Einstein–de Sitter model.

22   Derive (4.81) by direct differentiation of $D_1(z)$ with respect to z.

23   Derive (4.83) from first principles and use it to show that $z_m \to \infty$ as $q_0 \to 0$.

24   If in a Friedmann universe we have a fixed number of sources in a unit comoving coordinate volume, and each source emits a line radiation of fixed total intensity L_0 at frequency $\bar{v}$, show that the radiation background produced by such sources at the present epoch will have the frequency spectrum $S(v)\,dv$, where $S(v) = 0$ for $v > \bar{v}$, while for $v < \bar{v}$,

$$S(v) = \frac{c}{H_0}\,\bar{n}_0 L_0\,\frac{v^{3/2}}{[2q_0\bar{v} + (2q_0 - 1)v]^{1/2}},$$

where $n_0 = $ proper number density of sources at the present epoch.

25   Discuss Olbers's calculation on the darkness of the night sky in the Euclidean universe.

26   Show that a finite answer can be obtained in an Olbers-type calculation if the universe is finite in extent or finite in age.

27   Show that a finite answer can be obtained in an Olbers-type calculation by assuming that the sources are finite in size and therefore that nearby sources tend to block radiation from the more distant sources.

28   Show that the Olbers paradox can be resolved by assuming that a typical source can radiate only for a finite interval of time because of its finite reservoir of energy.

29 Review all the possible means of resolving the Olbers paradox and state your own preference for any particular resolution.

30 Derive (4.93) and (4.94) for the λ-cosmologies and deduce the conservation law from them.

31 Given that objects during the quasistationary phase of the Eddington–Lemaître cosmology are now seen with the redshift $z = 2$, what can you say about the value of λ?

32 Deduce that the scale factor in the λ-cosmology with $k = 1$ satisfies the differential equation

$$\dot{S}^2 = c^2 \left(\tfrac{1}{3}\lambda S^2 - 1 + \frac{\gamma}{S} \right),$$

where

$$\gamma = \frac{2q_0 + \dfrac{2}{3}\dfrac{\lambda c^2}{H_0^2}}{2q_0 - 1 + \dfrac{\lambda c^2}{H_0^2}} \left(\frac{c}{H_0} \right).$$

33 Write down an integral that gives the age of a big bang universe for $\lambda \neq 0$. Discuss qualitatively how the λ-term may be used to increase the age of the universe.

34 In λ-cosmology, what is the lower limit on the value of λ, given the value of q_0?

35 Compute the invariants R, $R_{ik}R^{ik}$ and $R_{iklm}R^{iklm}$ for the Friedmann models and show that they all diverge as $S \to 0$. Is there an exceptional case?

36 Repeat Exercise 35 for λ-cosmologies and show that the same conclusions follow for the models with $S \to 0$.

37 Give a general argument to show that for sufficiently small S, the λ-force is ineffective in preventing the spacetime singularity.

38 Let us return to Newtonian cosmology. Suppose, continuing in the vein of Exercises 1 to 3 of Chapter 3, we have a cosmic velocity field $v = H(t)r$. Use the continuity equation in fluid dynamics to show that the density ρ of cosmic fluid satisfies the conservation law $\rho S^3 = $ constant.

39 Use Euler's equations of motion in Exercise 38 and deduce from them the differential equation

$$\ddot{S} = \frac{4\pi G\rho}{3} S.$$

In Exercise 38 assume that the force F per unit mass on the cosmic fluid satisfies Gauss's theorem for Newtonian gravitation: $\operatorname{div} F = -4\pi G\rho$.

40   Integrate the differential equation of Exercise 39 and relate the constant of integration to the curvature parameter k in the Friedmann models.

41   Show that the force per unit mass F that acts on the cosmic fluid in Exercise 39 may be written as

$$\mathbf{F} = -\frac{4\pi G\rho}{3}\,\mathbf{r}.$$

Interpret what this expression means and comment on the status of boundary conditions at infinity in Newtonian cosmology.

42   Show how the result of Exercise 39 is modified if we introduce the analogue of the λ-term in Newtonian cosmology.

5

Relics of the big bang

5.1 The early universe

In Chapter 4 we saw that all Friedmann models have an epoch in the past when the scale factor S was zero. We referred to this epoch as the big bang epoch. To mathematicians the big bang implies a breakdown of the concept of spacetime geometry, and they have come to recognize it as an inevitable feature of Einstein's general relativity. It is a feature that prevents the physicist from investigating what happened at $S = 0$ or prior to it. To some physicists this abrupt termination of the past signifies an incompleteness of the theory of relativity. To them a more complete theory of the future may show a way of avoiding the catastrophic nature of the $S = 0$ epoch. A universe that has been expanding forever or that has been oscillating between maximum and minimum (but finite) values of S might result from such a theory.

In this chapter we will continue to put our faith in the validity of general relativity and push our investigations into the past of the universe as close as possible to the $S = 0$ epoch. The purpose of such investigations will be to find out whether we can point to any present-day evidence that the universe indeed had a past epoch when S was close to zero. In short, we will be looking for relics of the big bang.

Pioneering work in this field was done by George Gamow in the mid-1940s. Gamow was concerned with the problem of the origin of elements. Starting from the (then available) basic building blocks of neutrons and protons, Gamow attempted to describe the formation of nuclei of deuterium, helium, and so on. The process envisaged by him involved nuclear fusion, that is, a process in which nuclei are formed by bringing together neutrons and protons. Astrophysicists were already sure by the 1940s that such processes operate inside stars, where the necessary conditions of high temperature and density were known to exist. Gamow pointed out that similar conditions must have existed in a typical Friedmann universe soon after the big bang.

We know from (4.22) that the density ρ was very high at small values of S. What about temperature? A simple calculation shows how the temperature also might have been high. This calculation requires the assumption that at present we have a radiation density u_0 that is a relic of an early hot era. With this assumption, the radiation energy density at a past epoch S is given by (4.27):

$$u = u_0 \frac{S_0^4}{S^4}. \tag{5.1}$$

We also saw in Chapter 4 that at a critical value of the scale factor the contribution of radiation energy density equals that of matter energy density, and that prior to this epoch the former was more dominant. Gamow therefore assumed that in the early epochs the dynamics of expansion were determined by radiant energy rather than by matter in the form of dust.

If we wish to make a simplified calculation, we can assume that the radiation was in blackbody form with temperature T, so that

$$u = aT^4, \tag{5.2}$$

where a is the radiation constant. This means that in the early stages of the big bang universe

$$T_0^0 = aT^4, \quad T_1^1 = T_2^2 = T_3^3 = \tfrac{1}{3}aT^4. \tag{5.3}$$

We also anticipate the the space curvature parameter k will not affect the dynamics of the early universe significantly, and set it equal to zero. Thus from (4.12)

$$\frac{\dot{S}^2}{S^2} = \frac{8\pi Ga}{3c^2}T^4. \tag{5.4}$$

Further, from (5.1) and (5.2) we get

$$T = \frac{A}{S}, \quad A = \text{constant}. \tag{5.5}$$

Substituting (5.5) into (5.4) gives a differential equation for S that can be easily solved. Setting $t = 0$ at $S = 0$ we get

$$S = A \left(\frac{3c^2}{32\pi Ga}\right)^{-1/4} t^{1/2}, \tag{5.6}$$

and, more importantly,

$$T = \left(\frac{3c^2}{32\pi Ga}\right)^{1/4} t^{-1/2}. \tag{5.7}$$

Notice that all the quantities inside the parentheses on the right-hand side of the above equation are known physical quantities. Thus we can express the above result in the following form:

$$T_{kelvin} = 1.52 \times 10^{10} t_{seconds}^{-1/2}. \tag{5.8}$$

In other words, about one second after the big bang the radiation temperature of the universe was 1.52×10^{10} K. The universe at this stage was certainly hot enough to facilitate nucleosynthesis, as Gamow supposed.

The idea of a hot big bang, as the above picture is called, depends therefore on the assumption that there is relic radiation present today and the microwave background discovered in 1965 by Arno Penzias and Robert Wilson is that relic radiation. For the present we will accept this evidence as confirming Gamow's notion of the hot big bang and proceed further.

5.2 Thermodynamics of the early universe

Considerable progress has been made in our understanding of the properties of particles and their basic interactions since the days when Gamow and his colleagues R. A. Alpher and R. Herman did their calculations of primordial nucleosynthesis. In the following pages we will briefly outline the basic principles on which the modern calculations are usually based.

First it is necessary to specify the building blocks from which nuclei were constructed in the early epochs. The physicist would naturally like to imagine that the universe started with the simplest possible composition (whatever that may be!) and that more complex structures were built out of simpler ones by physical interactions. Thus the cosmologist is forced to take stock of the knowledge of particle physics. While Gamow and his colleagues took the existence of particles like protons, neutrons, electrons, and so on for granted, modern particle physicists believe that a more basic framework accounts for the creation or existence of these particles.

Here we take up the story from the stage when baryons (neutrons and protons), leptons (electrons, muons, neutrinos, and their antiparticles), and photons (the particles of light) are already in existence and are in thermodynamic equilibrium as particles of an ideal gas. In Chapter 6 we will consider the more speculative and earlier epochs and discuss how these particles came into existence.

Before proceeding with calculations we must clarify what is meant by 'thermodynamic equilibrium' and 'ideal gas'. We have already mentioned that in these early epochs the dominant form of energy was in these particles moving relativistically. The question arises therefore whether these particles were interacting with one another or whether they were mostly moving freely. Such particles would interact and collide, of course, but these instances are assumed to have occupied very brief time spans, and their effects on motions may be otherwise neglected. We will shortly express this idea in a quantitive manner.

The collisions and scatterings of the particles would, however, have helped to redistribute their energies and momenta. If these redistributions occurred frequently enough then the system of particles as a whole would have reached a state of thermodynamic equilibrium. In this case, for each species of particles there is a definite rule governing the number of particles in a given range of momentum. For thermodynamic equilibrium to be reached, the time scales between successive scatterings should be small compared with the expansion time scale for the universe. Again, we will express this idea quantitively in a short while.

5.2.1 Distribution functions

Assuming ideal gas approximation and thermodynamic equilibrium, it is then possible to write down the distribution functions of any given species of particles. Let us use the symbol A to denote typical species $(A = 1, 2, \ldots)$. Thus $n_A(P)\,dP$ denotes the number density of species in the momentum range $(P, P + dP)$, where

$$n_A(P) = \frac{g_A}{2\pi^2\hbar^3} P^2 \left[\exp\left(\frac{E_A(P) - \mu_A}{kT}\right) \pm 1\right]^{-1}. \tag{5.9}$$

In the above formula, $T =$ the temperature of the distribution, $g_A =$ the number of spin states of the species, $k =$ the Boltzmann constant, and

$$E_A^2 = c^2 P^2 + m_A^2 c^4 \tag{5.10}$$

is the energy corresponding to rest mass m_A of a typical particle. Thus for the electron $g_A = 2$; for the neutrino $g_A = 1$, $m_A = 0$, and so on. The $+$ sign in (5.9) applies to particles obeying the Fermi–Dirac statistics (these particles are called *fermions*), while the $-$ sign applies to particles obeying the Bose–Einstein statistics (particles known as *bosons*). For example, electrons and neutrinos are fermions, and photons are bosons.

The quantity μ_A is the chemical potential of the species A. For a detailed discussion of chemical potentials, see any standard text on thermodynamics and statistical mechanics. We note here that in any reaction involving these particles the μ_A are conserved (just as electric charge, energy, spin, and so on are conserved). Because photons can be absorbed or emitted in any number in a typical reaction, we set $\mu_A = 0$ for photons. Since particles and antiparticles (such as electrons and positrons) annihilate in pairs and produce photons, their chemical potentials are equal and opposite.

Apart from the dynamic quantities and the electric charge, several other quantities are found to be conserved in the interactions of particles. These are the baryon number, the muon lepton number, and the electron lepton number. In computing these numbers, a value of $+1$ is assigned to a particle and -1 to its antiparticle. The electron lepton number counts electrons (e^-) and their neutrinos (ν_e), while the muon lepton number counts muons (μ^-) and their neutrinos (ν_μ). Under these conservation rules reactions such as

$$n \rightarrow p + e^- + \bar{\nu}_e, \qquad p + \bar{\nu}_\mu \rightarrow \mu^+ + n$$

are permitted, while a reaction like the following is not:

$$n \rightarrow p + e^- + \nu_e.$$

(In Chapter 6 we will consider the situation in which the baryon number is not conserved. However, at the epochs that we are concerned with here we may safely assume the conservation of the baryon number density to apply.)

Hence, if we assume that, in any reaction electric charge, the baryon number, the electron lepton number, and the muon lepton number are conserved, then we have only four independent chemical potentials – those corresponding to protons, electron neutrinos, and muon neutrinos. (In Chapter 6 we will consider the possibility of more species of leptons/neutrinos being present.) From (5.9) we see that the total number of particles per unit volume in each of these species is needed to determine the corresponding μ_A and that the number densities will be large for large $\mu_A > 0$. These number densities are not known with any degree of accuracy, except that (as we shall see shortly) the ratio

$$\frac{N_B}{N_\gamma} = \frac{\text{number density of baryons}}{\text{number density of photons}} \sim 10^{-8} - 10^{-10}$$

is small compared with 1.

The smallness of the baryon number density suggests that the number densities of leptons may also be small compared with N_γ, and it is usually assumed that this hypothesis provides a good justification for taking $\mu_A = 0$ for all species. We will assume that $\mu_A = 0$ for all species in our calculations to follow.

We then get the following integrals for the particle number density (N_A), the energy density (ε_A), pressure (p_A), and entropy density (s_A):

$$N_A = \frac{g_A}{2\pi^2 \hbar^3} \int_0^\infty \frac{P^2 \, dP}{\exp[E_A(P)/kT] \pm 1}; \tag{5.11}$$

$$\varepsilon_A = \frac{g_A}{2\pi^2 \hbar^3} \int_0^\infty \frac{P^2 E_A(P) \, dP}{\exp[E_A(P)/kT] \pm 1}; \tag{5.12}$$

$$p_A = \frac{g_A}{6\pi^2 \hbar^3} \int_0^\infty \frac{c^2 P^4 [E_A(P)]^{-1} \, dP}{\exp[E_A(P)/kT] \pm 1}; \tag{5.13}$$

$$s_A = (p_A + \varepsilon_A)/T. \tag{5.14}$$

5.2.2 High- and low-temperature approximations

The above expressions become simplified for particles moving relativistically. In this case,

$$T \gg \frac{m_A c^2}{k} \equiv T_A. \tag{5.15}$$

The details are given in Table 5.1 for the different species of interest. The numbers are expressed in units of the quantities for the photon ($g_A = 2$; the symbol for a photon is γ):

$$N_\gamma = \frac{2.404}{\pi^2} \left(\frac{kT}{c\hbar}\right)^3, \quad \varepsilon_\gamma = \frac{\pi^2 (kT)^4}{15\hbar^3 c^3} = 3p_\gamma, \quad s_\gamma = \frac{4\pi^2 k}{45} \left(\frac{kT}{c\hbar}\right)^3. \tag{5.16}$$

In this approximation consider the electrical potential energy of any two electrons separated by distance r. This is given by

$$U = \frac{e^2}{r}.$$

Now the average interelectron distance is given by $N_e^{-1/3} \sim c\hbar/kT$. Thus average interaction energy is

$$(U) \sim \frac{e^2}{\hbar c} kT.$$

However, kT measures the energy of motion of electrons. Thus the interaction energy is $e^2/\hbar c \sim \frac{1}{137}$ of the energy of motion. Since the fraction is small, we are justified in treating the electrons as free gas.

Table 5.1 *Thermodynamic quantities for various particle species at* $T \gg T_A$

Particle species A	Symbol	T_A (K)	g_A	N_A/N_γ	$\varepsilon_A/\varepsilon_\gamma$	S_A/S_γ
Electron	e^-	5.93×10^6	2	$\frac{3}{4}$	$\frac{7}{8}$	$\frac{7}{8}$
Positron	e^+		2	$\frac{3}{4}$	$\frac{7}{8}$	$\frac{7}{8}$
Muon	μ^-	1.22×10^{12}	2	$\frac{3}{4}$	$\frac{7}{8}$	$\frac{7}{8}$
Antimuon	μ^+		2	$\frac{3}{4}$	$\frac{7}{8}$	$\frac{7}{8}$
Muon, electron	ν_μ, ν_e	0	1	$\frac{3}{8}$	$\frac{7}{16}$	$\frac{7}{16}$
neutrinos and their	$\bar{\nu}_\mu, \bar{\nu}_e$		1	$\frac{3}{8}$	$\frac{7}{16}$	$\frac{7}{16}$
antineutrinos						
Pions	π^+		1	$\frac{1}{2}$	$\frac{1}{2}$	$\frac{1}{2}$
	π^-	1.6×10^{12}	1	$\frac{1}{2}$	$\frac{1}{2}$	$\frac{1}{2}$
	π^0		1	$\frac{1}{2}$	$\frac{1}{2}$	$\frac{1}{2}$
Proton	p	10^{13}	2	$\frac{3}{4}$	$\frac{7}{8}$	$\frac{7}{8}$
Neutron	n	$T_n - T_p \sim 1.5 \times 10^{10}$	2	$\frac{3}{4}$	$\frac{7}{8}$	$\frac{7}{8}$

By contrast, at low temperatures $T \leqslant T_A$ we have for all species with $m_A \neq 0$

$$N_A = \frac{g_A}{h^3} \left(\frac{m_A kT}{2\pi} \right)^3 \exp\left(-\frac{T_A}{T} \right),$$

$$\varepsilon_A = m_A N_A, \quad p_A = N_A kT, \quad s_A = \frac{m_A N_A}{T} c^2. \tag{5.17}$$

Notice that with fall in temperature all these quantities drop off rapidly. We will often refer to this limit as the nonrelativistic approximation. (For the photon and a zero-rest-mass neutrino $T_A = 0$, and this approximation never applies.)

5.2.3 The behaviour of entropy

We now recall the conservation law satisfied by ε and p in the early stages of the expanding universe, the law given by (4.15),

$$\frac{d}{dS} (\varepsilon S^3) + 3p S^2 = 0, \tag{5.18}$$

and use it in conjuction with the second law of thermodynamics. This law tells us that the entropy in a given volume S^3 stays constant as the volume expands adiabatically. From (5.14) we therefore get

$$\frac{d}{dt}(S^3 s) = \frac{d}{dt}\left[\frac{S^3}{T}(p + \varepsilon)\right] = 0, \tag{5.19}$$

where $s = \sum_A s_A$ is the total entropy of all the particles in the expanding volume.

Rewriting (5.19) with the help of (5.18), we get

$$0 = \frac{d}{dt}\left(\frac{S^3 p}{T}\right) + \frac{1}{T}\frac{d}{dt}(S^3 \varepsilon) + S^3 \varepsilon \frac{d}{dt}\left(\frac{1}{T}\right)$$

$$= \frac{d}{dt}\left(\frac{S^3 p}{T}\right) - \frac{3pS^2}{T}\dot{S} + S^3 \varepsilon \frac{d}{dt}\left(\frac{1}{T}\right),$$

that is,

$$\frac{dp}{dT} = \frac{1}{T}(p + \varepsilon). \tag{5.20}$$

The relation can be directly derived from (5.12) and (5.13) by a simple manipulation of the integrals. Then, starting from (5.20) we can derive (5.19). We will use the constancy of

$$\sigma = \frac{S^3}{T}(p + \varepsilon) \tag{5.21}$$

in our later calculations.

In the high-temperature approximation we get $p = \varepsilon/3 \propto S^{-4}$ from (5.18). Hence from the constancy of σ we recover the relation (5.5):

$$T \propto 1/S.$$

A simple relation like this does not hold if the high-temperature approximation is not valid.

5.3 Primordial neutrinos

From Table 5.1 we see that for $T < 1.5 \times 10^{12}$ K, the only particles that can be present with appreciable number densities in thermal equilibrium are $\mu^\pm$, $e^\pm$, ν_e, $\bar{\nu}_e$, ν_μ, $\bar{\nu}_\mu$, and γ. The baryons (p and n) and pions $(\pi^\pm, \pi^0)$ will be cooled below their critical temperatures T_A, so that for them the low-temperature approximation holds. The photons, $e^\pm$ and $\mu^\pm$, follow their respective distributions of the type (5.9). The neutrinos, however, require some attention, since this phase happens to be crucial in determining the extent of their survival.

The neutrinos are absorbed, emitted, or scattered in reactions such as the following:

$$e^- + \mu^+ \leftrightarrow \nu_e + \bar{\nu}_\mu, \qquad e^+ + \mu^- \leftrightarrow \bar{\nu}_e + \nu_\mu, \qquad \nu_e + \mu^- \leftrightarrow \nu_\mu + e^-,$$

$$\bar{\nu}_e + \mu^+ \leftrightarrow \bar{\nu}_\mu + e^+, \qquad \nu_e + e^- \leftrightarrow \nu_e + e^-, \qquad \bar{\nu}_e + e^+ \leftrightarrow \bar{\nu}_e + e^+,$$

These are all examples of weak interactions. For $T \leqslant T_\mu$ the cross-section of a typical reaction is of the order

$$\Sigma = \mathcal{G}^2 \hbar^{-4} (kT)^2 c^{-4}, \tag{5.22}$$

where $\mathcal{G} = 1.4 \times 10^{-49} \text{ erg cm}^{-3}$ is the weak interaction coupling constant. From (5.16) and Table 5.1 we see that the number densities of participating particles $e^\pm$ is of the order

$$(kT/c\hbar)^3,$$

while for muons we should take account of (5.17) and introduce an exponential damping factor of

$$\exp\left(-\frac{T_\mu}{T}\right).$$

Thus typical neutrino reaction rate is

$$\eta = c\Sigma \cdot \left(\frac{kT}{c\hbar}\right)^3 \exp\left(-\frac{T_\mu}{T}\right) = \mathcal{G}^2 \hbar^{-7} c^{-6} (kT)^5 \exp\left(-\frac{T_\mu}{T}\right). \tag{5.23}$$

We must now take note of the other rate that is relevant to the maintenance of equilibrium of neutrinos – the rate at which a typical volume enclosing them expands. From Einstein's equations we get

$$H^2 = \frac{\dot{S}^2}{S} = \frac{8\pi G}{3c^2} \varepsilon \approx \frac{16\pi^3 G}{90\hbar^3 c^5} (kT)^4. \tag{5.24}$$

H, the Hubble constant at the particular epoch, measures the rate of expansion of the volume in question. Thus the ratio of the reaction rate to the expansion rate is given by

$$\frac{\eta}{H} \sim G^{-1/2} \hbar^{-11/2} \mathcal{G}^2 c^{-7/2} (kT)^3 \exp\left(-\frac{T_\mu}{T}\right)$$

$$\sim \left(\frac{T}{10^{10} \text{ K}}\right)^3 \exp\left(-\frac{10^{12} \text{ K}}{T}\right) \tag{5.25}$$

$$= T_{10}^3 \exp\left(-\frac{1}{T_{12}}\right).$$

Here we have substituted the values of G, $\hbar$, $\mathcal{G}$, c, k, and T_μ and arrived at the above numerical expression. Further, we have written the temperature using the notation T_{10}, T_{12}, and so on. In general T_n indicates temperature expressed in units of 10^n K.

What does (5.25) tell us? As the temperature drops below 10^{12} K, the exponential decreases rapidly. This means that the reactions involving neutrinos run at slower rate compared to the expansion rate of the universe. The neutrinos then cease to interact with the rest of the matter and therefore drop out of thermal equilibrium as temperatures fall appreciably below $T_{12} = 1$. How far below?

The original theory of weak interactions suggested that this temperature may be around $T_{11} = 1.3$. In the late 1960s and early 1970s successful attempts to unify the weak interaction with the electromagnetic interaction led to additional (neutral current) reactions that keep neutrinos interacting with other matter at even lower temperatures. The outcome of these investigations is that the neutrinos can remain in thermal equilibrium down to temperatures of the order $T_{10} = 1$.

However, even though neutrinos decouple themselves from the rest of the matter, their distribution function still retains its original form with the temperature dropping as $T \propto S^{-1}$. This is because as the universe expands the momentum and energy of each neutrino falls as S^{-1} and the number density of neutrinos falls as S^{-3}. Since the temperature of the rest of the mixture also drops as S^{-1} and since the two temperatures were equal when the neutrinos were coupled with the rest of the matter, they continue to remain equal even though neutrinos and the rest of the matter are no longer in interaction with one another. These remarks apply to neutrinos of all four species ν_e, $\bar{\nu}_e$, ν_μ, $\bar{\nu}_\mu$.

There is, however, another epoch when the neutrino temperature begins to differ from the temperature of the rest of the matter. First consider the universe in the temperature range $T_{12} = 1$ to $T_{10} = 1$. In this phase we have the neutrinos, the electron–positron pairs, and the photons, each with distribution functions of type (5.9) in the high-temperature approximation (see Table 5.1). Thus

$$\varepsilon = \varepsilon_{\nu_e} + \varepsilon_{\bar{\nu}_e} + \varepsilon_{\nu_\mu} + \varepsilon_{\bar{\nu}_\mu} + \varepsilon_{e-} + \varepsilon_{e+} + \varepsilon_\gamma.$$

Counting the various g-factors from Table 5.1, we get

$$\varepsilon = \tfrac{9}{2} a T^4. \tag{5.26}$$

Thus in this period the expansion equation is modified from our simplified formula (5.4) to

$$\frac{\dot{S}^2}{S^2} = \frac{12\pi Ga}{c^2} T^4 \tag{5.27}$$

and the relation (5.7) is changed to

$$T = \left(\frac{c^2}{48\pi Ga}\right)^{1/4} t^{-1/2}, \tag{5.28}$$

which we may rewrite as

$$T_{10} = 1.04 t_{\text{seconds}}^{-1/2}. \tag{5.29}$$

However, in the next phase the situation becomes complicated, as the $e^{\pm}$ pairs are no longer relativistic. Thus the high-temperature approximation is no longer valid for them and we have to use the full formulae (5.12) and (5.13) to determine the ε and p and the expansion rate of the universe. We will not go into details of this phase but instead jump across to its end, when the pairs have annihilated, leaving only photons:

$$e^- + e^+ \rightarrow \gamma + \gamma. \tag{5.30}$$

Thus the energy, originally in $e^{\pm}$ and photons, has now vested only in photons, raising their number and temperature. How can we evaluate this change? It is here that (5.21), telling us of the constancy of σ, comes to our help.

In the relativistic phase $(T_9 > 5)$ of $e^{\pm}$ we have

$$\sigma = \frac{4S^3}{3T} (\varepsilon_{e-} + \varepsilon_{e+} + \varepsilon_\gamma) = \tfrac{11}{3} a (ST)^3. \tag{5.31}$$

When the $e^{\pm}$ have annihilated and left only photons, we have the photon temperature T_γ given by

$$\sigma = \frac{4}{3} \frac{S^3}{T_\gamma} \varepsilon_\gamma = \tfrac{4}{3} a (ST_\gamma)^3. \tag{5.32}$$

We now use the result that the neutrino temperature always changes as S^{-1}. Let us write it as

$$T_\nu = B/S, \qquad B = \text{constant}. \tag{5.33}$$

Then (5.31) gives

$$\sigma = \tfrac{11}{3} a B^3 \left(\frac{T}{T_\nu} \right)^3. \tag{5.34}$$

Similarly, (5.32) gives

$$\sigma = \tfrac{4}{3} a B^3 \left(\frac{T_\gamma}{T_\nu} \right)^3. \tag{5.35}$$

Now in the preannihilation era, $T = T_\nu$, so that (5.34) tells us $\sigma = \tfrac{11}{3} a B^3$. After annihilation σ must have the same value, so we may equate it to the value given by (5.35). Thus we arrive at the conclusion that the photon temperature at the end of $e^{\pm}$ annihilation has risen *above* the neutrino temperature by the factor

$$\frac{T_\gamma}{T_\nu} = \left(\tfrac{11}{4} \right)^{1/3} \approx 1.4. \tag{5.36}$$

So the present-day neutrino temperature is *lower* than the photon temperature by the factor $(1.4)^{-1}$. If we take the latter ~ 3 K, the former is ~ 2 K.

5.4 The neutron/proton ratio

We have so far developed a picture of the early universe that is best expressed in the form of a time–temperature table of events, as shown in Table 5.2 (see also Figure 5.1).

In our discussion so far we have not paid much attention to baryons – the protons and neutrons that are also present in the mixture. In our approximation of setting the chemical potentials to zero we took the baryon number to be zero. The validity of the approximation depended on the baryon number density being several orders (8 to 10) of magnitude smaller than the photon density. Nevertheless, we must now take note of the existence of baryons, however small their number density; for we need them in order to consider Gamow's idea of nucleosynthesis in the hot universe.

First notice that the temperatures T_n and T_p of Table 5.1 are very high, so that the neutron and proton distribution functions follow the non-relativistic approximations of (5.17). Thus we get

Table 5.2 *A time–temperature table of events preceding nucleosynthesis in the early universe*

Time since big bang (s)	Temperature (K)	Events
$\leqslant 10^{-4}$	$> 10^{12}$	Baryons, mesons, leptons, and photons in thermal equilibrium.
10^{-4}–10^{-2}	10^{12}–10^{11}	$\mu^{\pm}$ begin to annihilate and disappear from the mixture. Neutrinos begin to form rest of matter.
10^{-2}–1	10^{11}–10^{10}	Neutrinos decouple completely. $e^{\pm}$ pairs still relativistic.
1–180	10^{10}–10^{9}	The pairs of $e^{\pm}$ annihilate and disappear, raising the photon gas temperature to ~ 1.4 times the temperature of neutrinos.

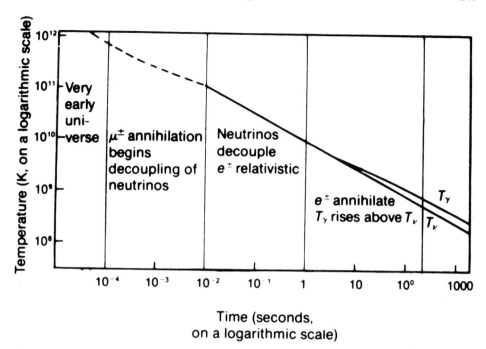

Fig. 5.1 The time–temperature plot of the early universe. The dotted portion does not describe the (t, T) relationship accurately, because the particles (especially pions and μ^+) interact and are not really free. A more reliable picture emerges for $t \geqslant 10^{-2}$ s. Notice the difference in temperatures of neutrinos and photons for $t \geqslant 10$ s. See the text for details.

$$N_{\mathrm{p}} = \frac{2}{\hbar^3} \left(\frac{m_{\mathrm{p}} kT}{2\pi} \right)^{3/2} \exp\left(-\frac{T_{\mathrm{p}}}{T} \right),$$

$$N_{\mathrm{n}} = \frac{2}{\hbar^3} \left(\frac{m_{\mathrm{n}} kT}{2\pi} \right)^{3/2} \exp\left(-\frac{T_{\mathrm{n}}}{T} \right). \tag{5.37}$$

In this approximation the neutron-to-proton number ratio is given by

$$\frac{N_{\mathrm{n}}}{N_{\mathrm{p}}} \approx \exp\left(\frac{T_{\mathrm{p}} - T_{\mathrm{n}}}{T} \right) = \exp\left(-\frac{1.5}{T_{10}} \right). \tag{5.38}$$

The ratio therefore drops with temperature, from near $1 : 1$ at $T \geqslant 10^{12}$ K to about $5 : 6$ at $T = 10^{11}$ K, and to 3.5 at 3×10^{10} K.

For thermodynamic equilibrium to be maintained, the reactions that convert neutrons to protons and vice versa have to be rapid enough compared with the rate at which the universe expands. These interactions are none other than the weak interactions considered earlier when we discussed the decoupling of neutrinos from the rest of the primordial brew (see section 5.3). There is one difference, however. In discussing the decoupling of neutrinos we were concerned mainly with the reaction of a neutrino with leptons like $e^{\pm}$, $\mu^{\pm}$, and the cross-section Σ given by (5.22) was determined for such interactions. Similarly, the reaction rate η given by (5.23) was obtained by multiplying by the number densities of participating leptons.

In the present case the cross-section for a typical reaction like

$$\nu_e + n \leftrightarrow e^- + p$$

is larger than that for the pure leptonic reaction like

$$\nu_\mu + e \leftrightarrow \nu_e + \mu.$$

Also, the lepton densities used in (5.23) were considerably higher than the nucleon densities we are considering now. So the probability of a given nucleon interacting with any neutrino is higher than the probability of a given neutrino interacting with any nucleon. The result is that the effective temperature at which n and p cease to be in thermodynamic equilibrium is lower than the effective temperature for neutrino decoupling determined earlier.

Quantitatively, instead of $\Sigma \propto T^2$ as in (5.22), the cross-section in the present case goes as $\propto T$, and the effective decoupling temperature T_* at which the reaction rate is just about equal to H is $<10^{10}$ K. Note that if the universe was expanding faster then T_* would be higher and the ratio N_n/N_p at decoupling as given by (5.38) would be higher.

Once the thermodynamic equilibrium ceases to be maintained, the N_n/N_p ratio is not given by (5.38) but by detailed consideration of specific reactions involving the nucleons.

As the universe cooled further, this ratio was therefore determined by the reactions that change protons to neutrons and vice versa. These are essentially weak interactions of the type

$$n + \nu_e \leftrightarrow p + e^-, \qquad n \leftrightarrow p + e^- + \bar{\nu}_e.$$

The reaction rates are therefore determined by the cross-sections computed according to the weak interaction theory. Until the electro-weak gauge theory became established in the late 1970s, the $V - A$ theory of weak interaction was used for these computations. We will not go into details of the calculation here, the purpose of which is to come up with a differential equation for the ratio

$$X_n = \frac{N_n}{N_p + N_n}. \tag{5.39}$$

If $\lambda(n \to p)$ denotes the rate at which neutrons are converted to protons and $\lambda(p \to n)$ the corresponding rate for protons changing to neutrons then clearly X_n satisfies the equation

$$\frac{dX_n}{dt} = (1 - X_n) \cdot \lambda(p \to n) - X_n \cdot \lambda(n \to p). \tag{5.40}$$

The rates λ depend on distribution functions of leptons, which in turn depend on the temperature, which is related to the scale factor of the expanding universe. The integration of (5.40) has to be done numerically, and it is continued until all $e^{\pm}$ pairs have dropped out of the mixture – which happens at $T \geqslant 10^9$ K.

When all $e^{\pm}$ have disappeared it is still possible for the neutrons to decay via the reaction

$$n \to p = e^- + \bar{\nu}_e,$$

with a characteristic time $\tau = 1013$ s. So from the time the pairs disappear to the onset of nucleosynthesis the neutron ratio X_n will decrease by the exponential factor $\exp(-t/\tau)$.

Thus the ratio of neutrons to protons is uniquely determined at the time nucleosynthesis begins, once we know all the parameters of the weak interaction process. This is one good aspect of primordial nucleosynthesis theory, which we will now proceed to discuss.

5.5 The synthesis of helium and other nuclei

A typical nucleus Q is described by two quantities, A = atomic mass and Z = atomic number, and is written

$$^A_Z Q.$$

This nucleus has Z protons and $(A - Z)$ neutrons. If m_Q is the mass of the nucleus, its binding energy is given by

$$B_Q = [Zm_p + (A - Z)m_n - m_Q]c^2. \tag{5.41}$$

Let us now consider a unit volume of cosmological medium containing N_N nucleons, bound or free. Since the masses of protons and neutrons are nearly equal, we may denote the typical nucleon mass by m. Thus $m_n \approx m_p = m$. If there are N_n free neutrons and N_p free protons in the mixture

$$X_n = \frac{N_n}{N_N}, \qquad X_p = \frac{N_p}{N_N} \tag{5.42}$$

will denote the fractions by weight of free neutrons and free protons. If a typical bound nucleus Q has atomic mass A and there are N_Q of them in our unit volume, we may denote the weight fraction of Q by

$$X_Q = \frac{N_Q A}{N_N}. \tag{5.43}$$

Now at very high temperatures ($T \gg 10^{10}$ K), the nuclei are expected to be in thermal equilibrium. However, even at these temperatures $T \ll T_Q$ and (5.17) holds. Further, since we are now concerned with relative number densities, we can no longer ignore the chemical potentials. Thus

$$N_Q = g_Q \left(\frac{m_Q kT}{2\pi\hbar^2}\right)^{3/2} \exp\left(\frac{\mu_Q - m_Q c^2}{kT}\right) \tag{5.44}$$

where we have reinstated the chemical potentials μ_Q. Since chemical potentials are conserved in nuclear reactions,

$$\mu_Q = Z\mu_p + (A - Z)\mu_n, \tag{5.45}$$

assuming that the nuclei were built out of neutrons and protons by nuclear reactions.

The unknown chemical potentials can be eliminated between (5.44) and similar relations for N_p and N_n. The result is expressed in this form:

$$X_Q = \tfrac{1}{2} g_Q A^{5/2} X_p^Z X_n^{A-Z} \xi^{A-1} \exp\left(\frac{B_Q}{kT}\right) \tag{5.46}$$

where

$$\xi = \tfrac{1}{2} N_N \left(\frac{mkT}{2\pi\hbar^2}\right)^{-3/2}. \tag{5.47}$$

For an appreciable buildup of complex nuclei, T must drop to a low enough value to make $\exp(B_Q/kT)$ large enough to compensate for the smallness of ξ^{A-1}. This happens for nucleus Q when T has dropped down to

$$T_Q \sim \frac{B_Q}{k(A - 1)|\ln \xi|}. \tag{5.48}$$

Let us consider what happens when we apply the above formula to the nucleus of ^{4}He. The binding energy of this nucleus is $\approx 4.3 \times 10^{-5}$ erg. If we substitute this value in (5.48) and estimate N_N from the presently observed value of nucleon density of around 10^{-6} cm^{-3}, we find that T_Q is as low as $\sim 3 \times 10^9$ K (see Exercise 23). However, at this low temperature the number densities of participating nucleons are so low that four-body encounters leading to the formation of ^{4}He are extremely rare. Thus the underlying assumption of thermodynamic equilibrium (which requires frequent collisions) leading to (5.48) becomes invalid. We therefore need to proceed in a less ambitious fashion in order to describe the buildup of complex nuclei.

Hence we try using two-body collisions (which are not so rare) to describe the buildup of heavier nuclei. Thus deuterium (d), tritium (^{3}H), and helium (^{3}He, ^{4}He) are formed via reactions like

$$p + n \leftrightarrow d + \gamma,$$
$$d + d \leftrightarrow {}^3\text{He} + n \leftrightarrow {}^3\text{H} + p, \tag{5.49}$$
$${}^3\text{H} + d \leftrightarrow {}^4\text{He} + n.$$

since formation of deterium involves only two-body collisions, it quickly reaches its equilibrium abundance as given by

$$X_{\text{d}} = \frac{3}{2^{1/2}} X_{\text{p}} X_{\text{n}} \xi \exp\left(\frac{B_{\text{d}}}{kT}\right). \tag{5.50}$$

However, the binding energy B_{d} of deuterium is low so that unless T drops to less than 10^9 K, X_{d} is not high enough to start further reactions leading to ^{3}H, ^{3}He, and ^{4}He. In fact the reactions given in (5.49), with the exception of the first one, do not proceed fast enough until the temperature has dropped to $\sim 8 \times 10^8$ K.

Although at such temperatures nucleosynthesis does proceed rapidly enough, it cannot go beyond ^{4}He. This is because there are no stable nuclei with $A = 5$ to 8, and nuclei heavier than ^{4}He. So the process terminates there. Detailed calculations by several authors have now established this result quite firmly.

So, starting with primordial neutrons and protons, we end up finally with ^{4}He nuclei and free protons. All neutrons have been gobbled up by helium nuclei. Thus if we consider the fraction by weight of primordial helium, it is very simply related to the quantity X_n – the neutron concentration before nucleosynthesis began. Denoting this fraction by weight by the symbol Y, we get

$$Y = 2X_{\text{n}}. \tag{5.51}$$

In Figure 5.2 the cosmic weight fractions of ^{4}He, ^{3}He, and ^{2}H and so on are plotted against a parameter η defined by

$$\eta = \left(\frac{\rho_0}{2.7 \times 10^{-26} \text{ g cm}^{-3}}\right)\left(\frac{3}{T_0}\right)^3. \tag{5.52}$$

Thus η essentially measures the nucleon density in the early universe through the formula

$$\rho = \eta T_9^3, \qquad T_9 < 3. \tag{5.53}$$

Note that the ^{4}He weight fraction is close to $\sim \frac{1}{4}$ and insensitive to the

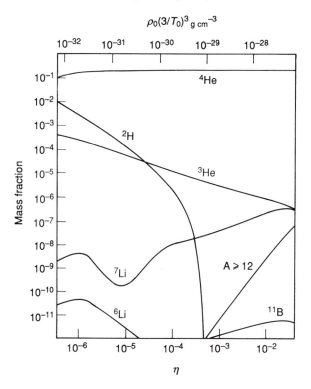

Fig. 5.2 Primordial abundances of light nuclei as functions of the present density of matter in the universe. The relation between ρ_0 and η is given by (5.52). (After R. V. Wagoner, The early universe. In R. Balian, J. Audouze, & D. N. Schramm, eds, *Physical Cosmology*, Les Houches Lectures Session XXXII, p. 395 (Amsterdam: North Holland, 1979).)

parameter η. This is because, as we saw just now, it depends only on X_n, which in turn depends more critically on the epoch when the weak interaction rate fell below the expansion rate. If we go back to (5.38) we see that in the very early stages the neutron/proton ratio depends on temperature T_*. A faster expansion rate implies that the ratio becomes frozen at a higher temperature and so is higher, thus leading to a higher ^{4}He abundance.

To see the effect quantitatively, recall from (5.38) that there was a 'last epoch' of temperature T_* when the neutron/proton ratio was determined from considerations of thermodynamic equilibrium:

$$x = \frac{N_n}{N_p} = \exp\left(-\frac{1.5}{T_{*10}}\right).$$

The temperature T_* was determined by equating the Hubble constant H to the reaction rate η for n $\leftrightarrow$ p conversions. Now

$$H \propto g^{1/2} T_*{}^2 \qquad \text{and} \qquad \eta \propto T_*{}^4,$$

so that

$$T_*{}^2 \propto g^{1/2}.$$

Writing $Y = 2N_n/(N_n + N_p)$ for the weight fraction of helium, we can estimate the change in Y due to a change in g (from increasing the neutrino flavours) as follows. As $g \to g + \delta g$, the above relations imply $x \to x + \delta x$, $Y \to Y + \delta Y$, where

$$\delta Y = \frac{x \ln x}{(1 + x)^2} \frac{\delta g}{g}.$$

If there are l lepton families,

$$g = \tfrac{7}{8}(4 + 2l) + 2.$$

Hence an increase from $l = 2$ to $l = 3$ gives $\delta g/g \sim \tfrac{1}{5}$. For $Y = \tfrac{1}{4}$, $x = \tfrac{1}{7}$ we get $\delta Y \approx 0.02$.

This result is relevant to the question of how many different types of neutrinos exist primordially. In the GUT formalism described in Chapter 6 there are three neutrino types, ν_e, ν_μ, ν_γ. Other formalisms may permit even more types of neutrinos to exist, thereby forcing the value of Y upwards. When we look at observations we will discover that the present estimates of helium abundance rule out more than three neutrino types. It is also interesting that the particle accelerator experiments appear to lead to the same conclusion.

In contrast to the behaviour of Y, which does not sensitively depend on η, the abundances of other nuclei do depend on η. These abundances are very small compared with Y. Only nuclei heavier than ^{4}He eventually survive; ^{3}H (tritium) decays to ^{3}He. Of nuclei heavier than ^{4}He, only ^{7}Li (lithium) appears with any appreciable quantities, although smaller than ^{3}He. The most interesting situation exists for deuterium, whose abundance sharply drops as η rises above 10^{-4}. For $T_0 = 3$ K, this corresponds to

$$\rho_0 \sim 2.7 \times 10^{-30} \text{ g cm}^{-3}. \tag{5.54}$$

Comparing this with (4.35), we see that for $h_0 = 1$, $\Omega_0 \leqslant 0.12$ and hence $q_0 \leqslant 0.06$. Therefore, if even a small amount of deuterium believed to be primordial in origin was found, Friedmann models of the closed variety would be ruled out. There is, however, a loophole in this argument: we can still accommodate nonbaryonic matter in the universe.

Figure 5.2 shows that the primordial production of heavy nuclei $(A \geqslant 12)$ is very little, and in no way can it account for their observed abundances. The main reason for this is that there are no stable nuclei with $A = 5$ and 8. Thus any attempt to synthesize heavier nuclei by adding to ^{4}He fails, whether we add a proton or another ^{4}He. In fact, to cross this gap and reach stable heavy nuclei like ^{12}C, ^{16}O, we need an altogether different scenario. Deep interiors of stars on their way to becoming red giants are suitable sites for making such nuclei.

We can sum up by saying that Gamow's expectation that the early hot universe would synthesize all types of nuclei has been only partially fulfilled. The idea works for light nuclei like D, ^{4}He, etc. To obtain complex nuclei heavier than ^{4}He (and possibly ^{7}Li), astrophysicists have to look to other sources: the stars.

5.6 The microwave background

Gamow and his colleagues Alpher and Herman made another prediction, however, that appears to have received confirmation. This is the prediction that the photons of the early hot era would have cooled down to provide a thermal radiation background in the microwaves at present. As mentioned earlier, such radiation was first detected in 1965 by Penzias and Wilson. To see how this background forms we have to follow our history of the early universe to stages subsequent to nucleosynthesis.

The era of nucleosynthesis took place when the temperature was around 10^9 K. The universe in subsequent phases cooled as it expanded, with the radiation temperature dropping as S^{-1}. The presence of nuclei, free protons, and electrons did not have much effect on the dynamics of the universe, which was still radiation-dominated. However, these particles, especially the lightest of them, the electrons, acted as scattering centres for the ambient radiation and kept it thermalized. The universe was therefore quite opaque to start with.

However, as the universe cooled, the electron–proton electrical attraction began to assert itself. In detailed calculations performed by P. J. E. Peebles, the mixture of electrons and protons and of hydrogen atoms was studied at varying temperatures. Because of Coulomb attraction between the electron and the proton, the hydrogen atom has a certain binding energy B. The problem of determining the relative number densities of free electrons, free protons (that is, ions), and neutral H-atoms in thermal equilibrium is therefore analogous to that we considered earlier in deriving (5.46) in section 5.5 for the mixture of free and bound nucleons. The only difference is that the binding to be considered now is electrostatic and not nuclear. Following the same method, we arrive at the formula relating the number densities of electrons (N_e), protons ($N_p = N_e$), and H-atoms (N_H) at a given temperature T:

$$\frac{N_e^2}{N_H} = \left(\frac{m_e kT}{2\pi\hbar^2}\right)^{3/2} \exp\left(-\frac{B}{kT}\right), \tag{5.55}$$

where m_e = electron mass. This equation is a particular case of *Saha's ionization equation*.

Writing N_B for the total baryon number density, we may express the fraction of ionization by the ratio

$$x = \frac{N_e}{N_B}.$$

Then, since $N_H = N_B - N_e$, we get from (5.55)

$$\frac{x^2}{1-x} = \frac{1}{N_B}\left(\frac{m_e kT}{2\pi\hbar^2}\right)^{3/2} \exp\left(-\frac{B}{kT}\right). \tag{5.56}$$

For the H-atom, $B = 13.59\,\text{eV}$. Substituting for various quantities on the right-hand side of (5.56), we can solve for x as a function of T. The results show that x drops sharply from 1 to near zero in the temperature range of $\sim 5000\,\text{K}$ to $2500\,\text{K}$, depending on the value of N_B, that is, on the parameter $\Omega_0 h_0^2$ (see Chapter 4). For example, for $\Omega_0 h_0^2 = 0.1$, $x = 0.003$ at $T = 3000\,\text{K}$.

Thus by this time most of the free electrons have been removed from the cosmological brew, and as a result the main agent responsible for the scattering of radiation disappears from the scene. The universe now becomes effectively transparent to radiation. This is called the recombination epoch.

The transparency of the universe means a light photon can go a long way ($\sim c/H$) without being absorbed or scattered. Therefore this epoch signifies the beginning of the new phase when matter and radiation become decoupled. This phase has lasted up to the present epoch. During this phase, the frequency of each photon is redshifted according to the rule

$$\nu \propto \frac{1}{S},$$

while the number density of photons falls as

$$N_\gamma \propto \frac{1}{S^3}.$$

It is easy to see that under these conditions the photon distribution function preserves the Planckian form, with the temperature dropping as

$$T \propto \frac{1}{S}.$$

A present background temperature of $\sim 3\,\mathrm{K}$ therefore means that the epoch when matter decoupled from radiation corresponds to a redshift of $\sim 10^3$. However, in section 4.2 we also saw that the universe changed over from being radiation-dominated to being matter-dominated around the same epoch. Why the transition from opaqueness to transparency and from radiation domination to matter domination should take place around the same time is at present unexplained and must be considered a coincidence.

Another result as yet unexplained by early universe physics is the observed ratio of photons to baryons:

$$\frac{N_\gamma}{N_B} = 4.57 \times 10^7 (\Omega_0 h_0^2)^{-1} \left(\frac{T_0}{3}\right)^3. \tag{5.57}$$

This ratio has been conserved since the time the universe became essentially transparent, although both N_γ and N_B can be studied theoretically at even earlier epochs. Why the above ratio and no other? Many physicists feel that ideas from particle physics are needed to throw light on this mystery.

The important signature of the relic radiation is, however, its spectrum. There may be small perturbations of the radiation background caused by formation of discrete structures. But these apart, we should find the background spectrum to be very close to the Planckian form. We will recall this prediction when taking stock of the observations of the microwave background.

5.7 Concluding remarks

Our investigations of the early universe in this chapter started at the epoch when the universe was very hot and barely $10^{-4}\,\mathrm{s}$ old. They concluded at epoch of redshift $\sim 10^3$ when the universe became transparent. We now have two ways to go: backwards from $10^{-4}\,\mathrm{s}$ or forwards from the 10^3 redshift epoch. In Chapter 6 we go backwards, and in Chapter 7 forwards.

The opaqueness of the universe prevents us from 'seeing' directly the redshifts of $\geqslant 10^3$. Thus any evidence of the big bang or hot universe must come indirectly. In this sense the abundances of light nuclei and the detailed observations of the microwave background provide us with the only means of checking the early history of the universe.

Exercises

1   Give the arguments that led George Gamow to the concept of the hot big bang.

2   Substitute the values of c, G, and a in (5.7) and verify the numerical coefficient in (5.8).

3   Plot a graph of T against t as given by (5.8) on a log–log scale to show the variations of temperature with time. Use this graph to read off the age of the universe when the temperature was equal to (a) 10^{12} K, (b) 10^{11} K, and (c) 10^9 K.

4   From a textbook on statistical mechanics, find the arguments that lead to the distribution functions (5.9).

5   From the reactions

$$e^- + \mu^+ \rightarrow v_e + \bar{v}_\mu, \qquad e^- + p \rightarrow v_e + n, \qquad \mu^- + p \rightarrow v_\mu + n$$

deduce that the corresponding chemical potentials satisfy the relations

$$\mu_{e^-} - \mu_{v_e} = \mu_{\mu^-} - \mu_{v_\mu} = \mu_n - \mu_p.$$

6   Give arguments to show that there are just four independent conserved quantum numbers in nuclear reactions of the type shown in Exercise 5.

7   Derive the relations (5.11) to (5.14) from the formula (5.9) in the approximation in which the chemical potentials are neglected.

8   Obtain (5.16) from (5.11) to (5.14) in the high-temperature approximation.

9   Using the Table of Constants at the end of this book, compute T_A for the species given in Table 5.1 and verify the numbers given in that table.

10   Deduce (5.17) from (5.11) to (5.14) in the low-temperature approximation.

11   Deduce (5.20) directly from (5.13) and (5.12).

12   Reverse the arguments given in the text to deduce the constancy of σ from energy conservation and the relation (5.20).

13   Write down all possible reactions involving the electron, the muon, their respective neutrinos, and the antiparticles of all of them.

14   Why does the neutrino have the degeneracy factor $g_v = 1$?

15   Use the Table of Constants at the end of this book to derive (5.25).

16   Use (5.23) to deduce that if the universe were to expand faster then the neutron/proton ratio would be frozen at a higher temperature.

17   Give arguments to show that the neutrino temperature drops as
      S^{-1} after they decouple from the rest of the matter.

18   Why is the present neutrino temperature expected to be lower than
      the photon temperature? Derive the ratio of the two temperatures
      from considerations of the early universe.

19   Using the formulae (5.12) to (5.14), deduce that during the phase
      in which the pairs e^+ and e^- are annihilating and producing
      photons the constancy of σ tells us that the photon temperature T
      changes with S according to the law

$$ST\xi(T) = \text{constant},$$

where

$$[\xi(x)]^3 = 1 + \frac{15}{2\pi^4} \int_0^\infty \frac{y^2(3x^2 + 4y^2)\,dy}{(x^2 + y^2)^{1/2}[\exp(x^2 + y^2)^{1/2} + 1]}.$$

20   A primordial mixture of relativistic bosons and fermions in the
      early universe of temperature T has the total density given by the
      formula

$$\varepsilon = \frac{\pi^2}{30\hbar^3 c^3} g_*(kT)^4.$$

Show that $g_* = g_b + \frac{7}{8}g_f$, where g_b = total spin degeneracy of all
bosons and g_f = total spin degeneracy of all fermions.

21   Discuss, with the help of Exercise 20, how the rate of expansion of
      the universe is affected by the number of species of relativistic
      particles that are present in it. How is the time–temperature
      relation affected?

22   From relations of the type (5.44), express μ_Q, μ_n, and μ_p in terms
      of the remaining quantities. Then use (5.45) to derive (5.46).

23   The binding energy of the ^{4}He nucleus is $B \approx 4.3 \times 10^{-5}$ erg.
      Show that for this nucleus $B/k(A - 1) \approx 10^{11}$ K. Next assume that
      the present value of radiation temperature is 3 K, and that of the
      nucleon density 10^{-6} cm^{-3}. Using the result that $N_N T^3 = $ constant,
      show that (5.48) givens T_Q for ^{4}He as $\sim 3.2 \times 10^9$ K.

24   Give arguments to show why the primordial helium abundance is
      insensitive to the number density of baryons in the universe.

25   The abundances of which nuclei are likely to provide a sensitive
      test of the baryonic density of the universe?

26   Can you think of a loophole that would allow $\Omega_0 = 1$ and still
      permit deuterium to be formed primordially in a standard hot
      universe model?

27 If m is the mass of a nucleon and if Ω_0 is the density parameter, show that the present number density of baryons is $3H_0^2\Omega_0/8\pi Gm$. Use this formula and the present microwave background temperature $T_0 = 3\,\mathrm{K}$ to estimate N_B in (5.56). Solve the Saha equation for $\Omega_0 = 0.1$, $h_0 = 1$ to show that at 3000 K, $x = 0.003$.

28 Derive (5.57) using the blackbody spectrum and the Friedmann cosmology. What is the corresponding ratio N_v/N_B for neutrinos?

29 Show that the form of the blackbody spectrum is preserved as the universe expands, with the effective temperature declining as S^{-1}.

30 Why is it not possible to observe the past of the universe beyond the redshift of $\sim 10^3$?

31 What could be considered possible candidates for relics of the big bang?

32 If the spectrum of the microwave background turns out to be markedly different from the Planckian form, what implication will it have for the hot big bang?

33 Show that the space curvature parameter k or the cosmological constant are unlikely to affect the calculations for the early universe.

34 Using the Thomson scattering cross-section for the electrons, show that the optical depth of the universe at the present epoch would be given by $0.08\Omega_0 h_0$ if all the electrons in the universe were free and equal in number to the baryons.

35 Assuming that in the past the electron number density increased as $(1 + z)^3$, use the analysis of Exercise 34 to estimate the smallest redshift at which the Einstein–de Sitter universe was opaque to radiation. (Take $h_0 = 1$). Comment on the fact that your answer comes out very much lower than $z \sim 1000$.

6

The very early universe

6.1 Cosmology and particle physics

In Chapter 5 we discussed the properties of the big universe starting from the epoch in which it was $\sim 10^{-4}$ second old, when a mixture of baryons, mesons, leptons, and photons was in thermodynamic equilibrium with a temperature of $\sim 10^{12}$ K. We discussed how this hot primordial gas evolved as the universe expanded and cooled down. We ended our story with the formation of the helium nucleus, by which time the universe was ~ 3 minutes old.

In the 1960s the above range of epochs would have been considered as describing the early universe. Today the interest has shifted to the 'very early universe': the era preceding the above phase, when matter was in an even more elementary form than that considered above. The reason for this shift lies less in any development in cosmology than in particle physics. The remarkable developments in particle physics, which signify progress towards a unification of the basic interactions of physics, have found their echoes in cosmology.

So far, physicists have relied on the use of powerful accelerators to study the interaction of particles at high energy. From elementary quantum theory, it follows that in order to be able to probe smaller and smaller distances, higher and higher momenta must be achieved. Thus high-energy accelerators are required in order to probe the structure of particles like the proton or the pion. The present accelerators achieve energies of the order of a few tens or hundreds of GeV($1 \text{ GeV} \equiv 10^9$ eV). The 'supercollider' of the next generation may reach energies of the order of 4×10^4 GeV. These values may be compared with the energies $\sim 10^{16}$ GeV, at which interesting unification phenomena are predicted by particle physicists. Energy ranging as high as this value is far beyond what could be achieved by the technology of the foreseeable future.

It is against this background that particle physicists have turned to cosmology in the realization that the very early hot universe is the poor man's high-energy accelerator. This is not the first time physicists have turned to astronomy in order to study the behaviour of physical processes under conditions unattainable in a terrestrial laboratory. Even before thermal fusion could be achieved on Earth, physicists were studying the process inside stars. And to go even further back in history, it was astronomy of the solar system that provided the real testing ground for the law of gravitation.

Naturally, the interplay of cosmology and particle physics that we plan to discuss in this chapter is highly speculative on both fronts. It depends on the validity of the cosmological model and on the viability of (as yet fluid) ideas of particle physics. The best that can be claimed is consistency between the two. The reader should bear this in mind throughout the various calculations given here.

Let us first consider what particles might exist in the early universe, out of which the baryons and mesons are formed. This information is supplied by particle physics and is listed in Table 6.1. Note that the quarks are listed according to their six 'flavours': up, down, strange, charm, truth, and beauty. Each quark comes in three 'colours': red, white (sometimes called green), and blue. These are constituents of baryons and mesons, three quarks making a baryon and a quark–antiquark pair making a meson. The quarks interact with each other by exchanging gluons, just as electrons interact with each other by exchanging photons.

Table 6.1 also lists six leptons, which come in pairs. Two pairs, (e, v_e) and (μ, v_μ), we have already encountered in Chapter 5. A third pair (τ, v_τ) is now known. The list of bosons includes the graviton, the photon, the eight gluons, as well as the charged particles $W^\pm$ and the neutral particle Z^0. Do these numbers have any special significance? Why six quarks? Why six leptons? Why eight gluons? Particle physicists have found it useful to describe the framework of all these particles in the abstract language of group theory (see section 6.3).

The masses in Table 6.1 are listed in the unit of $MeV(1\,MeV \equiv 10^6\,eV)$. We have so far not introduced this unit. It is convenient to do so now, since we shall be using many ideas from particle physics, where this unit is commonly used. Thus for each mass m expressed in grams, mc^2 is energy expressed in ergs. We then use the following conversion scale:

$$1\,MeV = 1.602\,191\,7 \times 10^{-6}\,erg.$$

Further, since we are going to describe the hot universe, it is also convenient to express the temperature in the same unit. Thus for T

Table 6.1 Elementary particles in the early universe

Particle		Mass (MeV)*	Spin, h	Electric charge, e	Interaction
Quarks	u	? + 4		$\frac{2}{3}$	
	d	? + 8		$-\frac{1}{3}$	G, W
	c	? + 1150	$\frac{1}{2}$	$\frac{2}{3}$	E, C
	s	? + 150		$-\frac{1}{3}$	
	t	? + $\geqslant$ 5000		$\frac{2}{3}$	
	b	? + 4500		$-\frac{1}{3}$	
Leptons	ν_e	$< 6 \times 10^{-6}$?		0	G, W
	e	0.5110		-1	G, W, E
	ν_μ	< 0.65	$\frac{1}{2}$	0	G, W
	μ^-	105.66		-1	G, W, E
	ν_τ	< 250		0	G, W
	τ^-	< 1780		-1	G, W, E
Bosons	graviton	$\leqslant 10^{-36}$	2	0	G
	γ	$\leqslant 7 \times 10^{-22}$	1	0	G, E
	gluons (8)	$\leqslant 100$	1	0	G, C
	$W^\pm$	$\sim 8 \times 10^4$	1	-1	G, W, E
	Z^0	$\sim 9 \times 10^4$	1	0	G, W

[a] Quark masses are not uniquely determined, since free quarks have not yet been found. There is some indication that the mass of ν_e may exceed the value given here.

[b] G: gravitation; W: weak interaction; E: eletromagnetism; C: chromodynamics.

Source: Based on R. V. Wagoner, The early universe. In R. Balian, J. Audouze, and D. N. Schramm, eds, *Physical Cosmology*, Les Houches Lectures Session XXXII, p. 395 (Amsterdam: North Holland, 1979).

expressed in kelvin, kT is energy expressed in ergs, which can be written in units of MeV or GeV. Therefore we have

$$1 \text{ gram} \sim 5.618 \times 10^{28} \text{ MeV} = 5.618 \times 10^{25} \text{ GeV},$$

$$1 \text{ kelvin} \sim 8.617 \times 10^{-11} \text{ MeV} = 8.617 \times 10^{-14} \text{ GeV}.$$

Although these conversion factors involve many powers of 10, they show why these are good units for the early universe. For example, a temperature of the order of 10^{12} K is a few MeV. Similarly, Table 6.1 shows that masses of the listed particles are given by moderate numbers when expressed in MeV. For higher energies we may use GeV.

We now recall from Chapter 5 the result that relates the temperature of the universe to its age as given by the Einstein equation

$$\frac{\dot{S}^2}{S^2} = \frac{8\pi G}{3}\rho. \tag{6.1}$$

If there are bosons with a total g_b of g-factors, and fermions with a total g_f of g-factors, then

$$\rho c^2 = \tfrac{1}{2}gaT^4, \tag{6.2}$$

with

$$g = g_b + \tfrac{7}{8}g_f. \tag{6.3}$$

Thus we have for $g = $ constant

$$S \propto t^{1/2}, \tag{6.4}$$

with

$$t = \left(\frac{3c^2}{16\pi Ga}\right)^{1/2} g^{-1/2}T^{-2}. \tag{6.5}$$

This relation can be expressed as

$$t_{\text{second}} = 2.4g^{-1/2}T_{\text{MeV}}^{-2} = 2.4 \times 10^{-6}g^{-1/2}T_{\text{GeV}}^{-2}. \tag{6.6}$$

6.2 Survival of massive particles

We will begin with a simple extrapolation of the approach adopted in Chapter 5. We will assume in this section that quarks have combined to form particles (and antiparticles) and investigate the criteria that determine the survival of a particular species of particles. In the ideal gas approximation, we will assume the distribution functions to be those given by (5.9). In the relativistic (high-temperature) approximation of section 5.2, we have the following formula for the number density of particles of species A:

$$N_A = \eta g_A N_\gamma = \eta g_A \frac{2.4}{\pi^2}\left(\frac{kT}{ch}\right)^3, \tag{6.7}$$

where N_λ is the number density of photons and $\eta = \tfrac{1}{2}$ for bosons and $\tfrac{3}{8}$ for fermions. In the nonrelativistic approximation we get

$$N_A = \frac{g_A}{h^3}\left(\frac{m_A kT}{2\pi}\right)^{3/2} \exp\left(-\frac{m_A c^2}{kT}\right). \tag{6.8}$$

The assumption leading to (6.7) or (6.8) is that the species is in thermodynamic equilibrium with the rest of the particles. For (6.7) to hold we used $T \gg T_A \equiv m_A c^2/k$, while for (6.8) to hold we should have $T \ll T_A$. Exactly similar results must hold if the species A has antiparticles $\bar{A}$. To fix ideas (since we are eventually going to use these formulae for baryons-protons and neutrons) we will assume A to be a fermion. Thus $\eta = \frac{3}{8}$.

In general A and $\bar{A}$ may annihilate if they are brought together. In a typical reaction, two photons will be produced:

$$A + \bar{A} \rightarrow \gamma + \gamma. \tag{6.9}$$

In the reverse reaction pairs $(A, \bar{A})$ are produced. The question we wish to answer is, how does the interchange affect the number density N_A or $N_{\bar{A}}$?

To start with, suppose $N_A = N_{\bar{A}}$, and consider the particles (and antiparticles) in a comoving volume V_0. The corresponding proper volume is $V_0 S^3(t)$. Define

$$\mathcal{N}_A = N_A V_0 S^3(t), \qquad \mathcal{N}_{\bar{A}} = N_{\bar{A}} V_0 S^3(t). \tag{6.10}$$

Let $\psi(T)$ denote the production rate per unit volume and $\beta(T)$ the annihilation rate coefficient. Both ψ and β will depend on the temperature T and

$$\beta = \langle v\sigma \rangle, \tag{6.11}$$

where σ is the annihilation cross-section and v the velocity of particles. Accelerator experiments on nucleon–antinucleon cross-sections give us $\beta \sim 10^{-15} \text{ cm}^3 \text{ s}^{-1}$, in the energy range 0.4 to 7 GeV. Thus it is convenient to write

$$\beta = 10^{-15} \bar{\beta} (\text{cm}^3 \text{ s}^{-1}) \tag{6.12}$$

and anticipate that $\bar{\beta} \sim 1$. It is also worth noting that if we consider the Compton wavelength for a particle of mass m and define $\sigma = \pi(h/mc)^2$, then for a proton or a neutron with $v = c$, (6.11) gives $\beta \sim 4 \times 10^{-17}$. Thus we may set

$$\beta = \zeta \frac{\pi h^2}{m^2 c} \tag{6.13}$$

and expect $\zeta \sim 100$ for a proton or neutron.

The rate of change of $\mathcal{N}_A(\mathcal{N}_{\bar{A}})$ is then given by

$$\frac{d\mathcal{N}_A}{dt} = \frac{d\mathcal{N}_{\bar{A}}}{dt} = [\psi(T) - \beta(T)N_A^2(T)]V_0 S^3(t). \tag{6.14}$$

Frequent collisions are necessary to establish equilibrium. The collision rate is given by

$$\Gamma(T) = N_A(T)\beta(T) \propto T^3 \beta(T). \tag{6.15}$$

In general, $\beta(T)$ does not decrease as T increases. Hence in the very early stages $\Gamma(T)$ was so large that it exceeded the expansion rate of the volume, given by

$$3H(t) = \frac{3}{2t} \propto T^2. \tag{6.16}$$

Thus initially

$$t\Gamma(T) \gg 1, \tag{6.17}$$

guaranteeing frequent collisions. Under such circumstances an equilibrium is reached with detailed balancing between the creation and annihilation processes. In equilibrium $\psi(T) = \beta(T)N_{A0}^2(T)$, where $N_{A0}(T)$ denotes the equilibrium value of $N_A(T)$. Thus (6.14) becomes for either A or $\bar{A}$,

$$\frac{d\mathcal{N}}{dt} = \beta(\mathcal{N}_0 + \mathcal{N})(\mathcal{N}_0 - \mathcal{N}). \tag{6.18}$$

If we now refer back to (6.7) we see that in the relativistic approximation $N \propto T^3 \propto S^{-3}$, so that $\mathcal{N} \propto NS^3 = $ constant. Thus if particles are relativistic, then $\mathcal{N} = $ constant $= \mathcal{N}_0$ is a solution of (6.18). If on top of this we suppose that the relativistic regime lasted long enough for $t\Gamma(T)$ to drop below 1, then we encounter the situation in which $\mathcal{N}_0$ *is preserved for subsequent epochs*. This is because the rarity of collisions makes it unlikely that production or annihilations will significantly alter $\mathcal{N}$ once $t\Gamma(T)$ drops well below 1. We will now follow the analysis of G. Steigman.

For massless particles the relativistic regime lasts forever. Hence for these particles the above always holds. Indeed we encountered an example of this reasoning in the context of massless neutrinos in Chapter 5. The presentday neutrino distribution could be traced back to the epoch when they decoupled from the rest of the matter – when the weak interaction processes become slower than the rate of expansion of the universe.

It may, however, happen that the particles are massive and are in equilibrium even when they become nonrelativistic. In that case (6.8) applies and the number $\mathcal{N}_A$ drops rapidly as T decreases. At some stage, with $\mathcal{N}_A \ll \mathcal{N}_\gamma$, the collision rate drops ($t\Gamma \ll 1$), so that further changes in $\mathcal{N}_A$ through creation and annihilation are not possible. Let us denote this epoch by t_* and the corresponding temperature by T_*. The value of $\mathcal{N}_A$ at this epoch then becomes frozen; that is, unaltered for subsequent epochs. This number would survive as a relic of the hot universe.

Figure 6.1 shows how $\mathcal{N}_A$ depends on the mass of the species A. For massless particles like the photon (and the neutrino if it has $m = 0$), $\mathcal{N}_A$ is unchanged. The neutral leptons become frozen at the next lower value. The charged leptons can interact longer through the electromagnetic force and hence they decouple later and at lower values of $\mathcal{N}_A$ than do the neutral leptons. The lowest are the hadrons (mesons, neutrons, protons,

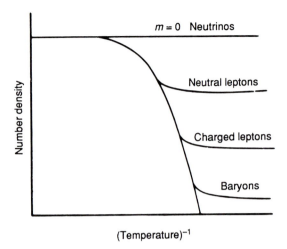

Fig. 6.1 Schematic description of how the surviving number density of a particle depends on its mass and on how strongly it interacts. This number is highest for neutrinos (with zero rest mass and weak interaction) and lowest for baryons (massive and strongly interacting).

and so on), which have strong interactions to hold them together and the largest masses.

Let us try to estimate this effect quantitatively. At t_* we have for species A in the nonrelativistic regime

$$N_A = \frac{g_A}{\hbar^3} \left(\frac{m_A k T_*}{2\pi} \right)^{3/2} \exp\left(-\frac{m_A c^2}{k T_*} \right). \tag{6.19}$$

Applying the condition $t_* \Gamma(T_*) = 1$ and using (6.15) and (6.19), we get

$$t_* \beta \frac{g_A}{\hbar^3} \left(\frac{m_A k T_*}{2\pi} \right)^{3/2} \exp\left(-\frac{m_A c^2}{k T_*} \right) = 1. \tag{6.20}$$

Define

$$x_* = \frac{m_A c^2}{k T_*} \tag{6.21}$$

and express masses and temperatures in MeV units. Thus in these units $x_* = m_A / T_*$. Then from (6.6) we have

$$2.4 g^{-1/2} T_*^{-2} \beta \frac{g_A}{\hbar^3 c^3} \left(\frac{m_A T_*}{2\pi} \right)^{3/2} e^{-x_*} = 1,$$

that is,

$$2.4 g^{-1/2} \cdot \frac{g_A}{\hbar^3 c^3} \frac{m_A}{(2\pi)^{3/2}} x_*^{1/2} e^{-x_*} = 1. \tag{6.22}$$

We may express this relation in the following form:

$$x_*^{-1/2} e^{x_*} = \Lambda g_A Z \tag{6.23}$$

where

$$Z = m_A \bar{\beta} g^{-1/2}. \tag{6.24}$$

Using (6.12), (6.24), and (6.23), we get

$$\Lambda = \frac{2.4 \times 10^{-15}}{(2\pi)^{3/2}} (\hbar c)^{-3} \approx 2 \times 10^{16}. \tag{6.25}$$

(*Caution*: We are using here the units seconds, centimetres, and MeV. Thus $\hbar c$ must be expressed in these units.)

Let us now apply these results to nucleons: to neutrons and protons together with their antiparticles. Then $g_A = 8$, and with $g_\gamma = 2$ for the photons (6.3) gives

$$g = 2 + 7 = 9. \tag{6.26}$$

The nucleon mass $m_A \simeq 940\,\text{MeV}$, and from (6.12), $\bar{\beta} \approx 1$. Thus $Z \approx 313$. With $g_A = 8$ we then have from (6.23) and (6.25)

$$x_*^{-1/2} e^{x_*} \approx 5 \times 10^{19} \tag{6.27}$$

and hence $x_* \simeq 47$. Thus

$$T_* \approx 20\,\text{MeV}, \qquad t_* \approx 0.002s. \tag{6.28}$$

We can also use the above calculation to compute the nucleon/photon ratio at the present epoch. Assuming that the $A-\bar{A}$ annihilation is the main source of photons, we get the photon number per unit comoving volume effectively frozen at the value it acquired at the epoch t_*. So the present value of N_A/N_γ will be the same as it was at $t = t_*$. Using (6.8) for N_A and (6.7) with $\eta = \frac{1}{2}$, $g_A = 2$ for the photons, we get

$$\frac{N_A}{N_\gamma} = \frac{g_A \pi^2}{2.4(2\pi)^{3/2}} \left(\frac{m_A c^2}{kT}\right)^{3/2} \exp\left(-\frac{m_A c^2}{kT}\right). \tag{6.29}$$

With $g_A = 8$ and x_* defined by (6.21), we get

$$\frac{N_A}{N_\gamma} \simeq 2x_*^{3/2} e^{-x_*}. \tag{6.30}$$

Now use (6.27) and $x_* \simeq 47$ to get

$$\frac{N_A}{N_\gamma} \simeq 2 \times 10^{-18}. \tag{6.31}$$

In Chapter 5 (5.57) gave the estimated present value of N_A/N_γ. In our present notation, this is given by

$$\frac{N_A}{N_\gamma} \simeq 2 \times 10^{-8} (\Omega_0 h_0^2) \left(\frac{T_0}{3}\right)^{-3}. \tag{6.32}$$

Since $T_0 \sim 3$, and $\Omega_0 h_0^2$ is not expected to be lower than $\sim 10^{-3}$ under the most extreme case, we have a large discrepancy to account for. There is one further point of criticism. If we are sure that the universe is made up predominantly of matter, than $N_A \gg N_{\bar{A}}$ and the formula (6.32) applies to $N_A (\approx N_A - N_{\bar{A}} = $ baryon number density). However, our analysis so far is symmetric between matter and antimatter and so leads to $N_A = N_{\bar{A}}$. Clearly new inputs are necessary in the discussion given above if we are to understand why $N_A \gg N_{\bar{A}}$ and why N_A/N_γ is as high as is indicated by (6.32).

We note that the ratio N_A/N_γ as given by (6.31) is a small number. In deriving it we have lost sight of the fundamental constants that went into it. It is instructive to see what (6.31) looks like in terms of c, $\hbar$, G, and m_A. Substituting $a = \pi^2 k^4/15c^3\hbar^3$ and using (6.13), (6.5), (6.20), and (6.29), we get

$$\frac{N_A}{N_\gamma} = \frac{\pi g^{1/2}}{7.2} \frac{x_*}{\zeta} \left(\frac{2Gm_A^2}{c\hbar} \right)^{1/2}. \tag{6.33}$$

We have already seen that $x_*/\zeta \sim \frac{1}{2}$ and $g^{1/2} \sim 3$, so that the coefficient in front of the expression in parentheses is of the order unity. So the smallness of N_A/N_γ is directly related to the ratio of the strengths of the gravitational interaction and the strong interaction. Denoting this ratio by

$$\alpha_G = \frac{Gm_A^2}{c\hbar} \sim 6 \times 10^{-39}, \tag{6.34}$$

we have

$$N_A/N_\gamma \sim \alpha_G^{1/2}. \tag{6.35}$$

The strength of the electromagnetic interaction is measured by the fine-structure constant $\alpha = e^2/\hbar c \sim \frac{1}{137}$. Notice how weak the gravitational interaction is by comparison. Had G been considerably higher than it is, we could have ended with a larger value of N_A/N_γ.

6.3 Grand unified theories and baryon asymmetry

Our simplified calculations of the previous section having led us into difficulties, it is evident that something more sophisticated is needed to understand (1) the present predominance of baryons over antibaryons, and (2) the baryon/photon ratio in the neighbourhood of 10^{-9}. Since our calculation assumed thermodynamic equilibrium and particle–antiparticle symmetry, any new input is expected to question these two assumptions. In this section we outline one of the ways in which this problem is being solved.

The solution is via the so-called grand unified theories (GUTs) – theories that seek to bring together three of the four basic interactions of physics into a single framework. The use of the plural shows that as yet there is no single theory that is universally accepted. In the present section we will follow the $SU(5)$ framework purely as an illustrative example and study its implications for the early universe. To understand what is involved, let us first have a superficial look at the three basic interactions from the group-theoretic point of view.

6.3.1 Electrodynamics

Let us begin with the simplest and the best understood interaction: the electromagnetic interaction. This describes how charged leptons (the $e^{\pm}$, $\mu^{\pm}$, $\tau^{\pm}$) interact through the exchange of photons. When an electron is shaken it emits photons. When a photon strikes an electron it accelerates. The information needed for studying this interaction requires a spinor wave function ψ for the lepton and a vector field A_i (the electromagnetic 4-potential) for the photon. The two physical effects described above are given by the following two equations (written in flat Minkowski space-time):

$$(A^{k,i} - A^{i,k})_{,k} \equiv F^{ik}_{,k} = \frac{4\pi e}{c} \bar{\psi}\gamma^i\psi, \tag{6.36}$$

$$\gamma^i \left(\nabla_i - \frac{e}{\hbar c} A_i\right)\psi - \frac{mc}{\hbar}\psi = 0. \tag{6.37}$$

γ_i are the 4×4 Dirac matrices, and e is the electric charge. ∇_i denotes differentiation with respect to spacetime coordinates.

It is easy to see that these equations are invariant under the transformation

$$A_i \rightarrow A_i + \Theta_{,i}, \qquad \psi \rightarrow \psi \exp\left(-\frac{ie\Theta}{\hbar c} Q\right) \tag{6.38}$$

where Q is the integer 1. The transformation of ψ is a unitary transformation, and since the exponent is a number (that is, a 1×1 matrix), these transformations form a *unitary group of one dimension*, denoted by $U(1)$.

6.3.2 The weak interaction

This weak interaction concerns both the charged and uncharged leptons in pairs: (e, ν_e), (μ, ν_μ), (τ, ν_τ), and so on. In a typical interaction the members of the pair are interchanged. This is the law of conservation of lepton numbers referred to in section 5.2. To describe the pair we therefore need two wave functions: for example, the combination

$$\Psi = \begin{pmatrix} \psi_e \\ \psi_v \end{pmatrix} \qquad (6.39)$$

describes the pair (e, v). From empirical considerations it is argued that the weak interaction is invariant under transformations of Ψ with 2×2 matrices that are unitary and have determinant 1. These transformations form a group denoted by $SU(2)$. Because of parity violation and the fact that neutrinos have only one spin state – they are *left*-handed – it is customary to write a subscript L in $SU(2)_L$. A typical member of the group is denoted by

$$U = \exp(-iH) \qquad (6.40)$$

where H is a 2×2 Hermitian matrix of zero trace. The most general such matrix is

$$\begin{pmatrix} a & b + ic \\ b - ic & -a \end{pmatrix} = a \begin{pmatrix} 1 & \\ & -1 \end{pmatrix} + b \begin{pmatrix} & 1 \\ 1 & \end{pmatrix} + c \begin{pmatrix} & i \\ -i & \end{pmatrix}. \qquad (6.41)$$

Thus instead of a single number Q in (6.38) we need three real numbers a, b, c; or rather three matrices. (These matrices are proportional to the well-known Pauli matrices.) The 'charges' in this case are three matrices, two of which are nondiagonal. The nondiagonal matrices permit an interchange of ψ_e and ψ_v in (6.39). This means physically that e and v are interchanged. In this process a charged boson W_1 is exchanged; for example,

$$e \rightarrow W_1 + v. \qquad (6.42)$$

Corresponding to the three matrices in (6.41) there are three W-particles, two with charges $\pm e$ and the third (W_3) neutral.

Although the weak interaction does not directly involve the electric charge, it still seems to demand the charged bosons W_1 and W_2. This circumstance prompted efforts to link it with the electromagnetic interaction. This link has been achieved via the $SU(2)_L \times U(1)$ framework originally proposed by A. Salam and S. Weinberg and sometimes called *the electro-weak interaction*. The link brings the photon (which is a boson) closer to the three particles W_1, W_2, and W_3. In this unified picture it is more convenient to talk of another neutral particle Z^0 instead of W_3. Z^0 has zero mass and charge, just like the photon. However, the photon does not interact with the neutrino, while the Z^0 does. The exchange of Z^0 does not alter the electric charge, and hence such an interaction is called a *neutral current interaction*. Thus in the electron neutrino scattering

$$e + \nu_e \rightarrow e + \nu_e$$

we have $e \rightarrow e$ and $\nu_e \rightarrow \nu_e$ in the neutral current interaction while $e \rightarrow \nu_e$, $\nu_e \rightarrow e$ in the charged current interaction.

The unification program makes use of the so-called *gauge theories*. The electromagnetic theory is a gauge theory in the sense that its equations are invariant under the gauge transformation of its potential. The transformation of A_i given by (6.38) is a gauge transformation, and in the Weinberg–Salam model similar gauge transformations play a pivotal role for the $SU(2)_L \times U(1)$ framework.

One reason for using the gauge theory is that it is 'renormalizable'. This is a technical term which gained currency in quantum electrodynamics (QED), which is a renormalizable and gauge theory. In QED the standard calculations of probability amplitudes, average values, energy levels etc., lead to infinities because the relevant integrals diverge at high energies. Renormalization is a technique of subtracting one infinity from another so as to arrive at a finite and physically meaningful answer. Although mathematicians would baulk at such an approach, the theoretical physicist has come to accept it, its merit being that it is unambiguous to operate. A discussion of this highly interesting topic will, however, take us too far from cosmology and into technical details of field theory. We simply mention that the accelerator experiments have measured the masses of the W and Z bosons and have found them in conformity with theoretical expectations.

6.3.3 *Quantum chromodynamics*

The third basic interaction of physics is the strong interaction described in the framework of quantum chromodynamics (QCD). This makes use of transformations under the $SU(3)$ group. The basic fields here are the quark fields, which are three-component vectors in an abstract space called the *colour space* with three 'dimensions': red, white, and blue. Again we have a relation like (6.40) in 3×3 matrices. The matrix H now has eight independent components, and so like (6.41) we have eight matrix charges, $T_1 \ldots T_8$, of which two (T_3 and T_8) are diagonal. Again the matrix character of (6.40) allows quarks to be exchanged. Corresponding to the three Ws in the $SU(2)$ framework we now have eight bosons $G_1 \ldots G_8$ that are called the gluons. No colour change takes place when the gluons G_3 and G_8 are exchanged.

The gluons generate an interquark force (just as the photon is responsible for the electromagnetic force between the charged particles).

This force is believed to be so large that quarks are expected to be in bound states of two or three. The states with two, a quark and an antiquark, form mesons (like π^+), while states with three quarks are baryons. Quarks have fractional charges. The u-quark has charge $2e/3$, while a d-quark has charge $-e/3$. Thus a proton is made of two u-quarks and one d-quark, while a neutron is made of two d-quarks and one u-quark.

6.3.4 GUT:SU(5)

In a typical unification attempt we expect the participating interactions to have comparable strengths. In normal laboratory energies the strong interaction (quantum chromodynamics) is the most powerful, followed by electrodynamics and then by the weak interaction. However, as the energy is increased the gap between the three narrows. At around 100 GeV, the last two are comparable in strength, thus making a unified 'electroweak' theory viable. Theoretical considerations suggest that if we extrapolate to considerably higher energies, the strong interaction reduces in strength, while the electroweak interaction gains. At around 10^{15} GeV these interations become comparable and their unification may seem natural. Figure 6.2 illustrates the changes in strengths of the three interactions with growing energy. Figure 6.2 also shows another landmark in energy at $\sim 10^{19}$ GeV. This is the Planck energy

$$E_P = \left(\frac{c^5\hbar}{G}\right)^{1/2} \approx 1.2 \times 10^{19} \text{ GeV.} \tag{6.43}$$

Clearly, with G and $\hbar$ in it this expression would have to do with quantum gravity. We shall consider it separately later. For the time being we exclude it from the unification attempts.

If we wished to unify all three interactions in a grand unification scheme, we could trivially combine the three into a structure

$$SU(3) \times SU(2)_L \times U(1).$$

However, it was realized that such a structure can form part of a single larger structure denoted by $SU(5)$. Again, if we go back to (6.40) and apply it to 5×5 matrices, the matrix H has 24 arbitrary constants. Thus there are 24 bosons that now mediate between the different basic entities. Of these we already have 4 from the combined electroweak interaction and 8 (gluons) from chromodynamics. Thus 12 more bosons are needed to make up the list of 24. For want of any specific designation, they are referred to simply as the X-bosons.

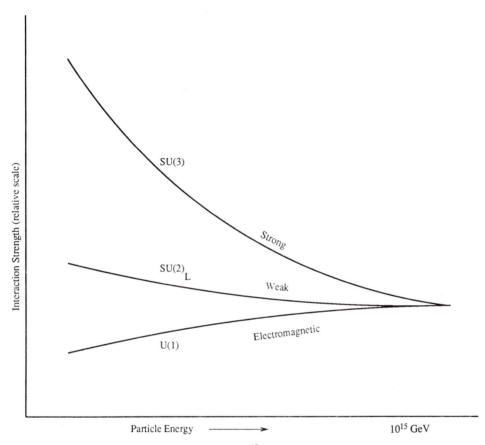

Fig. 6.2 At energies of the order of $\sim 10^{15}$ GeV the strengths of the strong, the weak and the electromagnetic interactions become comparable, thus suggesting this as a suitable energy for grand unification.

The X-bosons are expected to link the participants of chromodynamics (that is, the quarks) with the participants of electroweak interaction (that is, the leptons). In the $SU(5)$ theory, therefore, it is possible to change any of the six quarks (u, d, c, s, t, b) into any of the six leptons (e, μ, τ, ν_e, ν_μ, ν_τ) or vice versa by the exchange of the X-bosons. This is where it becomes possible to create or destroy baryons. Figure 6.3 outlines the scenario leading to the decay of a proton.

In Figure 6.3 an $\bar{X}$ (that is, an anti-X particle) is emitted and absorbed. Assuming that the mass of this particle is m_X, the probability amplitude for the above interaction will contain a factor m_X^{-2}. The proton decay lifetime τ_P will therefore vary as the fourth power of m_X. Since we expect the lifetime to contain the constants $\hbar$, c, and the proton mass m_P, from dimensional considerations we write

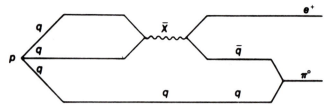

Fig. 6.3 The decay of the proton takes place through the mediation of the X-bosons, which change quarks into leptons and vice versa. This figure illustrates how this could come about. Two of the three quarks (q) in the proton (p) combine to form $\bar{X}$, which decays into positron and an antiquark. The latter combines with the third quark to form a π-meson.

$$\tau_P \sim \frac{\hbar m_X^4}{m_P^5 c^2}$$

$$\sim 2.87 \times 10^{-32} [m_X c^2 \text{ (GeV)}]^4 \text{ yr.}$$

(6.44)

The failure to observe the decays of protons in the laboratory experiments sets a lower limit to τ_P at $\sim 10^{29}$ years. Hence (6.44) suggests that

$$m_X c^2 \geq 10^{15} \text{ GeV}.$$

(6.45)

At a value of $\tau_P \sim 10^{30}$ years we have no hope of observing the decay of a particular proton. However, from a large population of protons a small fraction may decay. For example, in 1000 tons of matter about 50 protons are expected to decay every year if $\tau_P \sim 10^{30}$ years. Experiments during the 1980s failed to observe such decays in an unambiguous manner for $\tau_P \leq 10^{31}$ years. This led to the abandoning of the above simple $SU(5)$ theory in favour of more complex frameworks.

Exact dynamic theories are needed to quantify τ_P and m_X. However, while the proton decay experiment is barely feasible if $\tau_P \leq 10^{33}$ years, the full testing of the predictions of GUTs is clearly beyond the scope of present technology. It is worth mentioning one prediction that is commonly known as the *hypothesis of asymptotic freedom*. According to this hypothesis, at extremely high energies the particle interactions begin to lose their strength. However, even this hypothesis is still to be tested experimentally.

The other alternative, of course, is to use the hot universe for testing theoretical predictions. Even here, for a mass of 10^{15} GeV, the temperature ($= mc^2/k$) will be as high as $\sim 10^{28}$ K! A temperature of 10^{15} GeV gives, according to (6.6), the age of the universe as low as $\sim 10^{-36}$ second. We will refer to it as the GUT epoch. M. Yoshimura has suggested that under GUTs it is possible to produce a slight excess of baryons over antibaryons because the baryon number is not conserved.

However, further assumptions are needed to actually produce the result in accord with observations. The following scenario is that suggested by S. Weinberg and F. Wilczek.

6.3.5 *The production of baryon excess in the early universe*

Let us denote the mass of the X-boson (which causes baryon nonconservation) by m_X, and its coupling strength by α_X. The coupling strength may be 10^{-2} or 10^{-5}, depending on what type of particle X is. Let us denote by Γ_c the rate of collisions that do not conserve baryon number; that is, collisions in which the X-boson is involved. The X-boson itself does not last very long, its time scale being of the order of $\hbar/m_X c^2$. Denote the characteristic decay rate of the X-boson by Γ_X. We thus have three time scales to play with: Γ_X^{-1}, Γ_c^{-1}, and H^{-1}. The trick lies in adjusting these time scales suitably to produce the desired answer. The argument, qualitatively, goes like this.

At the earliest epochs, with temperature $\geq 10^{19}$ GeV, gravity was the strongest force between the various constituents of the universe. Other interactions (including the strongest of them, QCD) were unimportant under the hypothesis of asymptotic freedom. As the universe continued to expand and its temperature dropped there was a phase when gravity became weaker while the other interactions still remained unimportant. Thus for $T \leq 10^{19}$ GeV the particles remained essentially free for some time.

During this phase it becomes necessary to examine the nature of distribution functions that are given by the formula (5.9). There we saw that so long as $T \gg T_A$, that is, so long as we are in the relativistic regimes, the distribution function preserves its equilibrium form during free expansion with $T \propto S^{-1}$. However, if $T \leq T_A$ then the distribution function cannot preserve its form under free expansion. Thus it may get distorted from its equilibrium form.

Now of the various species in the early universe, the X-bosons are probably the most massive. Thus, provided they have a high enough value of T_X, there is a chance that the X-bosons will first drop out of equilibrium. For this to happen, however, it is also necessary that they have not all decayed by then. The decay rate of the X-boson is of the order

$$\Gamma_X \approx \alpha_X g m_X c^2/\hbar, \tag{6.46}$$

where g is the effective number of degrees of freedom for the various particle species (g may well lie between 100 and 200).

The expansion rate, on the other hand, is given by (6.1). The collision rate $\Gamma_c \approx \alpha_X \ll \Gamma_X$. A comparison of the three rates shows that $\Gamma_c < \Gamma_X < H$ soon after gravity became weak. Thus the universe was expanding at this stage with essentially no interaction between the species. The X-bosons began to decay when the age of the universe became comparable to Γ_X^{-1}. Using (6.1), (6.2), and (6.46) we get

$$T = \left(\frac{3g\alpha_X^2 m_X^2 c^4}{4\pi Ga\hbar^2}\right)^{1/4}. \tag{6.47}$$

By the time the universe had cooled to the above temperature the X-bosons would have begun to decay. The question is, were they in equilibrium till then?

As was seen above, this question is decided by a comparison of T with T_X. Two cases are of interest: (1) $T \gg T_X$ and (2) $T \leqslant T_X$. These are illustrated by Figures 6.4 and 6.5, respectively.

In case (1) the decays occured while the X-bosons still had their distribution functions in the equilibrium form. Under these circumstances the X-bosons could not have generated any net excess of baryons; for

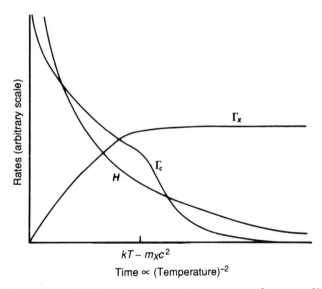

Fig. 6.4 The three rates H, Γ_c, and Γ_X for the case $m_X c^2 < \alpha_X 10^{20}\,\mathrm{GeV}$. When kT drops below $m_X c^2$, Γ_X exceeds the expansion rate H and the X-bosons decay exponentially in number while maintaining the equilibrium distribution. No net baryon excess is generated in this case. (After D. N. Schramm & M. S. Turner, *The origin of baryon number and related problems*. In R. Balian, J. Audouze, & D. N. Schramm, eds, *Physical Cosmology*, Les Houches Lectures Sessions XXXII, p. 501 (Amsterdam: North Holland, 1979).)

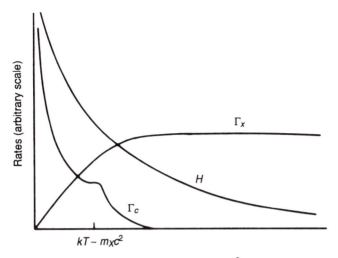

Fig. 6.5 Figure 6.4 redrawn for the case $m_X c^2 > \alpha_X 10^{20}$ GeV. When kT drops below $m_X c^2$, decays and annihilations are not effective, since both Γ_X and Γ_c are less than H. Until Γ_X exceeds H, the X-bosons do not come into equilibrium. At that stage X, $\bar{X}$ decay freely and may generate a net baryon excess. The excess stays since $\Gamma_c < H$. (After D. N. Schramm & M. S. Turner, The origin of baryon number and related problems. In R. Balian, J. Audouze, & D. N. Schramm, eds, *Physical Cosmology*, Les Houches Lectures Session XXXII, p. 501 (Amsterdam: North Holland, 1979).)

thermal equilibrium implies that any decays (like that in Figure 6.3) leading to destruction of baryon number would be compensated by inverse decays. In case (2), however, the distribution function of X-bosons was distorted from its equilibrium form and hence the detailed balancing between decays and inverse decays would not happen. The new input required into the early-universe scenario discussion in section 6.2 is therefore provided by case (2). By departing from thermodynamics equilibrium at the right time, the X-boson distribution has a chance of producing baryon asymmetry.

The condition that T_X exceeds the value of T given by (6.47) may be expressed as

$$m_X > \left(\frac{3g\alpha_X^2 k^4}{4\pi Gac^4}\right)^{1/2} = g^{1/2}\alpha_X m_p. \qquad (6.48)$$

Empirical considerations of the $SU(5)$ framework suggest that from the above inequality $m_X c^2$ should exceed $\sim 10^{16}$ GeV. This is consistent with our earlier estimate of the mass of the X-boson from the lower limit on the lifetime of the proton.

So far we have introduced the assumption of departure from thermo-dynamic equilibrium. We now introduce the other assumption of baryon–antibaryon asymmetry. Suppose an X-boson decays into two states with baryon numbers B_1 and B_2 with fractions r in state 1 and $1 - r$ in state 2. In a perfectly symmetrical situation, the $\bar{X}$-boson would decay into state 1 with baryon number $-B_1$ with fraction r and state 2 with baryon number $-B_2$ with fraction $1 - r$. However, if perfect symmetry does not exist, then the fractions would be $\bar{r}$ and $1 - \bar{r}$ respectively for the $\bar{X}$-decay ($\bar{r} \neq r$). The net baryon number generated by these processes is therefore

$$\Delta B = (r - \bar{r})(B_1 - B_2). \qquad (6.49)$$

Since the baryon nonconserving collision that could destroy ΔB are running at a smaller rate than $H (\Gamma_c < H)$, we expect ΔB to be preserved.

Thus, to account for the observed excess of baryons over antibaryons and to argue that the net baryon number density observed today is $\sim 10^9$ times the observed photon density, we have to make sure that the parameters of the GUT are such as to give appropriate quantitative expression to ΔB above. It is claimed that the reasonable values of the parameters of GUT do lead to a formula in agreement with (6.32).

Whether or not such claims turn out to be justified, the above argument illustrates how the early universe provides an interesting arena for the application of GUTs.

6.3.6 The spontaneous breakdown of symmetry

The change of a larger group of symmetries to the subgroup $SU(3) \times SU(2)_L \times U(1)$ is spontaneous. The actual mechanism involves a set of scalar fields called the Higgs fields ϕ that change over from their initial zero values to a set of finite values when this happens. Why and how this happens and the role the Higgs fields play in the process is a long story which would take us into the labyrinths of gauge field theories. The explanation given below skirts the problem and provides a superficial description.

We begin with the analogy of ferromagnetism and the crucial role of the Curie temperature (770 °C for iron). Above this temperature a bar of iron shows no magnetism in an external field. This is because its elementary nuclear magnets are randomly aligned with no resultant magnetization. Energetically, this is the lowest state for the bar and it chooses to remain in that state as the most stable one. Below the Curie temperature the state of lowest energy changes to that in which all the nuclei are aligned along the bar, which develops polarity at its ends. There are two states of the same lowest energy possible, depending on which (north or south) of the two poles falls at a given end. The ultimate choice of one state apparently breaks the symmetry although theoretically and inherently the symmetry is always there.

In the early universe something similar happens to the ϕ-field. Above a critical temperature T_c, the vacuum state, the state of lowest energy, is none other than $\phi = 0$. Below T_c the state of lowest energy changes. It now corresponds to a situation when ϕ has nonzero values. We will encounter explicit examples of this in section 6.5.

For the time being, let us suppose that there exist alternative values ϕ_i ($i = 1, 2, \ldots$) of the ϕ field, all corresponding to states of the same lowest energy which now acquire that status of vacuum. There is basic symmetry with respect to all ϕ_i, but in practice the system may spontaneously acquire one of them. This is again an apparent breakdown of symmetry.

The consequences of this for the very early universe are that it is divided into different domains, each with a different value of ϕ_i. In this way the universe acquires discontinuities along the domain walls. These translate into highly significant discontinuities of matter distribution. The fact that we do not see such discontinuities in actuality (say in the form of large sheets of matter) is hard to explain away. This difficulty is known as the *domain wall problem*.

The intersection of two domain walls is a linear structure known as 'cosmic string'. Such filamentary structures have been invoked in scenarios for galaxy formation (see Chapter 7).

6.4 Some problems of standard cosmology

It may appear from the above that by going over to the very early universe we have made progress in understanding some of the present features of the universe. In fact the situation is the exact opposite: we have acquired more problems than we managed to solve by this device. The domain wall problem is one of them. Other, more important problems are highlighted below.

6.4.1 The horizon problem

Let us suppose that the initial conditions for the universe were set fairly early on, at an epoch t in the radiation-dominated phase. From the considerations of Chapter 4 adapted to the scale factor $S \propto t^{1/2}$ we find that the proper radius of the particle horizon at that epoch was

$$R_L = 2ct. \tag{6.50}$$

Whatever physical processes operated at this epoch were limited in range by R_L. As such we do not expect the homogeneity of physical quantities to extend beyond the diameter $2R_L$, unless we make the somewhat contrived assumption that the universe was *created* homogeneous. In other words, the casual limitations tell us that no region larger than $2R_L$ in size should be homogeneous.

When the initial conditions were so set, this region grew to much larger size at the present epoch; the factor η by which it would grow is the ratio of scale factors

$$\eta = \frac{S(t_0)}{S(t)}$$

at the present and initial epochs. How do we estimate η?

The simplest method is to compare the temperatures at t and t_0, since (from considerations of Chapter 5) $S \propto T^{-1}$. Thus

$$\eta = \frac{T(t)}{T(t_0)}.$$

$T(t)$ is given by (6.6). It is convenient to express T_0 also in GeV:

$$T_0 \, (\text{GeV}) = 2.585 \times 10^{-13} \left(\frac{T_0}{3 \, \text{K}} \right). \tag{6.51}$$

Combining (6.6) and (6.51) and writing the value of c in (6.50) we get the present limit on a homogeneous region as

$$R_{\text{Hom}}(t_0) = 2ct \tag{6.52}$$
$$= 5.57 \times 10^{17} \times T_{\text{GeV}}^{-1} g^{-1/2} \times \left(\frac{3 \, \text{K}}{T_0} \right) \text{cm}.$$

For $T_{\text{GeV}} \approx 10^{15}$, $g \approx 100$, $T_0 \approx 3 \, \text{K}$ we get the surprisingly small value of 55 cm! In other words we have no reason to expect homogeneity on a scale larger than, say, 1 metre. The fact that the relic microwave background is homogeneous on the cosmological scale of $\sim 10^{28}$ cm tells us that there is something seriously wrong with our reasoning above. Yet, the standard model does not provide any loophole out of this so-called *horizon problem*. Notice also that the further we go back in the past (in our attempts to set the initial conditions) the larger will be T_{GeV} and the smaller will be the value of $R_{\text{Hom}}(t_0)$. Figure 6.6 illustrates the horizon problem.

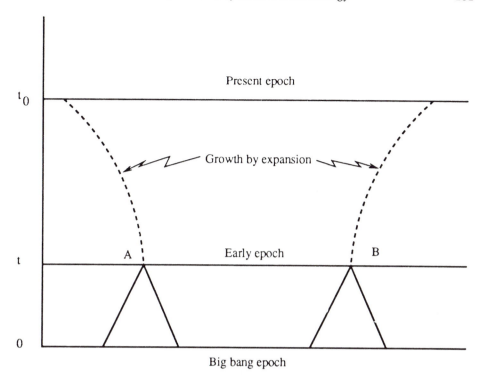

Fig. 6.6 At a very early epoch t, the observers A and B have non-overlapping particle horizons. Thus there is no *a priori* reason why A and B should have the same initial conditions. Yet the universe *as seen at present* is homogeneous over distance far larger than what AB would grow to (shown by dotted segment) at present.

6.4.2 The flatness problem

When discussing the early and the very early universe we ignored the kc^2/S^2 term in the field equations. Thus (6.1) should actually have been

$$\frac{\dot{S}}{S^2} + \frac{kc^2}{S^2} = \frac{8\pi G\rho}{3}. \tag{6.53}$$

Our justification in ignoring that term was that as $S \to 0$ $\dot{S}^2 \to \infty$ and, thus, far exceeds $|kc^2|$. This argument is, however, *scale-dependent*. Thus, if we write $S = At^{1/2}$, then $\dot{S}^2 = A^2/4t$. Whether $\dot{S}^2$ exceeds c^2 for $k = \pm 1$ would depend on A. A priori, we do not know A, unless we link it with the present size of the universe. It is more convenient to look at the density parameter Ω instead.

Writing $\rho = \Omega\rho_c$ as in (4.43), we have, at any general epoch when $S \propto t^{1/2}$,

$$\frac{kc^2}{S^2} = (\Omega - 1)\frac{\dot{S}^2}{S^2} = \frac{\Omega - 1}{4t^2}. \tag{6.54}$$

For the present epoch, on the other hand,

$$\frac{kc^2}{S_0^2} = (\Omega_0 - 1)H_0^2. \tag{6.55}$$

Dividing (6.54) by (6.55) and using $S \propto T^{-1}$, we get, for $k = \pm 1$,

$$\Omega - 1 = (\Omega_0 - 1) \cdot 4H_0^2 t^2 \cdot \frac{T^2}{T_0^2}.$$

Except for $(\Omega_0 - 1)$, all quantities on the right-hand side are known. Using (6.6) for t and (6.51) for T_0 we get

$$(\Omega - 1) \approx 3.5h_0^2 g^{-1} \times 10^{-21} T_{\text{GeV}}^{-2}\left(\frac{3\,\text{K}}{T_0}\right)^2 (\Omega_0 - 1). \tag{6.56}$$

For $T_{\text{GeV}} = 10^{15}$ and $g \approx 100$ we get for $T_0 \approx 3\,\text{K}$

$$\Omega - 1 \approx 3.5h_0^2 \times 10^{-53}(\Omega_0 - 1). \tag{6.57}$$

This expression epitomises what has come to be known as the *flatness problem*. Suppose that the initial conditions including the density parameter Ω were set at the GUT epoch when $T \approx 10^{15}$ GeV. Then the present value of $(\Omega_0 - 1)$ is given by (6.57). Or, to invert the chain of reasoning, suppose that the present observational uncertainty tells us that $|\Omega_0 - 1| \leqslant \mathbb{O}(1)$. Then from (6.57), at the GUT epoch Ω was differing from unity by a fraction of the order of 10^{-53}. In other words, the departure from the flat value of $\Omega(= 1)$ at this stage has to be extremely small. Any relaxation of this fine tuning would lead to a far wider range of Ω_0 at present than that permitted by observations.

So our neglect of the curvature term kc^2/S^2 is linked with an extremely fine tuning of the universe to the flat ($k = 0$) model. If this tuning were not there, the universe could either have gone into a collapse ($k = 1$) or an expansion to infinity ($k = -1$) in time scales of the order of 10^{-35} s that were characteristic of the GUT era.

Figure 6.7 illustrates this conundrum. The shaded region denotes the finely tuned set of Friedman models that end up today within the observed range $|\Omega_0 - 1| \leqslant \mathbb{O}(1)$. The curves at the top and bottom of the figure with time scales $\sim 10^{-35}$ s, should normally have operated at the GUT stage. What made the universe get into the shaded region instead?

This problem was first highlighted in 1979 by R. H. Dicke and P. J. E. Peebles, who discussed it not at the GUT epoch but at $t \sim 1$ s when the neutrinos had decoupled and pair ($e^{\pm}$) annihilation was to begin. Thus $T \sim 10^{-3}$ GeV, $g \sim 10$, and we get $\sim 10^{-16}$ instead of 10^{-53} as the

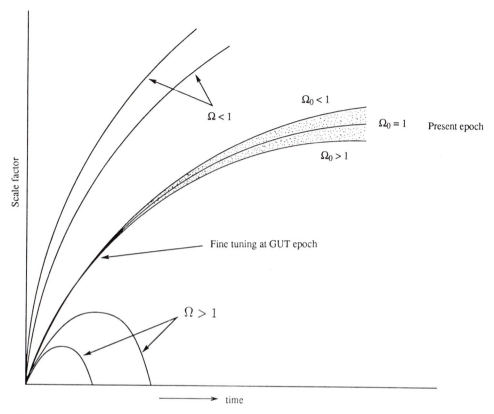

Fig. 6.7 The flatness problem, illustrated with the help of the expansion functions for the $k = 0, \pm 1$ models ($\Omega \geqslant, =, < 1$). The observable uncertainty extends over the range of curves in the shaded region, all of which were tightly bunched together at the GUT epoch, close to the $\Omega = 1$ curve.

coefficient of (6.57). It is clear that the further back in time and closer to $t = 0$ we go, the finer is the tuning required. (For the Planck epoch, we get 10^{-61} instead of 10^{-53}.)

6.4.3 The entropy problem

This is a restatement of the flatness problem and the horizon problem in a somewhat different form. The entropy in a given comoving volume stays constant in an adiabatic expansion (see section 5.2). The present photonic entropy in the observable universe of size $R \approx h_0^{-1} \cdot 10^{28}$ cm (radius) is given by the very large value

$$\Sigma = \frac{4\pi}{3k} \alpha T_0^3 R^3 \approx h_0^{-3} \times 6 \cdot 10^{87} \left(\frac{T_0}{3}\right)^3. \tag{6.58}$$

If the entropy was conserved then we would have $ST = $ constant. However, we found that in the flatness problem this hypothesis led to fine tuning while for the horizon problem it gave an extremely small size of homogeneity. It therefore appears that the trouble lies in $\Sigma = $ constant: it could be resolved if the adiabatic assumption were violated at some stage and Σ boosted to its present value by an enormously large factor.

6.4.4 The monopole problem

In a grand unified theory, whenever there is a breakdown of symmetry of a larger fundamental group like the $SU(5)$ to a subgroup like $SU(3) \times SU(2)_L \times U(1)$ which contains the $U(1)$ group, there inevitably arise particles that have the characteristics of a magnetic monopole. This is a rigorous mathematical conclusion in gauge field theories. Typically, the mass of the monopole (in energy units) is given by $\sim 10^{16}$ GeV. Monopoles are highly stable particles and once created they are not destructible. And so they would survive as relics to the present epoch.

At the GUT epoch t, the horizon size being $2ct$, we expect at least one monopole per horizon-size sphere, i.e. a monopole mass density of

$$\frac{10^{16} \text{ GeV}/c^2}{\dfrac{4\pi}{3}(2ct)^3}.$$

At present this is diluted by the factor $(T_0/T)^3$. For T_0 (GeV) given by (6.51) and $T = 10^{15}$ GeV we get the present monopole density as

$$\rho_M \approx 2 \times 10^{-13} \left(\frac{T_0}{3\,\text{K}}\right)^3 \text{g cm}^{-3}. \tag{6.59}$$

This is far in excess of the closure density $\sim 10^{-29}$ g cm^{-3}, thus making it a very awkward problem for the standard model to solve. Again, as in the earlier cases, the discrepancy grows if, instead of the GUT epoch, we use an even earlier epoch.

6.5 The inflationary universe

Thus difficulties of the standard big bang model seem to require a new input at or around the GUT epoch, an input that would change the dynamics of the universe, at least for a temporary period. In 1981 Alan Guth proposed the so-called *inflationary phase* as the solution to these problems. The word 'inflation' is supposed to indicate a rapid expansion. Thus we envisage the following sequence:

$t < t_1$: scale factor $S(t) \propto t^{1/2}$.

$t_1 < t < t_2$: scale factor $S(t) \propto \exp(at)$, $a =$ constant.

$t_2 < t$: scale factor $S(t) \propto t^{1/2}$. (6.60)

In short, we have inserted a phase of rapid exponential expansion during $[t_1, t_2]$. What is this time range? How do we set the value of the time constant a? To answer these questions let us first look at Guth's method.

6.5.1 Guth's inflationary model

As we saw earlier, the breakdown of GUT symmetry to $SU(3) \times SU(2)_L \times U(1)$ leads to a phase transition in which the vacuum state (i.e., the state of lowest energy) of the Higgs field ϕ changes. The original vacuum with $\phi = 0$ is no longer the true vacuum. The inflationary stage arises, however, if the true vacuum is not immediately attained.

An analogy will illustrate the scenario. Suppose steam is being cooled through the phase transition temperature of $100\,°C$. Normally we expect the stream to condense to water at this temperature. However, it is possible to supercool the steam to temperatures below $100\,°C$, although it is then in an unstable state. The instability sets in when certain parts of the steam condense to droplets of water which then coalesce, and eventually the condensation is complete. In the supercooled state the steam still remains its latent heat, which is released as the droplets form.

Suppose that similar supercooling takes place past the GUT phase transition temperature. What happens then is somewhat similar to the steam–water analogy. Its details depend on the potential energy function $V(\phi, T)$ which we consider next.

Consider the action principle defining the dynamics of the ϕ-field by

$$\mathcal{A}[\phi] = \int [\tfrac{1}{2}\phi_i\phi^i - V(\phi)]\, d^4x, \qquad (6.61)$$

where $\phi_i = \partial\phi/\partial x^i$ and $V(\phi)$ is given as follows. First, ϕ is a scalar gauge field, but it has internal degrees of freedom decided by the number of generators of the gauge group. Let the generating matrices be $\tau_A (A = 1, 2, \ldots, N)$. Then we write

$$\phi = \sum_A \phi^A \tau_A \qquad (6.62)$$

and consider the following *quartic* form for V:

$$V = -\tfrac{1}{2}\mu^2 Tr\phi^2 + \tfrac{1}{4}a(Tr\phi^2)^2 + \tfrac{1}{2}b(Tr\phi^4) + \tfrac{1}{3}c(Tr\phi^3), \qquad (6.63)$$

where μ, a, b, c are coupling constants.

In a typical symmetry breaking of the kind

$$SU(5) \rightarrow SU(3) \times SU(2)_L \times U(1)$$

we have

$$\langle \phi \rangle = \Phi \, \text{diag} \, (1, \, 1, \, 1, \, -\tfrac{3}{2}, \, -\tfrac{3}{2}) \tag{6.64}$$

where Φ is an ordinary scalar. Alternatively, if

$$SU(5) \rightarrow SU(4) \times U(1) \tag{6.65}$$

then

$$\langle \phi \rangle = \sigma \, \text{diag} \, (1, \, 1, \, 1, \, 1, \, -4), \qquad \sigma \text{ a scalar.} \tag{6.66}$$

In each case, $\text{Tr} \, \phi = 0$. If the basic GUT symmetry group is different, we will of course have different representations of ϕ. For our cosmological purpose we need to know how $V(\phi)$ will affect the spacetime geometry via the Einstein equations. For this we need to average over the quantum fluctuations of the ϕ-field, which gives us an 'effective' average potential:

$$V_{\text{eff}}(\phi) = \alpha\phi^2 - \beta\phi^4 + \gamma\phi^4 \ln(\phi/\sigma^2) \tag{6.67}$$

where α, β, γ, σ are parameters from particle physics.

Since this analysis is to be carried out in the hot early universe, there will be thermal fluctuations also. Their inclusion leads to an addition to $V_{\text{eff}}(\phi)$ of a thermal component to give a total potential

$$V(\phi, T) = V_{\text{eff}}(\phi) + \frac{18T^4}{\pi^2} \int_0^\infty x^2 \ln\left[1 - \exp\left\{-\left(x^2 + a\frac{\phi^2}{T^2}\right)^{1/2}\right\}\right] dx. \tag{6.68}$$

Here a is a constant. Figure 6.8 plots $V(\phi, T)$ as a function of ϕ for a range of values of ϕ.

Notice that for a critical value of $T = T_c$, the $V(\phi, T)$ curve touches the ϕ-axis at two points, $\phi = 0$ and $\phi = \phi_0$, both of which points are local minima for $V(\phi, T)$. For $T \gg T_c$, there is only one minimum, at $\phi = 0$. As T is lowered, a second minimum appears at a higher level, while as T goes below T_c this minimum sinks to a *lower* level. In other words, for $T < T_c$ the state of lowest energy of the ϕ-field resides not at $\phi = 0$ but at a value of $\phi > 0$. This is where we have the supercooled steam situation.

Imagine the universe being cooled through the critical value T_c. As T drops below T_c the state of lowest energy shifts in a discrete fashion, signalling a phase transion. However, if the universe is supercooled, it stays in the 'false' vacuum at $\phi = 0$ until at some stage the ϕ-field tunnels across the $V(\phi) > 0$ barrier and falls down the $V(\phi)$ slope to its 'true' vacuum. Let us denote by ε_0 the difference between the energies of the

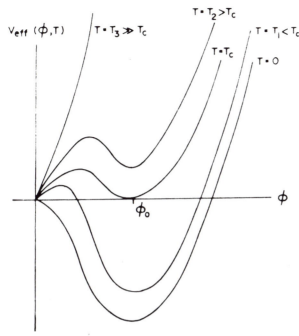

Fig. 6.8 $V_{eff}(\phi, T)$, the effective potential of the Higgs field at different temperatures. For $T < T_c$ the true vacuum, i.e. the state of lowest energy is no longer at $\phi = 0$.

two vacua. Until the tunnelling has taken place the universe has an extra energy density ε_0 at its disposal which must have dynamical effects via the Einstein equation:

$$\frac{\dot{S}^2 + kc^2}{S^2} = \frac{8\pi G}{3c^2}(\varepsilon_0 + \varepsilon_r). \tag{6.69}$$

Here $\varepsilon_r \propto 1/S^4$ is the energy density of radiation and relativistic particles. Since it falls as the universe expands, while ε_0 stays constant, the latter clearly dominates. Hence we ignore ε_r and solve (6.69). For $k = +1$ we get, for example,

$$S = \left(\frac{3c^4}{8\pi G\varepsilon_0}\right)^{1/2} \cosh\left[\left(\frac{8\pi G\varepsilon_0}{3c^2}\right)^{1/2} t\right]. \tag{6.70}$$

For $k = -1$ we get a similar expression with 'cosh' replaced by 'sinh'. The main point to note is that for

$$t \gg \left(\frac{3c^2}{8\pi G\varepsilon_0}\right)^{1/2}, \tag{6.71}$$

either solution approaches closely the $k = 0$ (flat) solution

$$S \propto \exp at, \qquad a = \left(\frac{8\pi G \varepsilon_0}{3c^2}\right)^{1/2}. \qquad (6.72)$$

This exponential expansion is reminiscent of the de Sitter model. Indeed, the energy tensor of false vacuum simulates the λg_{ik} term of the Eintein equations.

This rapid expansion in an exponential fashion continues until the tunnelling takes place and ϕ attains its true vacuum value. The average time τ for the tunnelling to occur can be computed quantum mechanically. One finds that

$$a\tau \approx 67, \qquad Z = \exp 67 \approx 10^{29}. \qquad (6.73)$$

In other words, the exponential expansion or *inflation* lasts long enough for the scale factor to blow up by a large multiple, $Z \sim 10^{29}$. Thus if we had started with a curvature term (kc^2/S^2) comparable to the expansion term $(\dot{S}^2/S^2)$ prior to inflation we would have ended up by having the former reduced by $Z^2 \sim 10^{58}$ while the latter stayed constant. This large factor Z not only takes care of the fine tuning in the flatness problem but also resolves the horizon problem (by blowing up the homogeneous region by Z in linear dimensions) and the monopole problem (by reducing the monopole density by Z^3). Similarly the domain walls are blown apart so that the chance of one crossing the observable universe is negligible.

There was one serious drawback, however, which rendered the Guth model unworkable. This comes from the entropy. The entropy is also blown up by the factor $Z^3 = 10^{87}$, thus apparently explaining why the present universe has such a large value for Σ. However, how was this entropy to be dumped in the universe?

The expectation was that as the phase transition is completed in a bounded region, it switches over to the Friedmann radiation-dominated expansion phase, since it no longer has the energy ε_0 to draw on. The inflating region therefore breaks up into Friedmann bubbles which expand. Most of the excess energy resides on the surfaces of these bubbles, so that when two bubbles collide the energy is thermalized. This is how a wider and wider region undergoes phase transition and acquires thermalized energy and entropy.

The expectation was nullified by the fact that as the universe outside the bubbles expands exponentially, bubbles nucleated in different parts move away from one another so fast that they cannot collide. The above thermalizing mechanism therefore does not work.

6.5.2 *The new inflationary universe*

A revised version of the inflationary scenario was not long in coming. A. D. Linde in the USSR and A. Albrecht and P. J. Steinhardt in the United States independently proposed what came to be known as the *new inflationary universe*. The crucial difference between the new and the original Guth version was in the choice of V_{eff} (ϕ, T). In the new model the V_{eff} (ϕ, T) was taken from the work of S. Coleman and E. Weinberg:

$$V_{\text{eff}}(\phi, T) = \tfrac{25}{16}\alpha^2[\phi^4 \ln \frac{\phi^2}{\sigma^2} + \tfrac{1}{2}(\sigma^4 - \phi^4)]$$

$$+ \frac{18}{\pi^2} T^4 \int_0^\infty \ln[1 - \exp(-\{x^2 + \tfrac{5}{12}\phi^2 g^2/T^2\}^{1/2})] \, dx, \quad (6.74)$$

where α, σ, and g are constants. Figure 6.9 plots V_{eff} (ϕ, T) for a characteristic value of T.

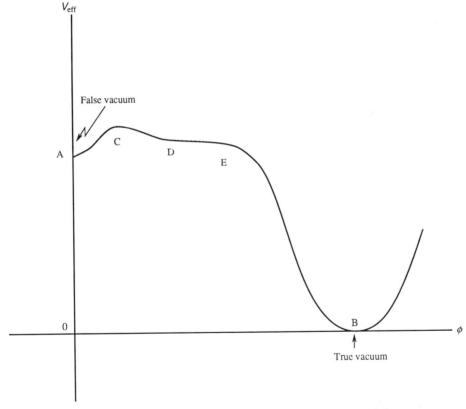

Fig. 6.9 $V_{\text{eff}}(\phi, T)$ used in the new inflationary model.

We have a false vacuum at A ($\phi = 0$) and the true vacuum at B ($\phi = \sigma$). There is a temperature-dependent bump C beyond A ($\phi > 0$) followed by a plateau portion DE which slopes down very gently before dropping down steeply from E to B. For inflation to take place for a sufficient time we need the system to remain in the upper part of the figure, i.e. during a time when it tunnels across the bump C and then rolls slowly over from D to E. Thereafter the system drops to B, but instead of staying there it executes damped oscillations during which energy is thermalized and entropy increased. The decay time scale for the ϕ field is Γ^{-1}.

The actual dynamics of the universe, as given by Einstein's equations, can be solved numerically. A 'satisfactory' solution is obtained by adjusting the parameters. Thus the following is a satisfactory solution:

$\phi = \phi_i \approx 0$ Roll-over down the plateau begins.

$t \leqslant 190 H^{-1}$ Roll-over time, which is also the duration of inflation with $S \propto \exp Ht$.

$H \approx 2 \times 10^{10}$ GeV

$Z \approx \exp 190 \approx 10^{50}$ Boost in linear size by inflation.

$\tau_{\mathrm{osc}} \approx \exp 4.8 \times 10^{-4}\ H^{-1}$ Time of oscillation before settling down at $\phi = \sigma \approx 2 \times 10^{15}$ GeV.

Notice that the model overcorrects the shortcomings of the standard model. For example, the horizon size is boosted by 10^{50}, so that the present size of a homogeneous region is greater than the observable universe by a factor $\sim 10^{23}$. Notice also that because a single region is large enough (more than enough!) to encompass the observable universe, there is no need for different bubbles to collide and coalesce. In the Guth version the bubbles were small so that their collision and coalescence was necessary to generate a large enough region – besides the requirement of reheating.

Since $\tau_{\mathrm{osc}} \ll H^{-1}$, the oscillations are quickly damped by the decay of ϕ into relativistic particles and radiation. With $\Gamma \approx 10^{13}$ GeV, the decay time Γ^{-1} is arranged to be $\ll H^{-1}$ to allow 'reheating' to take place. The temperature of the universe will rise again to $\approx 2 \times 10^{14}$ GeV.

The drawback of the new inflationary model is that it requires a fine tuning of ϕ_i/σ, where ϕ_i is the initial value from where the slow roll-over starts. We need $\phi_i/\sigma < 10^{-5}$ and the Higgs boson mass $m < 10^{-5}\sigma$. If $m \geqslant 10^9$ GeV then the model does not work. Since the whole concept of inflation was brought in to avoid fine tuning, this requirement is like breaking the ground rules.

6.5.3 Chaotic inflation

The original Guth model invoked a strongly first-order phase transition, while the second model may be considered as requiring a weakly first-order or even a second-order phase transition. Can we construct an inflationary model which has *no* phase transition involved?

Yes! Such a model was proposed by Linde under the title 'chaotic inflation'. The $V(\phi)$ function here has a simple form

$$V(\phi) = \lambda \phi^4. \tag{6.75}$$

Inflation results because of a rather slow motion of ϕ from some initial ϕ_0 towards the minimum. This initial value is believed to be due to chaotic initial conditions. While one can produce sufficient inflation this way, it is necessary to ensure that the initial kinetic energy of the ϕ-field is small compared with the potential energy. Detailed calculations show that this requires the field to be uniform over sizes bigger than the Hubble radius! Also, the value of λ has to be fine-tuned near 4×10^{-14} to get the correct density perturbations.

6.5.4 Inflation: drawbacks and epicycles

The role of inflation in growing density perturbations is an important subject, which we will discuss in the next chapter. It is probably the main issue on which the concept of inflation will survive or perish!

Nevertheless, the fluid state which the very-high-energy particle physics finds itself in today has its echoes in this branch of cosmology. Thus several epicycles of inflation have appeared, some stillborn, others with a half-life of 6 months to 1 year, while some still survive. It is not possible to review them all, especially since they have not yet produced a result that a cosmologist would be interested in.

A major unexplained point relates to the λ-term. Although the point (to be described below) would have arisen regardless of inflation, it was highlighted more by that scenario. The mystery is the smallness of λ_0 as used by Einstein (cf. Chapter 4) when expressed as the dimensionless ratio

$$\alpha_\lambda = \frac{G\hbar\lambda_0}{c^3} \approx 10^{-126}. \tag{6.76}$$

The suffix zero on λ indicates that we are interested in its present value, which is of the order of H_0^2/c^2. We saw, however, that during inflation the energy momentum tensor of the false vacuum is that corresponding to a λ-term with $\lambda \equiv \lambda_{GUT}$, where

$$\frac{\lambda_0}{\lambda_{\text{GUT}}} \approx 10^{-108}. \qquad (6.77)$$

Thus, to start with, λ may be as high as $G\hbar/c^3$, later changing to λ_{GUT} and finally to λ_0. (In between there is another phase transition at the breakdown of the electroweak symmetry, wherein $\lambda_0/\lambda_{\text{EW}} \approx 10^{-57}$.) The question is, how does λ manage to change from such a large initial magnitude to $\approx 10^{-126}$ of its initial value? What kind of fine tuning is this? This is known as the *graceful exit problem* for the cosmological constant.

With all its epicycles, the inflationary model makes one clearcut prediction about the present state of the universe, viz., $\Omega_0 = 1$. This is because the closeness of Ω to unity is such as to lead to $\Omega_0 = 1$ or very close to 1 (for $k = \pm 1$). This prediction automatically implies that there is dark matter present, since, from Chapter 5, $\Omega_\text{B} h_0^2 \lesssim 0.12$.

6.6 Primordial black holes

We now depart from the discussion of particle interactions and GUTs to study a peculiar consequence of gravity. The study relates to black holes, briefly described in Chapter 2.

As the name 'black hole' implies, we do not expect any radiation to come out of such an object. For a spherical object of mass M, the black hole condition is reached when its surface area equals $4\pi R_\text{s}^2$, where R_s, the Schwarzschild radius, is given by

$$R_\text{s} = \frac{2GM}{c^2}. \qquad (6.78)$$

No material particle or a light signal emitted from $R \leqslant R_\text{s}$ can go into the region $R > R_\text{s}$ – at least, this is what classical general relativity tells us.

Nevertheless, in 1974 Stephen Hawking made the remarkable suggestion that a black hole can radiate. Hawking's calculation went beyond classical physics: it considered what happens when any field (for example, the electromagnetic field) is *quantized* in the spacetime exterior to a black hole. The quantum mechanical description of a vacuum is much more involved than the classical description, which simply states that a vacuum is empty. According to quantum field theory, the vacuum is seething with virtual particles and antiparticles whose presence cannot be detected directly. Their interference with physical processes in spacetime can, however, lead to detectable results. Hawking found that one such result when considered in the spacetime outside a black hole is that an observer at infinity sees a flux of particles coming out from the vicinity of a black hole. We will not go through the calculations leading to this result; we will simply study the consequences of such a process in the early universe. Figure 6.10 provides a qualitative description of how the Hawking process operates. Not all aspects of the Hawking process have been worked out yet. An important issue still unresolved, for example, is that of back reaction: how the emission of particles by the black hole affects and alters the geometry of spacetime outside, and what effect this change has on the process of radiation by the black hole.

The idea we will use here is that a spherical black hole of mass M ejects particles in a thermal spectrum of temperature T given by

$$kT = \frac{\hbar c^3}{8\pi GM} \sim 10^{26} \, M_g^{-1}, \tag{6.79}$$

where $M_g = M$ expressed in grams. The emission of particles by the black hole leads to a mass loss rate given by

$$\frac{dM_g}{dt} \sim -10^{26} \, M_g^{-2} \, s^{-1}. \tag{6.80}$$

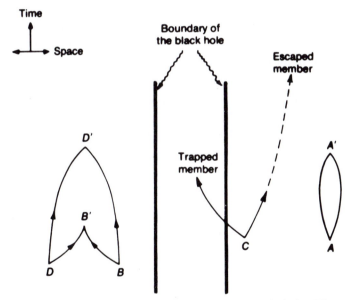

Fig. 6.10 The thick lines indicate the boundary of a black hole. The two arrows emerging from points A, B, C, ... indicate pair creation in vacuum fluctuations. Had there been no black hole, pairs would simply annihilate and disappear as at A', B'. The black hole may selectively attract only one member of a pair as at C and let the other member escape. A remote observer seeing this particle would conclude that the black hole has created this particle. Since virtual particles created may have negative energies, by absorbing a negative energy member of the pair the black hole loses part of its mass. The escaped particle carrying positive energy therefore describes energy emission by the black hole. This is the essence of the *Hawking process*.

The $\sim$ implies that a numerical constant of the order 1 appears on the right-hand side to take account of the number of particle species emitted. If we integrate (6.80), we find that the entire mass of the black hole is radiated away in a time τ given by

$$\tau \sim 3.10^{-27} M_g^3 \, s. \tag{6.81}$$

Thus a black hole created soon after the big bang with a mass $\sim 5 \times 10^{14}$ g would last until the present day.

The process described above is slow to start with, when a black hole is massive and cold. However, as M decreases T rises and the mass loss rate increases until finally it reaches a catastrophic level. This final stage is often called evaporation or explosion of a black hole. As seen above, a stellar mass black hole ($M_g \geqslant 10^{33}$) is hardly likely to explode in the lifetime of the universe! And since black holes considered in various astrophysical scenarios are at least as massive as $2M_\odot$, the Hawking process is only of academic interest with respect to them.

There are, however, scenarios in the very early universe that could lead to primordial black holes (PBH) of much lower masses than $M_\odot$. B. J. Carr in 1975 was the first to consider these at length. Carr investigated PBH formation and evaporation in order to see whether the presently observed nucleon density as well as the microwave background can be explained in terms of the emission of baryons, leptons, photons, and so on by low-mass black holes. Since these concepts are highly speculative, we will give only a brief survey of what PBHs could do.

It is argued that provided the equation of state is hard (that is, pressure/density of matter ~ 1), the size of a PBH at any given time is limited by the size of the particle horizon. This gives a simple relation

$$M_g \sim 10^{38} t_s \tag{6.82}$$

for a PBH of mass M_g formed at the epoch t_s. Thus to form a PBH of $M_g \leqslant 10^{14}$ we have to go as far back as 10^{-24} s. If we go as far back in time as $t_s = 10^{-43}$ s, when quantum gravity applies (see section 6.7), then $M_g \sim 10^{-5}$.

If the early universe contained density perturbations, then their growth could decide whether black holes would be formed and with what type of mass spectrum. Suppose the mass contrast at some initial time in a given region is

$$\delta_* = \frac{\Delta M}{M} = \varepsilon \left(\frac{M}{\bar{M}} \right)^{-n}, \tag{6.83}$$

where $\bar{M}$ = mass of the matter contained in the horizon at the initial instant $\bar{t}$. Carr found that only for $n = \frac{2}{3}$ is PBH formation over extended regions favoured. For $n > \frac{2}{3}$, PBH is not formed, except perhaps for $M = \bar{M}$. If $n < \frac{2}{3}$, PBH formation is favoured only when M is so large that it might split up into different universes casually disconnected from one another.

If $n = \frac{2}{3}$, and if we assume that the density excess in a region of mass M has a normal distribution with zero mean and standard deviation $= \delta_*$, then the present number density of PBHs in the mass range M, $M + \mathrm{d}M$ is given by $N(M)\,\mathrm{d}M$, where

$$N(M) \sim \bar{\rho} F(\bar{M})^{-2}\, \varepsilon \exp - \left[\frac{f}{2\varepsilon^2} \left(\frac{M}{\bar{M}} \right)^{-2(1+2f)/(1+f)} \right]. \qquad (6.84)$$

This formula ignores the degradation in mass of a PBH by evaporation. $\bar{\rho} = $ mean density of the universe at $\bar{t}$, $f = p/\rho$ at the time of PBH formation, $F = $ ratio of the number density now to what it would have been at $\bar{t}$. F goes as $(S_0/\bar{S})^{-3}$.

In determining how these black holes radiate, we have two alternative pictures. In the elementary particle picture the PBHs emit the so-called basic particles discussed earlier, such as quarks and gluons. Later these form particles such as baryons, mesons, and leptons. In the other composite particle picture, these latter particles are emitted directly. The PBH emission process does not take note of whether particles or antiparticles are being emitted. Thus it is not necessary that the process should conserve the baryon number and this possibility may therefore be used to account for the observed baryons-to-photons ratio.

The interesting aspect of this approach is that PBHs act as sources of various particles that have somehow got to be created in the universe. The suggestion that PBHs evaporating today might account for the observed γ-ray bursts, however, does not seem to be correct, since the spectrum of γ-rays emitted in such a process is not like the spectrum observed in burst events.

6.7 Quantum cosmology

In this chapter we finally push back our investigations into the past history of the universe down to the era $t \sim 10^{-43}$ s. Is it justified to put our faith in the standard big bang model when the universe was so young? One way to answer this question is to look for the limit where classical theory breaks down and quantum mechanics takes over. Beyond this limit we cannot trust the classical theory of gravity – that is, the general theory of relativity.

A look at the action principle (2.101) shows that the limit sought above can be obtained by equating the gravitational action

$$S_g = \frac{c^3}{16\pi G} \int_V R(-g)^{1/2}\, \mathrm{d}^4 x \qquad (6.85)$$

to Planck's constant. For $S_g \gg \hbar$ we can trust our classical description of spacetime geometry, while for $S_g \ll \hbar$ a quantum description of cosmology is indispensable. But to evaluate S_g we need V, the 4-volume of the spacetime manifold.

In the big bang model we take V as the spatial volume enclosed by the particle horizon and bounded by the time span of the universe. Thus at any epoch t for $k = 0$, $S \propto t^{1/2}$, the particle horizon is defined by

$$rS = 2ct.$$

For $S \propto t^{1/2}$, $R = 0$ and so $S_g = 0$. However, this happens because the trace of T^i_k is zero in the early universe. As an order-of-magnitude estimate we may take R^0_0 instead of R in the computation of S_g; R^0_0 gives us an idea of how the geometical part of the action changes with time. For $S \propto t^{1/2}$, $R^0_0 = \frac{3}{4}c^2 t^2$. Thus, up to the epoch t,

$$S_g \sim \frac{c^3}{16\pi G} \int_0^t \frac{3}{4c^2 t_1^2} \frac{4\pi}{3} (2ct_1)^3\, \mathrm{d}t_1 = \frac{c^5}{4G} t^2.$$

Equating S_g to $\hbar$, we get

$$t = 2t_P = 2\left(\frac{G\hbar}{c^5}\right)^{1/2} \approx 10^{-43}\ \mathrm{s}.$$

This time span is called the *Planck time. No classical discussion of standard big bang cosmology can be pushed to epochs $t < t_P$.* We already encountered this epoch in section 6.3, when the temperature was $E \sim 10^{19}$ GeV. This energy, as seen from (6.43), is simply $\hbar/t_P$.

Thus the present discussions of GUTs and cosmology already take us right up to the Planck epoch. Whether the universe did indeed have a spacetime singularity at $t = 0$ should be determined not by classical general relativity but by an appropriate theory of quantum gravity. At the present time the goal of having a working theory of quantum gravity seems far away. The different approaches that have been tried to quantize gravity do not agree on the answer to the question: did the universe have a singular epoch?

This concludes our discussion of the early universe, a discussion in which we have pushed back our incomplete knowledge as far as we possibly could. Figure 6.11 provides a schematic view of the events in the early universe that is built out of the present speculations. Our next investigation will relate to the formation of discrete structures in the universe, to the problem of evolving a successful theory that starts with 'seeds' of local fluctuations implanted in the very early universe and grows them to galaxies, clusters, superclusters, etc.

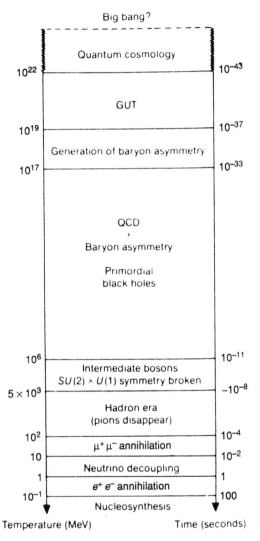

Fig. 6.11 The time axis going back to $-\infty$ on a logarithmic scale. The interactions and events that govern the state of the universe are shown in the relevant temperature sections. We have to remember that rising temperatures correspond to decreasing time by the formula (6.6). According to present speculations some brand of inflation took place around the GUT era.

Exercises

1   Explain what is meant by the remark: 'the early universe is the poor man's high-energy accelerator'.

2   Look up the mass and surface temperature of the Sun in an astronomy textbook and express both in MeV units.

3   Explain why MeV is a good unit to describe the masses and temperatures in the early universe.

4   Look up the values of the various fundamental constants appearing in (6.5) and verify the numerical coefficient in (6.6).

5   Assuming that the universe contains only those particles (as well as antiparticles of quarks and leptons) listed in Table 6.1, estimate the g-factor from (6.3).

6   Give qualitative arguments based on thermodynamic equilibrium and the survival of different particle species under various interactions to indicate why we expect hadrons to be the least abundant species in the universe.

7   Apply the arguments given in Exercise 6 to estimate quantitatively the ratio of baryons to photons at the present epoch. Comment on the smallness of this ratio.

8   Relate the smallness of the ratio of Exercise 7 to the relative strengths of the gravitational and the strong interactions. How much larger has the gravitational constant got to be in order that the calculated value of N_A/N_γ comes out as high as the observed value?

9   Show that in a theory having the symmetries of $SU(n)$ the number of 'charges' will be equal in n^2-1. What is the corresponding number of bosons in such a theory?

10 Illustrate the general result of Exercise 9 by specific examples of physical theories with $n = 2, 3$, and 5.

11 Distinguish between the natural current and the charge current components of the electroweak interaction. Give examples of each.

12 Show how baryons may be created or destroyed by suitable transformations in the $SU(5)$ version of GUTs. Show how a proton can decay, and indicate what the decay products could be in such an event.

13 Describe a laboratory experiment that might prove the existence of the X-bosons. Why can't these bosons be detected directly in a high-energy accelerator?

14   Explain how the hot universe can be a testing ground for the predictions of GUTs.

15   Show by a qualitative argument how a baryon excess can be

16   For the X-bosons estimate the time scale

$$\tau \sim \frac{\hbar}{m_X c^2}.$$

17   Explain why the horizon and flatness problems get more severe as we seek to set the initial conditions at earlier and earlier epochs.

18   Give *astrophysical* reasons why we do not expect free magnetic monopoles in significant quantities today.

19   In the Guth model of inflation let $\lambda(t_0)$ denote the rate at which bubbles form in a given proper volume and suppose that $p(t)$ denotes the probability that there are no bubbles engulfing a given point in space. Show that

$$p(t) = \exp\left[-\int_0^t \lambda(t_1) S^3(t_1) V(t, t_1) \, dt_1\right]$$

where

$$V(t, t_1) = \frac{4\pi}{3} \left[\int_{t_1}^t \frac{c \, dt_2}{S(t_2)}\right]^3.$$

20   If the nucleation rate may be approximated by a constant λ_0 in Exercise 19, show that

$$p(t) = \text{constant} \times \exp\left(-\frac{t}{\tau}\right),$$

where $\tau = 3a^3/4\pi\lambda_0$.

21   Show that in the Coleman–Weinberg case the evolution of ϕ during the slow roll-over is given approximately by

$$\ddot{\phi} + 3a\dot{\phi} + V'(\varphi) = 0,$$

where $\ddot{\phi}$ may be ignored and $a = 25\pi\alpha^2 G\sigma^4/24c^2$.

22   Show that even if we are able to explain why the variation in Ω, $\Delta\Omega = O(1)$ at this epoch, the future epoch will again have larger and larger $\Delta\Omega$.

23   Calculate the temperature in kelvins of a spherical black hole of mass equal to the mass of (a) a proton, (b) one ton, (c) the Earth, (d) the Sun, and (e) a star of mass $10 \, M_\odot$.

24   Substitute the values of $\hbar$, c, G, and k to verify the results (6.79) and (6.80).

25   Express in terms of the fundamental constants the time τ for which a black hole of mass M survives under its own radiation.

26   Taking $k = 0$, $S \propto t^{1/2}$ in the early universe, calculate the size of the particle horizon. By equating this to the Schwarzschild radius of a black hole, calculate the mass M of the black hole. Show that this mass is given by

$$M = \frac{c^3 t}{G}.$$

27   Discuss how the primordial black holes might act as sources of particles and radiation that are now found in the universe.

28   Compute $S_g(t)$ for the closed Friedmann model with given values of q_0 and h_0, taking the time interval as $(0, t)$ and the spatial extent covering the whole (spherical) space. Estimate the epoch at which $S_g = \hbar$. Why do you get an answer different from t_P?

29   Relate the Planck length associated with gravity to the Compton wavelength of the proton and to the strength factor α_G defined in (6.34). Show also that at the Planck epoch the Schwarzschild radius of a primordial black hole filling the particle horizon is of the same order as the Compton wavelength of the black hole.

30   Comment on the limit to which classical general relativity may be pushed in discussions of the early universe.

7

The formation of structures in the universe

7.1 A key problem in cosmology

Chapter 5 narrated the success story of the hot big-bang model, of how the particles combined to form light nuclei at temperatures of 10^8–10^{10} K, and how the relic of that hot era is today seen as the radiation background in microwaves. Encouraged by these achievements, the big bang cosmologists pushed their investigations further back in time, to epochs of very-high-energy particles. These investigations, outlined in Chapter 6, brought cosmologists in contact with the very-high-energy particle theorists, leading to a variety of new inputs to classical cosmology such as inflation, baryon nonconservation, etc.

Exciting though these investigations are, we must not lose sight of the fact that cosmology is a branch of physics, and as such requires hard facts to support these speculations. One important fact is the existence of discrete structures in the universe, ranging from galaxies to superclusters. How did these structures come about? Why are they distributed in an inhomogeneous fashion, when their radiation counterpart is so smooth? This key problem of cosmology must surely have a solution buried in the early history of the universe.

In this chapter we review some attempts to come to grips with this problem. If the big bang scenario is correct then the solution should incorporate some or all of the following epochs:

1. The Planck epoch.
2. The GUTs/inflation epoch.
3. The epoch when radiation decoupled from matter.
4. The epoch when the universe switched over from being radiation-dominated to being matter-dominated.
5. The epoch of redshift ~ 5, when galaxies and QSOs may have begun to form.

This last epoch is fixed by the observation that the numbers of these discrete objects seem to taper off as we approach redshifts of this order.

The strategy is to consider small fluctuations of density in the very early epochs (1) or (2) and work out their growth through the successive later epochs. Since the physics of the universe changes drastically, the techniques of working out the solution also change. We will consider first the epochs from (3) and (4) onward, which, historically, were the first to be tackled and which involve the less speculative parts of cosmology.

7.2 The Jeans mass in the expanding universe

As early as 1902, Sir James Jeans considered the problem of formation of galaxies in the universe as a process involving the interplay of gravitational attraction and the pressure force acting on a mass of nonrelativistic fluid. Jeans's treatment used Newtonian physics and assumed a static universe. However, his ideas can be adapted to suit our problem, at any rate part of it.

7.2.1 The basic equations

Consider the universe as filled with fluid of density ρ, pressure p, velocity field v, and the gravitational force field $\mathbf{F}$. We will assume Newtonian physics to hold for gravity as well as for fluid dynamics. Thus the continuity equation

$$\frac{\partial \rho}{\partial t} + \nabla \cdot (\rho \mathbf{v}) = 0 \tag{7.1}$$

and the Euler equation

$$\frac{\partial \mathbf{v}}{\partial t} + (\mathbf{v} \cdot \nabla)\mathbf{v} = -\frac{1}{\rho} \nabla p + \mathbf{F} \tag{7.2}$$

hold for fluid motion, while the Poisson equation holds for $\mathbf{F}$:

$$\nabla \times \mathbf{F} = 0, \qquad \nabla \cdot \mathbf{F} = -4\pi G \rho. \tag{7.3}$$

In the unperturbed situation of homogeneity and isotropy we get the following simple solution of the above equations (see Exercise 2 at the end of Chapter 3, and Exercises 38 and 41 at the end of Chapter 4):

$$p = 0, \qquad \rho \propto \frac{1}{S^3(t)}, \qquad \mathbf{v} = \mathbf{r}\frac{\dot{S}(t)}{S(t)}, \qquad \mathbf{F} = -\frac{4\pi G \rho}{3}\mathbf{r}. \tag{7.4}$$

The scale factor $S(t)$ satisfies the differential equation

$$\frac{\ddot{S}}{S} = -\frac{4\pi G\rho}{3}. \tag{7.5}$$

We now consider perturbations of this simple solution. Our aim in doing so is to see whether any initial clumpiness can grow in size by gravitational instability. Thus we consider small changes in ρ, $\mathbf{v}$, $\mathbf{F}$, and p in the above solution, denoting them by ρ_1, $\mathbf{v}_1$, $\mathbf{F}_1$, and p_1 respectively. To begin with, these perturbations are supposed to be small, so that the equations can be linearized. The linearized equations become

$$\dot{\rho}_1 + \frac{\dot{S}}{S}(\mathbf{r}\cdot\nabla)\rho_1 + 3\frac{\dot{S}}{S}\rho_1 + \rho\nabla\cdot\mathbf{v}_1 = 0, \tag{7.6}$$

$$\dot{\mathbf{v}}_1 + \frac{\dot{S}}{S}(\mathbf{r}\cdot\nabla)\mathbf{v}_1 + \frac{\dot{S}}{S}\mathbf{v}_1 + \frac{1}{\rho}\nabla p_1 - \mathbf{F}_1 = 0, \tag{7.7}$$

$$\nabla\times\mathbf{F}_1 = 0, \qquad \nabla\cdot\mathbf{F}_1 = -4\pi G\rho_1. \tag{7.8}$$

We also have for *adiabatic* fluctuations

$$p_1 = c_s^2\rho_1, \tag{7.9}$$

where c_s is the speed of sound $(=(\mathrm{d}p_1/\mathrm{d}\rho_1)^{1/2})$.

It is not difficult to see that plane–wave solutions of the following form exist for (7.6)–(7.8):

$$\rho_1(\mathbf{r},\, t) = \bar{\rho}_1(t)\mathrm{e}^{i\chi}, \qquad \mathbf{v}_1(\mathbf{r},\, t) = \bar{\mathbf{v}}_1(t)\mathrm{e}^{i\chi}, \tag{7.10}$$

with

$$\chi = \frac{\mathbf{r}\cdot\mathbf{k}}{S(t)} \tag{7.11}$$

and

$$\dot{\bar{\rho}}_1 + \frac{3\dot{S}}{S}\bar{\rho}_1 + \frac{i\mathbf{k}\cdot\bar{\mathbf{v}}_1}{S}\rho = 0, \tag{7.12}$$

$$\dot{\bar{\mathbf{v}}}_1 + \frac{\dot{S}}{S}\bar{\mathbf{v}}_1 + \frac{ic_s^2}{S\rho}\mathbf{k}\bar{\rho}_1 - \mathbf{F}_1 = 0, \tag{7.13}$$

$$\mathbf{k}\times\bar{\mathbf{F}}_1 = 0, \qquad i\mathbf{k}\cdot\mathbf{F}_1 = -4\pi G\bar{\rho}_1 S. \tag{7.14}$$

It is now convenient to split $\bar{\mathbf{v}}_1$ into two parts: along and perpendicular to the wave vector $\mathbf{k}$. Thus we write

$$\bar{\mathbf{v}}_1 = \frac{\mathbf{k}\cdot\bar{\mathbf{v}}_1}{k^2}\mathbf{k} + \frac{\mathbf{k}\times(\bar{\mathbf{v}}_1\times\mathbf{k})}{k^2}$$

$$= \mathbf{v}_\| + \mathbf{v}_\perp.$$

Taking the vector product of (7.13) with $\mathbf{k}$ we get

$$\left(\dot{\bar{\mathbf{v}}}_1 + \frac{\dot{S}}{S}\bar{\mathbf{v}}_1\right)\times\mathbf{k} = \mathbf{0},$$

from which our definition of $\mathbf{v}_\perp$ leads us to

$$\mathbf{v}_\perp S = \text{constant.} \tag{7.15}$$

Thus the transverse (or rotational) model tends to decrease in the expanding universe. What about the mode parallel to $\mathbf{k}$? Taking the scalar product of (7.13) with $\mathbf{k}$ gives the following relation:

$$\dot{v}_\| + \frac{\dot{S}}{S} v_\| + \frac{i}{k} \left(\frac{c_s^2 k^2}{S} - 4\pi GS \right) \frac{\bar{\rho}_1}{\rho} = 0. \tag{7.16}$$

We now define the density contrast parameter

$$\delta = \frac{\bar{\rho}_1}{\rho}. \tag{7.17}$$

Then, since $\rho \propto S^{-3}$, we get from (7.12)

$$\dot{\delta} = -\frac{ik}{S} v_\|. \tag{7.18}$$

Eliminating $v_\|$ between (7.16) and (7.18), we get

$$\ddot{\delta} + \frac{2\dot{S}}{S} \dot{\delta} + \left(\frac{c_s^2 k^2}{S^2} - 4\pi G\rho \right) \delta = 0. \tag{7.19}$$

This is the equation that tells us how or whether gravitational instability leads to the growth of condensations in the expanding universe.

We first consider (7.19) in the quasistatic approximation, wherein the expansion of the universe is neglected. Thus we set $S = \text{constant}$ and $\dot{S}/S = 0$. This brings us back to the original Jeans calculation of the static universe. We define $K = k/S$ as the effective wave number for the solution (7.10) and (7.11) and call

$$K_J = \left(\frac{4\pi G\rho}{c_s^2} \right)^{1/2} \tag{7.20}$$

the *Jeans wave number*. The equation (7.19) now looks like

$$\ddot{\delta} + c_s^2 (K^2 - K_J^2)\delta = 0. \tag{7.21}$$

In this approximation it is easy to see that (7.21) has sinusoidal (that is, oscillating) solutions for $K > K_J$ and exponential (growing as well as damped) solutions for $K < K_J$. If we write

$$\delta \propto e^{i\omega t} \tag{7.22}$$

Then

$$\omega^2 = c_s^2 (K^2 - K_J^2). \tag{7.23}$$

Notice first that for $K < K_J$ the growth rate $|\omega|$ is maximum when $K = 0$ and is given by

$$|\omega|_{\text{max}} = c_{\text{s}} K_{\text{J}}. \tag{7.24}$$

However, the expansion rate of the universe, which we have neglected so far, is also of this order. For, from Einstein's equations we get (for the $k = 0$ cosmology)

$$\frac{\dot{S}^2}{S^2} = \frac{8\pi G\rho}{3},$$

that is,

$$\frac{\dot{S}}{S} = \left(\frac{2}{3}\right)^{1/2} K_{\text{J}} c_{\text{s}}. \tag{7.25}$$

Thus we cannot legitimately neglect the expansion of the universe in the present problem.

Nevertheless, we can salvage something useful out of this analysis. If we set $K \gg K_{\text{J}}$ in (7.23) then we get sinusoidal disturbances that do not grow but simply propagate like sound waves. What does this mean? To understand the meaning of $K \gg K_{\text{J}}$, define a mass

$$M = \frac{4\pi n m_{\text{H}}}{3} \left(\frac{2\pi}{K}\right)^3. \tag{7.26}$$

M is the mass of a sphere of radius $2\pi/K$ containing a number density n of hydrogen atoms, each of mass m_{H}. As the universe expands, n decreases as S^{-3} and K decreases as S^{-1}. Thus M remains invariant. Taking $\rho \approx n m_{\text{H}}$ for the present, we see that the gravitational energy of this sphere is

$$\mathscr{E}_{\text{G}} \approx \frac{GM^2}{2\pi/K} \approx \frac{16\pi^2 G\rho^2}{9} \cdot \left(\frac{2\pi}{K}\right)^5.$$

The thermal energy of this sphere, on the other hand, is

$$\mathscr{E}_{\text{th}} = \frac{4\pi}{3} \rho c_{\text{s}}^2 \left(\frac{2\pi}{K}\right)^3.$$

Comparing the two expressions above, we see that

$$K \gg K_{\text{J}} \Rightarrow \mathscr{E}_{\text{th}} \gg \mathscr{E}_{\text{G}}. \tag{7.27}$$

Further, $K \gg K_{\text{J}}$ also gives us

$$|\omega| \gg \frac{\dot{S}}{S}. \tag{7.28}$$

Thus in the sound wave approximation the gravitational forces and the expansion of the universe may be neglected.

It is convenient to express the condition $K \gg K_J$ in the form

$$M \ll \frac{4\pi n m_H}{3}\left(\frac{2\pi}{K_J}\right)^3 \equiv M_J, \tag{7.29}$$

where M_J is called the *Jeans mass*. The above result therefore means that the only disturbances that have any prospects of growth are those whose mass exceeds the Jeans mass M_J.

7.2.2 The evolution of the Jeans mass

Let us try to follow the variation of M_J as the universe goes through different phases, starting with the era when e^+ and e^- annihilated (see Chapter 5). It is a good approximation to assume that until the electrons combined with protons to form hydrogen atoms, the universe is made largely of nonrelativistic ionized hydrogen in thermal equilibrium with the blackbody radiation at temperature T. In this era we may neglect the pressure and entropy of matter in comparison with that of radiation. Hence we have the density, pressure, and entropy density as given below:

$$\rho = n m_H + \frac{aT^4}{c^2}. \tag{7.30}$$

$$p = \tfrac{1}{3}aT^4, \tag{7.31}$$

$$s = \tfrac{4}{3}aT^3. \tag{7.32}$$

In adiabatic changes entropy of a comoving volume is constant, so $s/n =$ constant. Hence to evaluate c_s^2 we must calculate $dp/d\rho$ at constant s/n. A simple calculation gives

$$c_s^2 = \tfrac{1}{3}\cdot\left(\frac{Ts}{c^2 n m_H + Ts}\right)c^2. \tag{7.33}$$

In evaluating the Jeans mass from (7.29) we will replace ρ in (7.20) by $\rho + p/c^2$, without seriously altering any conclusions (which are order-of-magnitude anyway!). Thus a simple calculation gives

$$M_J = \frac{2\pi^{5/2}s^2}{9a^{1/2}n^2 m_H^2 G^{3/2}}\left(1 + \frac{Ts}{n m_H c^2}\right)^{-3}. \tag{7.34}$$

It is more convenient to use specific entropy

$$\sigma = \frac{s}{kn}, \qquad k = \text{Boltzmann's constant}. \tag{7.35}$$

Then

$$M_J = \frac{2\pi^{5/2}\sigma^2 k^2}{9a^{1/2}G^{3/2}m_H^2}\left(1 + \frac{\sigma kT}{m_H c^2}\right)^{-3}. \tag{7.36}$$

As the temperature drops to ~3000 K the recombination of electrons with ionized hydrogen is almost complete. This is the *recombination epoch*. If the present-day background temperature is taken as 3 K, the recombination era broadly corresponds to redshifts in the range $z \sim 1000$ to 1500. In making any numerical estimates we will take $z_R = 1000$ as the redshift at recombination.

After the recombination era, radiation pressure becomes unimportant and the gas (of H-atoms) behaves as a monoatomic gas is expected to behave: with $\gamma = \frac{5}{3}$ and

$$\rho = nm_H + \frac{3}{2}\frac{nkT}{c^2}, \tag{7.37}$$

$$p = nkT, \tag{7.38}$$

$$c_s^2 = \frac{5}{3}\frac{kT}{m_H}. \tag{7.39}$$

The Jeans mass then becomes

$$M_J = 4\left(\frac{\pi}{3}\right)^{5/2}\left(\frac{5kT}{G}\right)^{3/2}n^{-1/2}m_H^{-2}. \tag{7.40}$$

Just after recombination, the temperature T of matter is the same as the radiation temperature. So we can express our answer above in terms of σ by using (7.32):

$$M_J = \frac{2\pi^{5/2}5^{3/2}k^2\sigma^{1/2}}{9a^{1/2}G^{3/2}m_H^2}. \tag{7.41}$$

How does the matter temperature drop subsequently? In Chapter 4 we saw that random motions drop as S^{-1}, so that in this nonrelativistic era the temperature will fall as S^{-2}. Thus, starting from (7.41) at the recombination, M_J will drop according to (7.40), that is,

$$M_J \propto T^{3/2}n^{-1/2} \propto S^{-3/2}. \tag{7.42}$$

Figure 7.1 shows how M_J varies with the radiation temperature T_γ on the assumption that the present radiation temperature is ~3 K and is equal to that of the cosmic microwave background. The quantity that enters the expression (7.36) besides the temperature is σ, the specific entropy. In Chapter 5 we saw that σ, which is proportional to the photon/baryon ratio, is in the range 10^8 to 10^{10}. In Figure 7.1 we have taken $\sigma = 10^{10}$. It is convenient to express M_J in units of the solar mass $M_\odot$.

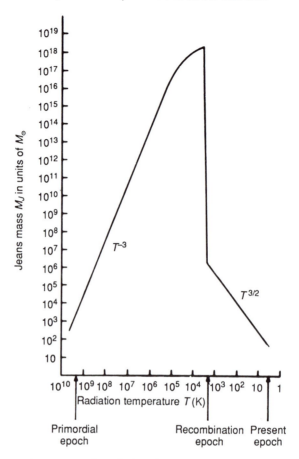

Fig. 7.1 An approximate graph of the Jeans mass M_J as a function of the radiation temperature in the universe for the entropy density $\sigma \approx 10^9$.

We see that at $T_\gamma \sim 10^9$ K, M_J was in the range $10^4 M_\odot$ to $10^5 M_\odot$. For $T \gg m_H c^2/\sigma k$, M_J increases as T_γ^{-3}. The increase continues until T_γ drops to a temperature of 10^4 to 10^5 K ($m_H c^2/\sigma k = 10^4$ for our chosen σ). The highest value reached by M_J is in the range of 10^{17} to $10^{19} M_\odot$. This is just before the recombination era, when M_J drops to the value of a few times $10^6 M_\odot$. This drop is a sharp one. But thereafter M_J drops further as $S^{-3/2}$, that is, as $T_\gamma^{3/2}$. This behaviour of M_J helps us understand the difficulties of forming galaxies in the expanding universe.

Suppose we are interested in forming a galaxy of typical mass $\sim 10^{11} M_\odot$. In terms of (7.36), M_J will be less than this value, until the temperature has dropped to $\sim 10^7$ K (see also Figure 7.1). From our crude theory of Jeans mass, we see that a fluctuation of mass $\sim 10^{11} M_\odot$ will have a chance to grow under its self-gravitation until the temperature drops to this value. The actual mode of growth must be calculated using the perturbation theory with the full general relativistic equations. This complicated problem was solved by E. Lifshitz in 1946. We will not go through the details here, but simply quote the result: *in the fastest growing normal models $\delta\rho/\rho$ increases as t.*

In the next phase, when $M_J > 10^{11} M_\odot$, our fluctuation cannot grow. It oscillates as a sound wave until the postrecombination era, when M_J has again dropped below $10^{11} M_\odot$ (see Figure 7.1). After the temperature has dropped to ~3000 K, growth is possible and we can use our simple Newtonian equations. We will proceed to solve this problem in the following section. It is clear, however, that so far, within the Jeans mass theory, the number $10^{11} M_\odot$ does not seem to emerge as having any particular significance. The typical mass at recombination is of the order of the mass of a globular cluster: it is much smaller than $10^{11} M_\odot$. This was pointed out by R. H. Dicke and P. J. E. Peebles in 1968, and has been the main difficulty in trying to understand why typical discrete units of $10^{11} M_\odot$ are found in the universe.

7.3 Growth in the postrecombination era

We now try to solve (7.19) in the framework of Friedmann models. Our purpose in doing so is to try to relate any present fluctuations in temperature or number density to those at the recombination epoch, with the hope that such a calculation may give us clues as to how galaxies may have formed in that era. We will consider the problem separately for the three types $k = 0$, $k = 1$, $k = -1$ of the Friedmann model. We will make one simplification in our calculation. We will neglect the term $c^2 k^2/S^2$ in comparison with $4\pi G\rho$. Physically, this means that we are neglecting random motions relative to the expanding substratum; that is, this approximation corresponds to neglecting M_J in comparison with the galactic mass. This neglect is valid, since in the postrecombination era M_J is as low as $10^6 M_\odot$ ($\approx 10^{-5} \times$ galactic mass).

7.3.1. The Einstein–de Sitter model

In this model (see Chapter 4)

$$S(t) = \left(\frac{t}{t_0}\right)^{2/3}, \qquad t_0 = \tfrac{2}{3}H_0^{-1}, \qquad \rho = \frac{1}{6\pi G t^2}. \qquad (7.43)$$

Therefore (7.19) becomes

$$\ddot{\delta} + \frac{4}{3t}\dot{\delta} - \frac{2}{3t^2}\delta = 0. \tag{7.44}$$

This equation has the general solution

$$\delta = At^{2/3} + Bt^{-1}. \tag{7.45}$$

Thus the growing mode is $\propto t^{2/3}$ and the damped mode $\propto t^{-1}$. If both modes are present in comparable form to start with, only the growing mode will be important eventually. Thus we will set $B = 0$.

At the recombination epoch let t_R and z_R denote the cosmic time and the redshift. Taking the temperature of the epoch as ~ 3000 K, we have

$$1 + z_R \approx 10^3, \tag{7.46}$$

since the radiation temperature increased in proportion to $(1 + z)$ in the past. Thus the density contrast δ should have grown by the factor

$$\Sigma = \frac{\delta(t_0)}{\delta(r_R)} = \left(\frac{t_0}{t_R}\right)^{2/3} = (1 + z_R) \sim 10^3. \tag{7.47}$$

7.3.2 The closed model ($k = 1$)

We use the relations (4.46)–(4.50) to write

$$ct = \tfrac{1}{2}\alpha(\Theta - \sin\Theta), \qquad S = \tfrac{1}{2}\alpha(1 - \cos\Theta),$$

$$\rho = \frac{3H_0^2}{4\pi G} \cdot \frac{q_0(1 - \cos\Theta_0)^3}{(1 - \cos\Theta)^3} = \frac{3H_0^2(2q_0 - 1)^3}{4\pi G q_0^2(1 - \cos\Theta)^3}, \tag{7.48}$$

$$\alpha = \frac{2q_0}{(2q_0 - 1)^{3/2}}\left(\frac{c}{H_0}\right).$$

Changing the independent variable from t to Θ in (7.19) we get

$$(1 - \cos\Theta)\frac{d^2\delta}{d\Theta^2} + \sin\Theta\frac{d\delta}{d\Theta} - 3\delta = 0. \tag{7.49}$$

This equation has the general solution

$$\delta = A\left(\frac{5 + \cos\Theta}{1 - \cos\Theta} - \frac{3\Theta\sin\Theta}{(1 - \cos\Theta)^2}\right) + B\frac{\sin\Theta}{(1 - \cos\Theta)^2}. \tag{7.50}$$

Again, the growing mode is that multiplying the constant A. Concentrating on this mode, we first note that for the recombination epoch z_R given by (7.46) Θ_R is small. Hence

$$1 + z_R = \frac{1 - \cos\Theta_0}{1 - \cos\Theta_R} \approx \frac{2}{\Theta_R^2}(1 - \cos\Theta_0),$$

that is,

$$\Theta_R = \left(\frac{2(1 - \cos \Theta_0)}{(1 + z_R)} \right)^{1/2}.$$ (7.51)

Thus the growth factor is given by

$$\Sigma = \frac{5(1 + z_R)[(5 + \cos \Theta_0)(1 - \cos \Theta_0) - 3\Theta_0 \sin \Theta_0]}{(1 - \cos \Theta_0)^3}$$

$$= \frac{5(1 + z_R)q_0}{(2q_0 - 1)^2} \left[(4q_0 + 1) - \frac{3q_0}{(2q_0 - 1)^{1/2}} \sin^{-1} \frac{(2q_0 - 1)^{1/2}}{q_0} \right],$$ (7.52)

where we have used (4.50) to express $\cos \Theta_0$ in terms of q_0.

7.3.3 The open model ($k = -1$)

Again using the relations (4.60)–(4.63), we write in terms of the parameter Ψ:

$$ct = \tfrac{1}{2}\beta(\sinh \Psi - \Psi), \qquad S = \tfrac{1}{2}\beta(\cosh \Psi - 1)$$

$$\rho = \frac{3H_0^2(1 - 2q_0)^3}{4\pi G q_0^2(\cosh \Psi - 1)^3},$$ (7.53)

$$\beta = \frac{2q_0}{(1 - 2q_0)^{3/2}} \left(\frac{c}{H_0} \right).$$

Proceeding exactly as for the closed model, we finally arrive at the growth factor

$$\Sigma = \frac{5(1 + z_R)q_0}{(1 - 2q_0)^2} \left\{ (1 + 4q_0) - \frac{3q_0}{(1 - 2q_0)^{1/2}} \sinh^{-1} \left[\frac{(1 - 2q_0)^{1/2}}{q_0} \right] \right\}.$$ (7.54)

Figure 7.2 plots Σ as a function of q_0 in the range $0 \leqslant q_0 \leqslant 5$. Notice that Σ increases up to $\sim 6 \times 10^3$. We have already seen that for $q_0 = \tfrac{1}{2}$, $\Sigma = 1 + z_R \sim 10^3$. What does Σ mean in terms of galaxy formation?

We have to admit that $\delta = \delta\rho/\rho$, representing the density contrast between galaxies and the surrounding medium, is considerably higher than 1, since density in a galaxy is higher than density of matter in a cluster of galaxies by at least a factor of 10. Moreover, the intracluster density is $\sim 10^{-28}$ g cm^{-3}, which is higher by another order of magnitude than the closure density of the universe ($\sim 10^{-29}$ g cm^{-3}). Thus, to apply our theory to galaxy formation we need $\delta\rho/\rho \gg 1$, and in this region linearization of the basic equations is not valid. So we cannot use our calculations in any exact sense.

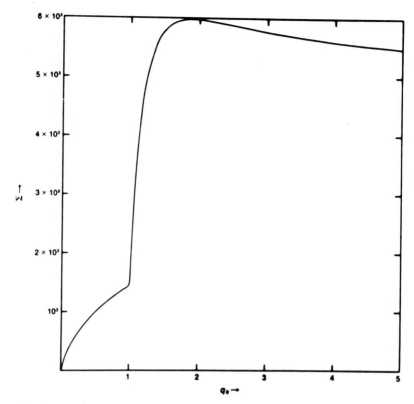

Fig. 7.2 The growth function Σ plotted against q_0 for $0 \leqslant q_0 \leqslant 5$, for $z_R \approx 1000$.

However, we can use the above analysis to demand that in order to form galaxies the density contrast must at least be unity at present. Thus if we set $\delta\rho/\rho \sim 1$ at present we may ask for $\delta\rho_R/\rho_R$ to be at least Σ^{-1} at the time of recombination. Only the full nonlinear theory can really tell us what $\delta\rho_R/\rho_R$ should have been in order to generate large density contrasts at present. However, even the lower limit Σ^{-1} is of the magnitude that could in principle be detected by accurate measurements of the microwave background, as we shall now see.

7.4 Observational constraints

The 'growth of fluctuations' idea outlined above encounters a problem when confronted with observations. We will not only briefly review this problem first but will also discuss certain other constraints that a successful theory of structure formation must satisfy. Although these constraints have posed severe difficulties for the standard big bang model, we should view them in a more positive way. For, they represent the remarkable progress that extragalactic observational astronomy has made in the 1980s, thanks to increasingly sophisticated observing techniques. To match these developments on the observational front, the cosmological theories have to be correspondingly more mature and less speculative.

We begin with the possible impact of the density fluctuations leading to galaxy formation on the microwave background.

7.4.1 Small-angle anisotropy

Let us estimate the effect of these fluctuations of $\delta\rho/\rho$, the density contrast at the recombination epoch on the radiation background. Assuming that the fluctuations are adiabatic, the particle number density will vary as the cube of the radiation temperature. Therefore

$$\left(\frac{\delta T}{T}\right)_R = \frac{1}{3}\left(\frac{\delta\rho}{\rho}\right)_R, \tag{7.55}$$

where the subscript R denotes the recombination epoch.

Since the universe is optically thin after this epoch, these fluctuations will be imprinted on the radiation background, and would be observed to this day. That is, if we sweep across the sky we should see ups and downs in the background temperature. What should be the order of magnitude of this fluctuation in temperature at the present epoch? Over what character-istic angular size should we observe these fluctuations?

Our calculations above have placed the value of $(\delta\rho/\rho)_R$ in the region of Σ^{-1}. For the different cosmological models (see Figure 7.2), Σ^{-1} lies in the range of $\sim 10^{-2}$ to 3×10^{-4}. Hence from (7.55) we should have present-day fluctuations of $\Delta T/T$ in the range $\sim 3 \times 10^{-4}$. This is of course true on the assumption of optical thinness mentioned earlier.

To fix the angular size of fluctuations we note that (7.26) relates the mass M of a typical fluctuation to the characteristic wavelength $2\pi/K$. What will be the angle subtended by a length $2\pi/K$ at the redshift of z_R? For this we need the formulae for angular size derived in Chapter 4. We recall the relevant formulae (4.72) and (4.76), and apply them in the limit of large redshift $(1 + z_R \approx 1000)$. Thus we get the angular size as

$$\Delta\theta = \frac{2\pi}{K} \cdot \frac{(1 + z_R)^2}{D_1},$$

where

$$D_1 \approx \frac{c}{H_0} \cdot \frac{z_R}{q_0} \approx \frac{c}{H_0} \frac{(1 + z_R)}{q_0}.$$

Hence from (7.26)

$$\Delta\theta = \frac{2\pi}{K} \cdot \frac{H_0 q_0}{c} (1 + z_R)$$

$$= \frac{H_0 q_0}{c} (1 + z_R) \cdot \left(\frac{3M}{4\pi n_R m}\right)^{1/3}.$$

Since $n_R = n_0(1 + z_R)^3$, where $n_0 =$ present number density, we get finally

$$\Delta\theta = \frac{H_0 q_0}{c} \left(\frac{3M}{4\pi n_0 m_H}\right)^{1/3}. \tag{7.56}$$

Using the result that $n_0 m_H = 3H_0^2 q_0/4\pi G$, we can express the above result in the following form:

$$(\Delta\theta) \approx 23\left(\frac{M}{10^{11} M_\odot}\right)(h_0 q_0^2)^{1/3} \text{ arc second}. \tag{7.57}$$

Thus galaxy formation should leave a characteristic patchiness of the angular size $\sim$20 arc second. However, observations (to be described in detail in Chapter 9) do not show any patchiness in spite of sensitivity that could detect $\Delta T/T$ as low as 2×10^{-5}. This may be called the 'smoothness problem'.

7.4.2 The horizon problem

The second difficulty is of a technical nature. Let us assume that at any epoch, t, $\bar{\rho}(t)$ denotes the smooth averaged-out density in the universe while $\rho(\mathbf{r}, t)$ denotes the actual density at any space-point with coordinate $\mathbf{r}$. To fix ideas, as well as to simplify matters, let us illustrate the problem for the $k = 0$ model. Define the 'density contrast' $\delta(\mathbf{r}, t)$ by

$$\delta(\mathbf{r}, t) = \frac{\rho(\mathbf{r}, t) - \bar{\rho}(t)}{\bar{\rho}(t)} = \int \delta_k(t) e^{i\mathbf{k}\cdot\mathbf{r}} \frac{d^3 k}{(2\pi)^3}. \tag{7.58}$$

If $S(t)$ is the scale factor then the proper length corresponding to $\mathbf{r}$ is $S(t)|\mathbf{r}|$. Hence the physical wave number for $\mathbf{k}$ is k/S.

The inhomogeneity denoted by $\delta(\mathbf{r}, t)$ is thus seen as a superposition of components of different wave numbers. A typical size $(2\pi/k)S(t)$ is stretched in an expanding universe. Besides the amplitude for a given $\mathbf{k}$ will grow due to gravitational instability. So an inhomogeneity of characteristic size λ_0 today would correspond to a proper length

$$\lambda(t) = \lambda_0 \frac{S(t)}{S(t_0)}. \tag{7.59}$$

With $S(t) \propto t^n$, say (viz. $n = \frac{1}{2}$ for the radiation-dominated phase and $n = \frac{2}{3}$ for the matter-dominated one), we find that $\lambda(t) \propto t^n$. However, the particle horizon size, as we saw earlier, is proportional to t (cf. equations (4.74) and (6.50)). Thus with $n < 1$, for sufficiently small t, $\lambda(t)$ would exceed the horizon size.

Since physical processes operate under the causality principle, it follows that any astrophysically relevant scale today demands seed fluctuations with scales not exceeding the horizon size at any earlier epoch. Thus there is manifestly a contradiction here. (For explicit numerical values of typical length scales see Exercises 20–21.)

7.4.3 The scale-invariant spectrum

First we consider the two-point correlation function $\xi(r)$ for galaxies defined by the probability δp of finding a galaxy in a given volume δV within a distance r from a given galaxy:

$$\delta p = \bar{n}\{1 + \xi(r)\}\, \delta V. \tag{7.60}$$

Here $\bar{n}$ = mean number density of galaxies. Detailed studies of galaxy counting indicate that $\xi(r)$ has the form

$$\xi(r) = \left(\frac{r}{r_0}\right)^{-\gamma}, \tag{7.61}$$

where $r_0 = 5h_0^{-1}\,\mathrm{Mpc}$ and $\gamma = 1.8$.

Now $\xi(r)$ is scale-invariant and is typical of fractals. Moreover, the galaxy, the cluster and the supercluster correlation functions have, surprisingly, the same functional form with the same γ. Thus, if we did not know what population was being described in a catalogue we would not be able to determine the answer from their correlation analysis. Figure 7.3 illustrates this commonality. In mathematical terms these correlation functions are adequately described by

$$\xi_L(r) = 0.3\left(\frac{r}{L}\right)^{-1.8}, \tag{7.62}$$

where $L = (\bar{n})^{-1/3}$. This scale-invariant spectrum has to be explained by a theory of galaxy formation.

7.4.4 Hierarchy of structures

The discrete structures range from galaxies on the scales of masses $\sim 10^{11} M_\odot$ and sizes $\sim 10\,\mathrm{kpc}$ to superclusters on the scales of masses

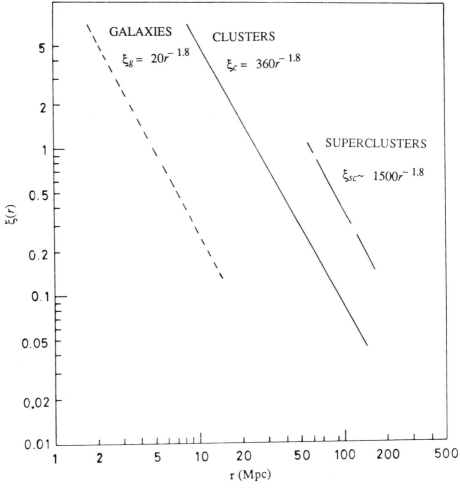

Fig. 7.3 The correlation function for clusters and galaxies as well as for superclusters is seen to be the same from this observed distribution of $\xi(r)$, although their characteristic scales are different. (After N. A. Bahcall, *Ann. Rev. Astron. Astrophys.*, **26**, 631, Fig. 10 (1988).)

$\sim 10^{14}$–$10^{15}\,M_{\odot}$ and sizes $\sim$50–100 Mpc. We also have to understand their large-scale filamentary structure interspersed with giant voids $\sim$100–200 Mpc in size. Two relatively nearby large-scale inhomogeneities are the so-called Great Wall and the Great Attractor.

A theory of structure formation may belong to one of two types: 'top-down' and 'bottom-up'. In the top-down scenario, the larger-scale structures form first and later fragment into smaller ones. The reverse is true in the bottom-up case, with smaller scale structures forming first and accreting together in groups to form the bigger ones.

7.4.5 Age distribution

Did all galaxies form more or less at the same epoch, or is the formation process a continuous and ongoing one? When did it begin? Was it related in its evolutionary sequence to the formation of QSOs?

Clues and possible checks can come from the redshift distributions of discrete objects, from the age estimates of galaxies and from their chemical evolution. Redshifts of QSOs indicate a tapering-off of their numbers beyond $z = 5$. Galaxies do seem to indicate a variety of ages, judging by the evolutionary stages of stars therein and by the abundances of heavy elements. These clues pose important constraints on structure formation theories.

7.5 Inputs from the inflationary phase

One of the attractive features of the inflationary models is that they hold out the possibility of generating seed fluctuations that can grow to form the large-scale structures with a scale-invariant spectrum. To illustrate this we first discuss a scenario that produces the observed structures from seed perturbations and decide what form of such perturbations is needed, then compute explicitly the nature of perturbations produced by inflation. We will then see to what extent the latter compare with the former.

7.5.1 Causal connections within the initial fluctuations

In the preceding section we saw how the physical wavelengths (of the present large-scale inhomogeneities) were larger than the horizon radius sufficiently early on in a Friedmann model, and thus could not be linked by causal interactions. This conclusion is altered if an inflationary phase is present. Let us see how this comes about, with an illustrative example.

Consider a wavelength λ_0 associated with a galactic mass M at the present epoch. With the mean density given by

$$\bar{\rho}_0 = \frac{3H_0^2}{8\pi G} \Omega_0$$

we have

$$M = \frac{4\pi}{3} \bar{\rho}_0 \lambda_0^3,$$

i.e.,

$$\lambda_0 = \left(\frac{2GM}{H_0^2 \Omega_0} \right)^{1/3}. \tag{7.63}$$

We now trace this length scale back to the epoch t_f when inflation had just ended. Since the scale factor varies as the reciprocal of the radiation temperature, the length scale at t_f was

$$\lambda_f \equiv \lambda(t_f) = \lambda_0 \frac{S(t_f)}{S(t_0)} = \left(\frac{2GM}{H_0^2 \Omega_0}\right)^{1/3} \frac{T_0}{T_f}. \tag{7.64}$$

Since during the inflationary phase $t_i \lesssim t \lesssim t_f$ the scale factor increased exponentially, the scale as t_i was

$$\lambda_i = \lambda_f \exp a(t_i - t_f) = \lambda_f Z^{-1}, \tag{7.65}$$

Z being the factor by which the universe inflated (see equation (6.73)). How does it compare now with the horizon size?

Assuming that the universe was in the de Sitter expansion mode during $t_i \lesssim t \lesssim t_f$, the nature of horizon changes. The de Sitter spacetime over its full timespan $-\infty < t < \infty$ does *not* have a particle horizon. It does have an event horizon c/a in radius if the expansion factor is $\exp at$. However, here we are dealing with a finite interval of the de Sitter expansion, and so the issue is somewhat vague. For causal connections which have developed through the past light cone one should, strictly, talk of the particle horizon. In the absence of a clearcut particle horizon, we may take c/a, the so-called 'Hubble radius', as a length scale up to which causal connections might get established.

Because of the largeness of the factor Z we expect that for most astrophysically relevant scales

$$\lambda_i < c/a. \tag{7.66}$$

In other words, the Hubble radius exceeds the length scale. Thus the original causality problem of standard cosmology is circumvented.

Figure 7.4 illustrates the revised situation. It shows the Hubble radius over the inflationary epoch, followed by the particle horizon radius in the Friedmann radiation-dominated expansion phase. The Hubble radius is constant at c/a for $t_i \lesssim t \lesssim t_f$. For $t > t_f$ the particle horizon grows at t. Compared with these scales, the typical length scale for a primordial fluctuation corresponding to the wave number k grows always in proportion to the scale factor of the universe, i.e., it grows during $t_i < t < t_f$ as $(2\pi/k) \exp at$. Thus it exceeds and crosses out of the Hubble radius at some time t_{exit} given by

$$\frac{2\pi}{k} \exp(at_{exit}) = \frac{c}{a}. \tag{7.67}$$

For $t_{exit} < t < t_{enter}$, the scale in question exceeds both c/a during the inflationary stage and the horizon size during the subsequent Friedmann

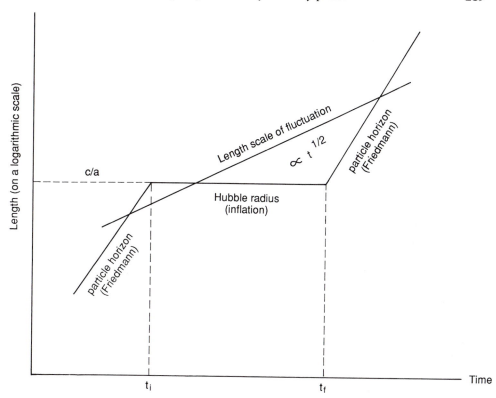

Fig. 7.4 The scale size of a fluctuation grows as $S(t)$. It exits from the Hubble radius of the inflationary model and later re-enters the particle horizon in the Friedmann phase.

stage. The instant t_{enter} is given by the epoch when the proper length of the fluctuation (which grows during the Friedmann regime in proportion to $t^{1/2}$) becomes equal to the horizon size (which grows as t). The suffix 'enter' indicates that, for $t > t_{enter}$, the length of the fluctuation will be *less* than the horizon size. Inflation therefore allows one to start with seed fluctuations at a very early epoch which then grow to present observed scales. It is during the interval $t_{exit} < t < t_{enter}$ that causal connections are lost and the correlations developed prior to t_{exit} are maintained intact. Notice also that both t_{exit} and t_{enter} depend on k and that this circumstance plays a key role in determining the spectrum of fluctuations.

7.5.2 *The scale-invariant spectrum*

Going by the above argument, we need to know the amplitude of a typical density perturbation at the time $t_{enter}(k)$ when it enters the horizon. For

$t > t_{\text{enter}}(k)$ we can then study its growth, first by linearization techniques until its magnitude $|\delta(\mathbf{k}, t)|^2$ becomes comparable to unity and then later by other methods. (The fact that $\Delta T/T$ in the microwave background radiation is $<10^{-5}$, implies that $|\delta(\mathbf{k}, t)|^2$ was $\ll 1$ in the radiation-dominated phase of expansion, and so linearization is justified.) Thus we need to know the function

$$F(\mathbf{k}) \equiv |\delta(\mathbf{k}, t)|^2_{t=t_{\text{enter}}(\mathbf{k})}. \tag{7.68}$$

Harrison in 1970 and Zeldovich in 1972 had argued independently, from theoretical considerations that at the time of entering the horizon, the perturbations should have the form $F(\mathbf{k}) \propto k^{-3}$. For it can be shown that the root-mean-square fluctuation of mass M as a fraction of average mass contained in a region of size R is proportional to $k^3 |\delta_k|^2$ at $k = R^{-1}$. Therefore, for the above $F(\mathbf{k})$, $\langle (\delta M/M)^2 \rangle$ will be independent of the scale R at $t = t_{\text{enter}}(\mathbf{k})$, thus giving equal power at all scales at the time they enter the horizon. As we saw in the proceding section, a scale-invariant spectrum is indicated by the distribution of discrete large-scale structures.

The inflationary model seems capable of producing this kind of spectrum, through fluctuations in the scalar field $\phi(\mathbf{r}, t)$ (see Chapter 6). We write the fluctuations as $f(\mathbf{r}, t)$ over a smoothed average value $\phi_0(t)$. Thus

$$\phi(\mathbf{r}, t) = \phi_0(t) + f(\mathbf{r}, t). \tag{7.69}$$

These fluctuations result in fluctuations of energy density.

Since the energy density of a scalar field is $\rho c^2 \approx \frac{1}{2}\dot{\phi}^2$, we get

$$\delta\rho(\mathbf{r}, t) \approx \dot{\phi}_0(t)\dot{f}(\mathbf{r}, t)/c^2 \tag{7.70}$$

for $|f| \ll |\phi_0|$. Writing

$$f(\mathbf{r}, t) = \int Q_k(t)e^{i\mathbf{k}\cdot\mathbf{r}} \frac{d^3k}{(2\pi)^3}, \tag{7.71}$$

we have

$$\delta\rho(\mathbf{k}, t)c^2 \approx \dot{\phi}_0(t)\dot{Q}_k(t). \tag{7.72}$$

The average energy density during inflation being dominated by the constant term V_0 (say) of the Coleman–Weinberg potential, we have the density contrast

$$\delta_k(t) = \frac{\delta\rho c^2}{V_0} = \frac{\dot{\phi}_0 \dot{Q}_k(t)}{V_0}. \tag{7.73}$$

For $\dot{\phi}_0$ we use the mean evolution of ϕ in the slow roll-over phase: but what is f? Now in actuality the fluctuations in ϕ are of quantum origin but here, in a classical approximation, we are using $f(\mathbf{r}, t)$ to mimic them classically. In quantum field theory the field would be an operator $\hat{\phi}(\mathbf{r}, t)$ whose Fourier coefficients $\hat{q}_\mathbf{k}(t)$ are also operators. In a quantum state specified by the wavefunction $\psi_\mathbf{k}$, the fluctuations of $\hat{q}_\mathbf{k}$ are given by the dispersion relation

$$\sigma_k^2(t) = \langle \psi_\mathbf{k} | q_k^2(t) | \psi_\mathbf{k} \rangle, \tag{7.74}$$

the mean value (in $\mathbf{k} \neq \mathbf{0}$ mode) of $\psi_\mathbf{k}$ being zero. This is because ϕ_0, the average of ϕ, is homogeneous. Since $\sigma_k^2(t)$ appears to be a good measure of quantum fluctuations, we may identify $Q_\mathbf{k}(t)$ with $\sigma_\mathbf{k}(t)$ and write

$$\delta_\mathbf{k}(t) = \frac{\dot{\phi}_0(t)}{V_0} \dot{\sigma}_\mathbf{k}(t). \tag{7.75}$$

Thus we have taken a semiclassical approximation to estimate the fluctuations in the energy density of the ϕ-field which act as the seed fluctuations of density during the inflationary phase $t_i < t < t_f$. For a comparison with observations we need the value of $\delta_\mathbf{k}(t)$ at $t = t_{\text{enter}}$. Several workers in inflation theory have found a way of relating $\delta_\mathbf{k}(t_{\text{enter}})$ to $\delta_\mathbf{k}(t_{\text{exit}})$ through an approximate conservation law,

$$\frac{\delta_\mathbf{k}(t_{\text{enter}})}{1 + W(t_{\text{enter}})} = \frac{\delta_\mathbf{k}(t_{\text{exit}})}{1 + W(t_{\text{exit}})}, \tag{7.76}$$

where $W(t)$ is the ratio of pressure $p(t)$ to density $\rho(t)$ of the average background field.

During the inflationary phase, with $\dot{\phi}_0^2 \ll V_0$,

$$p(t) = \tfrac{1}{2}\dot{\phi}_0^2 - V_0, \quad \rho(t) = \tfrac{1}{2}\dot{\phi}_0^2 + V_0, \quad 1 + W(t) \simeq \frac{\dot{\phi}_0^2}{V_0}. \tag{7.77}$$

In the radiation-dominated phase $1 + W = \tfrac{4}{3}$. Therefore,

$$\delta_\mathbf{k}(t_{\text{enter}}) = \frac{4}{3} \frac{V_0}{\dot{\phi}_0^2} \delta_\mathbf{k}(t_{\text{exit}}) \tag{7.78}$$

$$= \frac{4}{3} \frac{\dot{\sigma}_\mathbf{k}}{\dot{\phi}_0} \bigg|_{t=t_{\text{exit}}}.$$

For the Coleman–Weinberg potential, detailed calculations give $\sigma_\mathbf{k}(t)$ and $\phi_0(t)$. The final result is

$$\delta_\mathbf{k}(t_{\text{enter}}) \approx 10^2 k^{-3/2}. \tag{7.79}$$

In other words, the condition

$$k^3 |\delta_\mathbf{k}(t_{\text{enter}})|^2 = \text{constant} \tag{7.80}$$

required for a scale-invariant spectrum is satisfied. While this is undoubtedly a success for the inflationary model, the outcome is hardly satisfactory. For, after putting in numbers we find we have too high an amplitude for fluctuations! Instead of yielding values of the order $< 10^{-4}$, the equation (7.79) leads to an amplitude of $\sim 10^2$. Some unrealistic fine tuning of the parameters of the theory is needed to bring the amplitude down to the required level.

7.6 The role of dark matter

The presence of dark matter also plays a significant role in the formation of structures. We have seen how the smoothness of the microwave background limits the extent of $\delta\rho/\rho$ at the recombination epoch. The argument that was used in equating the $\delta\rho/\rho$ for matter and radiation depends on the matter being baryonic. Baryonic matter does interact with radiation, and so we cannot have large fluctuations $\delta\rho/\rho$ of such matter coexisting with much smaller fluctuations in the radiation background.

The argument, however, breaks down if the bulk of the matter is nonbaryonic and hence (possibly) does not interact with radiation. This would allow a large $\delta\rho/\rho$ of nonbaryonic matter at the recombination epoch. We may then arrange for the baryonic fluctuations (which were small at that epoch) to catch up with the larger fluctuations of the nonbaryonic matter at later epochs. For, the two kinds of matter interact gravitationally. Because nonbaryonic matter does not interact with radiation it is 'dark' for all astronomical purposes.

7.6.1 Types of nonbaryonic dark matter

Nonbaryonic matter can be broadly of two kinds, 'hot' or 'cold'. These adjectives indicate how fast a dark matter particle was moving when it decoupled from the rest of the (baryonic) matter in the universe. In Chapter 5 we saw that neutrinos decoupled from the rest of the matter at temperatures $\sim 10^{10}$ K. At temperature of this order an electron (with a rest mass of $\sim 0.5\,\text{MeV}/c^2$) would move relativistically. So, even if a neutrino has a rest mass of around 20–$40\,\text{eV}/c^2$, it would more relativistically at decoupling.

Neutrinos are therefore an example of 'hot dark matter' (HDM) particles. At the time of decoupling they were moving with relativistic speeds. On the contrary, particles whose velocities had dropped to values $\ll c$ when they decoupled, are called 'cold dark matter' (CDM) particles.

Table 7.1 gives a list of candidates for dark matter, baryonic as well as nonbaryonic. In the latter class the only familiar species are neutrinos which are HDM. All other particles are conjectured by the grand unified

Table 7.1. *Some dark-matter candidates*

Candidate/particle	Approximate mass	Predicted by	Astrophysical effects
Axion, majoron, goldstone boson	10^{-5} eV	QCD; symmetry breaking	Cold DM
Ordinary neutrino	10–100 eV	GUTs	Hot DM
Light higgsino, photino, gravitino, axino, sneutrino	10–100 eV	SUSY/SUGR[a]	Hot DM
Para-photon	20–400 eV	Modified QED	Hot/warm DM
Right-handed neutrino	500 eV	Superweak interaction	Warm DM
Gravitino, etc.	500 eV	SUSY/SUGR	Warm DM
Photino, gravitino, axino, mirror particle, Simpson neutrino	keV	SUSY/SUGR	Warm/cold DM
Photino, sneutrino, higgsino, gluino, heavy neutrino	MeV	SUSY/SUGR	Cold DM
Shadow matter	MeV	SUSY/SUGR	Hot-cold (like baryons)
Preon	20–200 TeV	Composite models	Cold DM
Monopoles	10^{16} GeV	GUTs	Cold DM
Pyrgon, maximon, perry pole, newtorities, Schwarschild	10^{19} GeV	Higher-dimension theories	Cold DM
Supersymmetric strings	10^{19} GeV	SUSY/SUGR	Cold DM
Quark nuggets, nuclearities	10^{15} g	QCD, GUTs	Cold DM

[a] SUGR ≡ Supergravity.

or supersymmetric (SUSY) particle theories. None have been detected in accelerator experiments. We therefore begin with a discussion of massive neutrinos. Of all those listed above, the only tangible particle so far is the neutrino. Although it is still uncertain as to whether the neutrino has a rest mass, it is worth while examining a few consequences of such a possibility.

7.6.2 Massive neutrinos

Experiments by F. Reines, H. W. Sobel, and E. Pasierb, as well as by V. A. Lyubimov *et al.* in 1980, suggested that neutrinos may indeed have a small rest mass. Subsequent experiments by different groups have been rather equivocal on this issue. However, this possibility opened up a number of interesting astrophysical consequences. As early as 1972, R.

Cowsik and J. McClelland had conjectured that the 'missing mass' in the universe (that is, the dark matter) may be accounted for by relic neutrinos. What can we say today about such a possibility?

Let us do the calculations, taking $g_\nu = 1$ even for massive neutrinos. If the rest mass of the neutrino is larger than $\sim 2 \times 10^{-4}$ eV then they will have small random velocities today. Since experiments suggest m of the order of a few electronvolts, we will write

$$m_\nu = M_\nu \, (\text{eV}).$$

From Table 5.1 we know that the number density of neutrinos is three-eighths of the number density of photons of the same temperature. We also know that the number density of photons goes as the cube of photon temperature. Since in the post-$e^+ - e^-$ annihilation phase

$$\left(\frac{T_\nu}{T_\gamma} \right)^3 = \tfrac{4}{11},$$

we get the present number density of neutrinos as

$$\left(\frac{N_\nu}{N_\gamma} \right)_0 = \tfrac{3}{22}. \tag{7.81}$$

Putting everything together, the mass density of neutrinos at present may be expressed as

$$\rho_\nu = \Sigma \Omega_\nu \rho_c, \tag{7.82}$$

where Σ denotes the sum over all types of neutrinos and

$$\Omega_\nu \simeq \frac{M_\nu}{150} \left(\frac{T_0}{3} \right)^3 h_0^{-2}. \tag{7.83}$$

A similar contribution to density will come from antineutrinos. If we consider all species of neutrinos (and their antineutrinos) together then we discover that their contribution to density becomes comparable to that of baryonic matter provided

$$\sum_{\text{all species}} m_\nu \geqslant 1.5 \, \text{eV}. \tag{7.84}$$

If neutrinos collapsed with the nucleons to form clusters, then we get a lower bound on the ratio of nonluminous to luminous (nucleonic) matter. This lower bound is

$$\frac{\Sigma \Omega_\nu}{\Omega_N} \geqslant \tfrac{2}{3} \Sigma M_\nu. \tag{7.85}$$

From cluster emission of X-rays it is estimated that the mass of hot gas is related to the total mass of the cluster by the formula

$$M_{HG} \approx 0.1(2h_0)^{-3/2} M_{Total}. \tag{7.86}$$

We may take M_{Total}/M_{HG} as an upper limit in (7.85). This gives

$$\Sigma M_v \leqslant 40 h_0^{3/2}. \tag{7.87}$$

Thus for h_0 in the range $\frac{1}{2}$ to 1, the upper limit on Σm_v lies in the range ~15 to 40 eV.

In 1979 S. Tremaine and J. Gunn pointed out another handle on neutrino masses. A massive neutrino will have this distribution function in the momentum space.

$$dn_v = \frac{g}{(2\pi\hbar)^3} \left[\exp\left(\frac{p_v c}{kT_v}\right) + 1 \right]^{-1} d^3 p_v \tag{7.88}$$

at the time decoupling. As they cool down $p_v \propto T_v$ and the neutrinos eventually become nonrelativistic. Slow-moving neutrinos would be susceptible to being trapped by the gravitational potential wells of massive systems that eventually form clusters or single galaxies. Trapping and collapse of neutrinos changes their distribution function from (7.88) to a Maxwellian distribution of an isothermal gas. This final distribution is given by

$$dn_v = \frac{\rho_v}{m_v^4} \frac{1}{(2\pi\sigma^2)^{3/2}} \exp\left(-\frac{v^2}{2\sigma^2}\right). \tag{7.89}$$

In order that (7.89) represents a gas trapped by the gravitational field of a mass M at a distance R, we need

$$\langle v^2 \rangle \equiv 3\sigma^2 \simeq \frac{GM}{R},$$

that is,

$$\langle \rho \rangle \equiv \frac{3M}{4\pi R^3} = \frac{9\sigma^2}{4\pi G R^2}. \tag{7.90}$$

Expressing M in terms of $M_\odot$, R in megaparsecs, and σ in units of $100\ km\ s^{-1} \equiv \sigma_{100}$, we get from above

$$\frac{M}{M_\odot} \simeq 7 \times 10^{12} \sigma_{100} R_{Mpc}, \quad \langle \rho \rangle \simeq 10^{-28} \left(\frac{\sigma_{100}}{R_{Mpc}}\right)^2 g\ cm^{-3}. \tag{7.91}$$

Now one feature of a collapse accompanied by rapid energy changes is that the maximum of phase space density decreases. (This happens because as the gas particles move, a mixing of states occurs in which the maximum of the original distribution function gets mixed up with lower-density parts of the distribution function.) Comparing the maxima of (7.88) and (7.89), we therefore get

$$\frac{g_v}{(2\pi\hbar)^3} > \frac{\rho_v}{m_v^4(2\pi\sigma^2)^{3/2}},$$

that is,

$$m_v > \left[\frac{\rho_v(2\pi)^{3/2}\hbar^3}{g_v\sigma^3}\right]^{1/4}. \tag{7.92}$$

Expressing this inequality in terms of (7.91), we get for $g_v = 1$

$$M_v \geqslant 4.5\sigma_{100}^{-1/4}R_{Mpc}^{-1/2}. \tag{7.93}$$

Relic neutrinos that are sufficiently heavy may therefore collapse and dominate the mass on the various scales given by (7.93). Tremaine and Gunn pointed out a curious aspect of this result. (Somewhat similar arguments were used by Cowsik and McClelland in 1973 to place lower limits on neutrino masses.) The larger the value of M_v, the larger the ratio in (7.85), that is, the unseen mass is larger compared to the luminous mass. Yet the ratio is known to be largest in clusters of galaxies and lowest in single galaxies. Thus it would appear that relic neutrinos don't seem to solve the missing mass problem. To resolve this contradiction, Schramm and Steigman have suggested that m_v may lie in the range 4 to 20 eV. Thus these neutrinos would not be massive enough to dominate gravitational clumping on the scale of a single galaxy, but may well be effective on the scale of clusters.

Very massive neutrinos will prove embarassing for big bang cosmology. If all neutrinos have on average a mass of ~ 25 eV, then $\Sigma\Omega_v$ is close to 1. Larger mass than this value and/or an increase in the number of relic neutrino species would increase $\Sigma\Omega_v$ and the overall Ω beyond the closure value $\Omega = 1$. As seen in Chapter 4, closed universes have shorter ages, and an overall age $< \sim 6 \times 10^9$ years may be embarassingly small. It has been suggested that under such circumstances λ-cosmologies might have to be invoked.

These calculations illustrate how astrophysics may provide valuable constraints on properties of elementary particles.

7.6.3 Dark matter and structure size

An interesting relation emerges between the mass of a nonbaryonic HDM particle and the mass of the large scale structure associated with it. The ideas is as follows. Suppose m_X is the mass of a particle X which moves in a collisionless fashion (i.e., it is noninteracting) with relativistic speed. Such a motion is called 'free streaming'. A population of such particles tends to wipe out any inhomogeneity. The limit on the size of the inhomogeneity is then placed by the size of the particle horizon. We estimate the effect as follows.

The particle will be relativistic until the ambient temperature drops to

$$T_X = \frac{m_X c^2}{k}. \tag{7.94}$$

The time–temperature relationship in the early universe will give the epoch as, cf. (6.5),

$$t_X = \left(\frac{3c^2}{16\pi Ga}\right)^{1/2} g^{-1/2} T_X^{-2}. \tag{7.95}$$

At this epoch the horizon size is

$$R_X = 2ct_X. \tag{7.96}$$

The energy density is given by

$$\varepsilon = \frac{\pi^2 (kT_X)^4}{15\hbar^3 c^3} \cdot \frac{g}{2}. \tag{7.97}$$

Therefore the total mass contained within the horizon sphere is given by putting together (7.94)–(7.97). After some manipulation we get its magnitude as

$$\mathcal{M} = \frac{4\pi}{3} R_X^3 (\varepsilon/c^2)$$
$$= \frac{3^{1/2}\pi^{3/2} g^{-1/2}}{60 G^{3/2} a^{3/2}} \cdot \frac{k^6}{\hbar^3 c^3 a^{3/2}} \cdot m_X^{-2}.$$

Writing the radiation constant and Planck mass as

$$a = \frac{\pi^2}{15} \cdot \frac{k^4}{\hbar^3 c^3}, \qquad m_P = \left(\frac{c\hbar}{G}\right)^{1/2}, \tag{7.98}$$

the above expression becomes

$$\mathcal{M} = \frac{3 \times 5^{1/2}}{4\pi^{3/2} g^{1/2}} \frac{m_P^3}{m_X^2} = \alpha \cdot \frac{m_P^3}{m_X^2}. \tag{7.99}$$

The constant α is of order unity. (We may take g between 10 and 100.)

Expressing neutrino mass in units of electronvolts and $\mathcal{M}$ in units of solar mass $M_\odot$, the above relation is

$$\mathcal{M} \approx 1.5\alpha \times 10^{15} \left(\frac{30eV}{m}\right)^2 M_\odot. \tag{7.100}$$

Thus, with massive electron neutrinos as HDM, we get the characteristic scale of supercluster-like inhomogeneities. Therefore, if we assume this type of HDM then we have the *top-down* scenario to think about.

For CDM, on the other hand, the particles hardly move after they have decoupled, and their masses are large. The resulting structures are therefore much smaller than for HDM and we are confronted with the *bottom-up* scenario.

Table 7.1 also lists intermediate mass particles like the gravitino which have mass $\sim 1\,\mathrm{keV}$ and may be considered 'warm'.

7.7 The nonlinear regime

The sequence of events leading to galaxy formation may be summarized as follows:

Stage 1: Quantum fluctuations in the primordial era were created, say, during the inflationary phase.

Stage 2: Fluctuations enter the horizon of the radiation-dominated universe and grow linearly until the recombination epoch.

Stage 3: In the post-recombination era the growth is strongly affected by the presence and nature of dark matter.

Stage 4: The fluctuations grow large enough, so that nonlinear processes become important. The end result of this stage is the large-scale structure we should be able to observe with telescopes.

We have discussed Stages 1–3 and will now consider the final stage.

7.7.1 The Zeldovich approximation

In 1970 Zeldovich gave a simplified picture of how the growing modes of density fluctuations would lead to a nonlinear regime. We briefly describe this approach.

Consider the cosmic material as made of fluid elements with trajectories given by

$$\mathbf{r} = S(t)\{\mathbf{q} - b(t)\nabla_{\mathbf{q}}\psi(\mathbf{q})\}. \tag{7.101}$$

Here $S(t)$ is the expansion factor, $\mathbf{q}$ the comoving coordinate of the fluid element, $b(t)$ describes the growth of fluctuation and ψ is the perturbation potential.

If ρ_0 is the density in comoving coordinates and $\rho(\mathbf{r}, t)$ the proper density then a simple mass-conservation relation gives

$$\rho(\mathbf{r}, t) = \frac{\rho_0}{S^3} \det \left\| \frac{\partial \mathbf{r}}{\partial \mathbf{q}} \right\|^{-1}. \tag{7.102}$$

The determinant is the Jacobian of transformation, the matrix of which will have eigenvalues λ_1, λ_2, λ_3 that are continuous random functions of coordinates $\mathbf{q}$. Thus the density becomes

$$\rho(\mathbf{r}, t) = \frac{\rho_0}{S^3} (1 - b\lambda_1)^{-1} (1 - b\lambda_2)^{-1} (1 - b\lambda_3)^{-1}. \qquad (7.103)$$

Without loss of generality we assume $\lambda_1 \geqslant \lambda_2 \geqslant \lambda_3$. Then as $b(t)$ grows the density becomes infinite as $b\lambda_1 \to 1$. The original volume element had a cubical shape that now flattens to a two-dimensional surface which Zeldovich called 'pancake'.

So far no gravity has been included. To make the picture self-consistent we need to satisfy the Poisson equation:

$$\nabla \cdot \ddot{\mathbf{r}} = -4\pi G\rho. \qquad (7.104)$$

To solve the equation write the three invariants of the transformation matrix

$$I_1 = \lambda_1 + \lambda_2 + \lambda_3, \qquad I_2 = \lambda_1\lambda_2 + \lambda_2\lambda_3 + \lambda_3\lambda_1,$$
$$I_3 = \lambda_1\lambda_2\lambda_3. \qquad (7.105)$$

Then the Poisson equation becomes

$$\left(3\frac{\ddot{S}}{S} + 4\pi G \frac{\rho_0}{S^3} \right) - [I_1 - 2bI_2 + 3b^2 I_3] \left\{ \ddot{b} + 2\frac{\ddot{S}}{S}\dot{b} + 3\frac{\ddot{S}}{S}b \right\}$$

$$+ 3\frac{\ddot{S}}{S} \{2b^3 I_3 - b^2 I_2\} = 0. \qquad (7.106)$$

The first term is zero by the cosmological expansion law. The second term is zero if

$$\ddot{b} + 2\frac{\ddot{S}}{S}\dot{b} + 3\frac{\ddot{S}}{S}b = 0. \qquad (7.107)$$

This is the growth equation for linear fluctuations. The last term can be related to a fractional error in density given by

$$\frac{\Delta\rho}{\rho} = \frac{2b^3 I_3 - b^2 I_2}{1 - bI_1 + b^2 I_2 - b^3 I_3}. \qquad (7.108)$$

For a planar collapse $\lambda_2 = \lambda_3 = 0$ and hence $\Delta\rho = 0$. This means that the Zeldovich approximation is exact. If $\lambda_1 > 0$ then the collapse occurs when $b = \lambda_1^{-1}$. For $b\lambda_1 \ll 1$ we are in the linear regime and (7.10) approximates to

$$\rho(\mathbf{r}, t) \approx \frac{\rho_0}{S^3} [1 + b(\lambda_1 + \lambda_2 + \lambda_3)],$$

i.e., the linearized over-density is

$$\delta(\mathbf{r}, t) \simeq bI, \qquad (7.109)$$

We thus have a simple picture of how transition occurs from the linear to the nonlinear regime. The approximation serves as a starting point for the more exact N-body simulations on a computer.

7.7.2 N-body simulations

A general scheme for numerical simulations may be as follows. We have N particles of (generally equal) masses $m_i (i = 1, \ldots, N)$. The force on particle i located at $\mathbf{r}_i$ is calculated as a modified inverse square law:

$$\mathbf{F}_i = Gm_i \sum_{j \neq i} \frac{m_j(\mathbf{r}_j - \mathbf{r}_i)}{[|\mathbf{r}_j - \mathbf{r}_i|^2 + \varepsilon^2]^{3/2}}. \tag{7.110}$$

The small number ε is used to avoid very large forces at close encounters. This force determines the acceleration of the ith particle. Given its position and velocity at one instant, they can then be calculated at a slightly later instant.

This method is direct but very time-consuming for large values of N. Faster approximate methods are therefore devised to make progress. However, the computer speeds fall far short of giving a realistic simulation of the actual problem. Statistical techniques are, however, useful as indicators of what is going on.

In a typical project $N \geqslant 10^5$ and the calculations begin at $\delta\rho/\rho \approx 0.2$. The Zeldovich approximation is used to work out the initial position and velocities in the growing mode perturbation. The free parameters of the calculation are H_0 and Ω_0, as well as the initial amplitude of the fluctuations, given in the shape of the spectrum. The spectrum is evolved by solving the linear fluctuation growth equations for each $\mathbf{k}$.

The end product for a typical HDM scenario is illustrated in Figure 7.5. Similar pictures are obtained for CDM also. The idea is to compare these diagrams with the actual redshift surveys that give a 3-D mapping of the universe. The large-scale motions are also compared with data.

It is fair to say that although such exercises have given us considerable insight into how nonlinear growth processes operate and how the different types of dark matter influence them differently, no successful simulation of the actual universe has yet been possible. An important question is: Do galaxies trace the mass distribution? The CDM scenario works if the galaxies do not trace the mass but form in some 'biased' regions. The biasing is introduced as follows. Consider a Gaussian distribution of fluctuations $\delta\rho$ of CDM, around an average ρ. If there were no biasing light would trace mass, i.e., galaxies would mimic $\delta\rho/\rho$. However, if

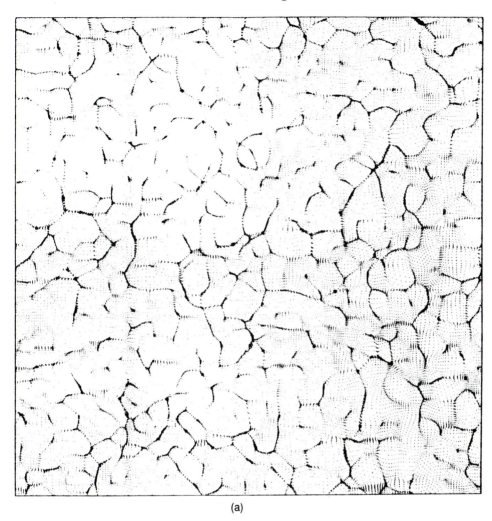

(a)

Fig. 7.5 Typical results of structures obtained by *N*-body simulations. In (a) we see an early stage with cellular structure and pancakes. In (b) the scale factor has increased by the factor 16, but the cellular structure is still apparent. (Source: A. L. Melott & S. F. Shandarin, *Ap. J.*, **343**, 26 (1989).)

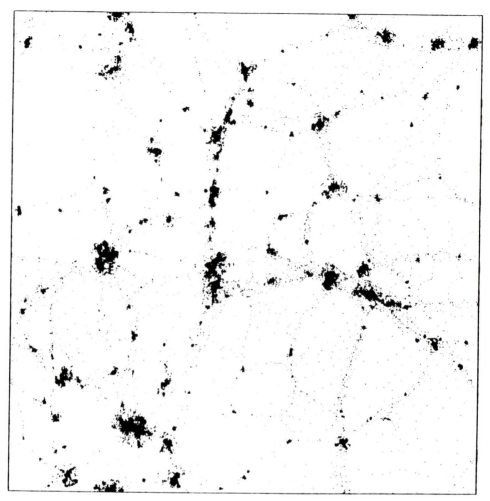

Fig. 7.5(b)

galaxies form preferentially at high values of $\delta\rho$ then there is biasing. The biasing parameter b is unity if there is no biasing and $b > 1$ for biasing. By restricting to $b > 1.5$, say, one is truncating the Gaussian to the tail where $\delta\rho$ is suitably high. The practical advantage of biasing is that it reduces $\Delta T/T$. For example, J. Silk and N. Vittorio have shown that at $4'.5$ and $1'.5$ angular scale

$$\frac{\Delta T}{T} \approx \frac{6 \times 10^{-6}}{\Omega_0 h_0 b}. \tag{7.111}$$

Thus for $\Omega_0 = 1$, $h_0 = 1$, $b = 1.5$ we expect $\Delta T/T$ to be about 4×10^{-6}. The present limits on $\Delta T/T$ permit $b > 0.4$.

Although the cold dark matter models survive on this count there are problems in explaining the very-large-scale streaming motions and the large structures like the Great Wall and the Great Attractor. The HDM models can explain large-scale structures but find it difficult to explain galaxy-size structures and the low values of $\Delta T/T$.

There are other scenarios besides the above CDM, HDM theories. In the cosmic strings hypothesis the linear discontinuities at the GUT phase transition (see Chapter 6) act as seeds for growth of fluctuations. The strings untangle as the universe expands, leaving a few long stretches and closed loops within the present Hubble radius. In the explosions model nongravitational processes like shock waves generated by the supernova explosions are called upon to trigger the process of structure formation. Neither approach can claim success with observations. Perhaps more daring ideas are needed! And so we leave this chapter with the problem of section 7.1 unsolved.

Exercises

1   Explain what is meant by the transverse or rotational modes of the velocity field in the first-order perturbation analysis in the expanding universe. Show that these modes decrease as the universe expands.

2   Derive (7.19) satisfied by the density contrast in the expanding universe.

3   Relate the longitudinal modes of the velocity field to the density contrast. Comment on the fact that the density contrast does not depend on the transverse modes.

4   What is the physical significance of the Jeans wave number? How is it related to the Jeans mass?

5   Show why we cannot neglect the expansion of the universe in a Jeans-type calculation.

6   Estimate the gravitational energy and the thermal energy of a typical spherical perturbation in the expanding universe. Relate the ratio of these energies to the ratio of the mass of the perturbation to the Jeans mass.

7   Explain the significance of the Jeans mass in relation to the perturbations that can or cannot grow in the expanding universe.

8   Using data on the Earth's atmosphere, estimate (a) the Jeans length and (b) the Jeans mass for air at normal temperature and pressure.

9   Show that with ρ and p given by

$$\rho = nm_{\mathrm{H}} + \frac{aT^4}{c^2}, \qquad p = \tfrac{1}{3}aT^4,$$

the speed of sound is given by

$$c_{\mathrm{s}}^2 = \tfrac{1}{3}c^2\left(1 + \frac{m_{\mathrm{H}}c^2}{\sigma kT}\right)^{-1},$$

where $k\sigma$ is the entropy per particle.

10   Show that after the recombination era the Jeans mass is given by

$$M_{\mathrm{J}} \simeq \frac{4\pi}{3}\left(\frac{5\pi kT}{3G}\right)^{3/2} n^{-1/2} m_{\mathrm{H}}^{-2},$$

where T is the matter temperature.

11   Assuming that in Exercise 10 the matter temperature equalled the radiation temperature at the recombination epoch, show that M_{J} at that epoch was given by

$$M_{\mathrm{J}} \simeq \frac{4\pi}{3}\left(\frac{5\pi kT_0}{3G}\right)^{3/2} n_0^{-1/2} m_{\mathrm{H}}^{-2},$$

where T_0 is the present temperature of the microwave background.

12   Evaluate M_{J} of Exercise 11 in a Friedmann universe of given (h_0, Ω_0) with $T_0 = 3K$. Show that

$$M_{\mathrm{J}} \simeq 2.54 \times 10^{39}(\Omega_0 h_0^2)^{-1/2}\,\mathrm{g}$$
$$\simeq 1.27 \times 10^6 (\Omega_0 h_0^2)^{-1/2}\,M_{\odot}.$$

13   Show that (7.41) gives at the recombination epoch

$$M_{\mathrm{J}} \simeq 100 M_{\odot}\sigma^{1/2}.$$

14   Follow the evolution of the Jeans mass in the expanding universe

and discuss qualitatively how a galaxy-size fluctuation is likely to behave in the pre- and post-recombination eras.

15   Show that in discussing the growth of a mass very much in excess of the Jeans mass in the postrecombination era the effect of pressure may be neglected. Is this a good assumption for studying the behavior of galaxy-size perturbations?

16   Discuss quantitatively the growth of fluctuations in the Friedmann models in the post-recombination era.

17   Solve from first principles the differential equation

$$(1 - \cos \Theta) \frac{d^2 \delta}{d\Theta^2} + \sin \Theta \frac{d\delta}{d\Theta} - 3\delta = 0$$

and relate its solutions to the behaviour of fluctuations in the postrecombination era of the closed Friedmann universe.

18   Verify by a suitable limiting process that as $q_0 \to \frac{1}{2}$ both (7.52) and (7.54) tend to (7.47). Plot Σ as a function of q_0 for $1 + z_R = 10^3$.

19   Review some of the attempts to understand the formation of galaxies.

20   Show that the mass associated with wavelength λ measured in Mpc is

$$M(\lambda) = 1.5 \times 10^{11} \Omega_0 h_0^2 \lambda^3 M_\odot.$$

21   In the previous exercise a mass of the order $10^{12} M_\odot$ corresponds to $\lambda \approx 1.88$ Mpc. Show that this wavelength was bigger than the horizon at all redshifts exceeding

$$z = 1.41 \times 10^5 (\Omega_0 h_0^2)^{1/3}.$$

22   Discuss in what way the inflationary phase helps in carrying forward the growth of primordial fluctuations.

23   Show that if the universe were dominated by three types of relic massive neutrinos at the present epoch, the average neutrino mass needed to close the universe would be

$$25 \left(\frac{T_0}{3} \right)^{-3} h_0^2 \, eV.$$

24   Suppose that the universe has enough nucleons to make $\Omega_N = 1$, and that it has in addition three species of neutrinos of average mass 25 eV. For $T_0 = 3$, $h_0 = 1$, calculate the age of the universe.

25   Discuss how the observation of neutrino mass affects the age of the universe. With the example given in Exercise 24, if the age of the universe comes out very low, can you think of a way out of the difficulty by using the λ-cosmologies of Chapter 4?

26 A primordial neutrino has rest mass 1 eV. Estimate its random velocity relative to the cosmological rest frame at the present epoch.

27 Describe how massive neutrinos might influence the condensation of matter into galaxies or larger structures. Is it possible to think of a consistent mass range of m_v that may account for the missing mass in galaxies and clusters of galaxies?

28 In equation (7.95) replace G by Planck mass and arrive at m.

29 Discuss why CDM scenarios lead to smaller structures than their HDM counterparts.

30 Outline the observational constraints that must be satisfied by theories of galaxy formation.

8

Alternative cosmologies

8.1 Alternatives to Friedmann cosmologies

In 1922–4, when Friedmann produced the expanding-universe solutions of
Einstein's equations, his work went largely unnoticed. Subsequent to
Hubble's discovery of nebular redshift, cosmologists came to regard these
models as the simplest starting point for discussing their subject. However,
the physicists considered these attempts as naive and speculative, and so
they did not pay as much attention to George Gamow's very seminal work
on the early universe. Eventually, the turning point for cosmology came in
1965 with the discovery of the microwave background radiation. The
MBR seemed to confirm the early universe scenario, and, taken together
with the extended validity of Hubble's law obtained by bigger and better
telescopes, laid a solid foundation for cosmology as a branch of physics.
By the mid-1970s a considerable body of physicists began to take the
Friedmann cosmology seriously.

Chapters 6 and 7 have given a glimpse of how this cosmology has
progressed with the inputs from particle physics. The question we will
properly address in the last chapter is to what extent Friedmann
cosmology is a correct theory of the origin and the large-scale structure of
the universe. While the majority of today's cosmologists would put their
money on the Friedmann models, there have been a few 'agnostics' who
were not satisfied with them. And out of their efforts have emerged
alternative theories to be compared with the standard big bang model.

These theories have not been worked through to the depth that
Friedmann cosmology can boast of. This is hardly surprising, considering
the much smaller brainpower that has worked on them. Nevertheless, they
contain different perspectives and may offer a resolution of some of the
outstanding problems that the Friedmann cosmology has been unable to
solve. In this chapter we describe a few such theories.

8.2 The steady state theory

In 1948, around the same time that George Gamow was initiating detailed studies of the physical properties of the universe close to the big bang epoch, three astronomers proposed an entirely new approach to cosmology. This model, now famous (or notorious!) as the steady state model, does not have a singular big bang type epoch; indeed, it does not have either a beginning or an end on the cosmic time axis. The cosmological scene was considerably enlivened for two decades after the inception of the steady state model by the observers' attempts to shoot this rival model down. What was the motivation that led Hermann Bondi, Thomas Gold, and Fred Hoyle to propose the steady state cosmology?

First of all, in 1948 the measured value of $T_0 \equiv H_0^{-1}$ was only $\sim 1.8 \times 10^9$ years. Consequently the age of a standard Friedmann model could not exceed T_0 – a value lower than even the geological age of the Earth! Thus a prima facie case existed for doubting the conclusion that the universe began ~ 1 to 1.8 billion years ago.

Secondly, if a model (like the Friedmann models) proposes that the universe began at $t = 0$, it should provide a physical discussion of the beginning. At least it should leave the question tractable for a future, more sophisticated physical theory. The spacetime singularity at the $t = 0$ epoch precludes any such discussion. For example, the question as to how the matter and radiation we see around us came into existence in the first place remains unanswered. Moreover, one may pose another philosophical question.

Have we any guarantee that the physical laws that we use here and now have always remained the same? We could have assumed this to be the case had the universe itself not changed considerably in the course of time. This, however, was not the case for the Friedmann universes. A typical standard model changes considerably in its physical content and properties from soon after $t = 0$ to the present day (see Chapters 5–7). So the assumption that the laws of physics have remained unchanged throughout the history of the standard models is more an article of faith than a verifiable fact.

Today, as we will see in Chapter 9, the age problem is still with us, although not in such a severe form as the low value of T_0 in 1948 implied. The questions of singularity and matter creation still remain with the standard models: the work discussed in Chapter 6 does not tell us what happened at $t = 0$. Hoyle's approach to the steady state theory was designed to attack the problem of primary creation of matter. However, his colleagues Bondi and Gold considered the assumption of constancy of physical laws as of paramount importance.

8.2.1 The perfect cosmological principle

Bondi and Gold argued that the cosmological principle (see Chapter 3) goes some way towards ensuring that the locally discovered laws of physics have universal validity; but it does not go far enough. This principle tells us that at any given cosmic time t, all fundamental observers see the same large-scale features of the universe. Thus we are justified in assuming no spatial variation in the basic physical laws at any given cosmic time. But there is no justification from the cosmological principle to assume that the laws remain unchanged with time.

To provide such a justification Bondi and Gold strengthened the cosmological principle to what they called the *perfect cosmological principle* (PCP). The PCP states that in addition to the symmetries implicit in the cosmological principle, the universe in the large is unchanging with time. Thus the geometrical and physical properties of the hypersurfaces $t = $ constant do not change with t.

It is important to emphasize the qualification 'in the large'. On a small enough scale the observed part of the universe *will* change. For example, stars in a galaxy will grow older, a small cluster of galaxies will evolve with time in shape and composition, and so on. However, according to the PCP the statistical properties of large-scale populations do not change.

For example, Hubble's constant should remain the same whether it is measured now or at any time past or present, since its accurate measurement involves the rate of expansion of the universe. This being a property of the large-scale structure of the universe, the constancy of H tells us immediately that

$$H = \frac{\dot{S}}{S} = \text{constant} = H_0, \quad \text{i.e.,} \quad S = \exp(H_0 t). \quad (8.1)$$

Further, the curvature of a $t = $ constant hypersurface is given by k/S^2. This could in principle be measured at different times and found to be changing unless $k = 0$. (See Exercise 4 for another argument leading to $k = 0$.) Thus the PCP leads us to the unique line element

$$ds^2 = c^2 dt^2 - e^{2H_0 t}[dr^2 + r^2(d\theta^2 + \sin^2\theta \, d\phi^2)]. \quad (8.2)$$

Notice that we have arrived at the line element of the steady state universe without having to solve *any* field equations, as we had to do to determine $S(t)$ and k in standard cosmology. Bondi and Gold cited this result as an example of the deductive power of the PCP. Two other examples of deductions from this principle are given.

Expansion of the universe

The line element (8.2) is completely characterized by H_0. It is possible to have $H_0 = 0$, $H_0 < 0$, or $H_0 > 0$, all consistent with the PCP. If, however, we take account of the local thermodynamic conditions, we are able to deduce that $H_0 > 0$. For, our observations show that the universe in our local neighbourhood is far from being in a state of thermodynamic equilibrium. Stars radiate; regions of high and low temperatures exist within the Galaxy and outside it. If $H_0 < 0$ then we would have a static, infinitely old Euclidean universe. Such a universe should have reached a thermodynamic equilibrium by now, as implied by the Olbers paradox (see Chapter 4). If $H_0 < 0$ then we would have a contracting universe in which radiation from distant objects would be blueshifted. Such radiation would lead to an infinite radiation background even worse than that indicated by the calculations of Olbers. Thus our local observations preclude $H_0 \leqslant 0$, leaving the case $H_0 > 0$, which is consistent with the finite and low night sky background (see Exercise 7). Hence the universe must expand: a conclusion arrived at without looking at any nearby galaxies!

Creation of matter

It is easily seen that a proper 3-volume V bounded by fixed (r, θ, ϕ) coordinates increases with time as

$$V \propto \exp 3 H_0 t,$$

$$\frac{\dot{V}}{V} = 3 H_0. \tag{8.3}$$

By the steady state hypothesis the density of the universe must remain constant at $\rho = \rho_0$. Then the amount of matter within V must increase in mass $M \equiv V\rho_0$ as

$$\dot{M} = 3 H_0 V \rho_0.$$

In other words,

$$Q = 3 H_0 \rho_0 \tag{8.4}$$

denotes the rate of creation of matter per unit volume. If we use c.g.s.

units we get

$$Q = 2 \times 10^{-46} \left(\frac{\rho_0}{\rho_c}\right) h_0^3 \, \mathrm{g\,cm^{-3}\,s^{-1}}, \tag{8.5}$$

where ρ_c and h_0 have been defined in Chapters 3 and 4.

The small value of Q shows that there is a very slow but continuous creation of matter going on, in contrast to the explosive creation at $t = 0$ of the standard models.

8.2.2 A field theory for creation

The creation field

Attractive though the above deductive approach is, it has its limitations. For example, we do not have a quantitative relation connecting H_0 to, say, the mean density ρ_0 as we have Friedmann cosmologies. Nor do we have any physical theory for such an important phenomenon as the continuous creation of matter. Is the sacrosanct law of conservation of matter and energy being violated in the process of matter creation? Bondi and Gold appreciated the fact that questions like these could be answered through a dynamical theory rather than from their deductive approach. However, they felt that, together, the PCP and local observations fix the large-scale properties of the universe in a form that can be tested by observations (see section 8.3). Therefore they attached a greater import-ance to testing the PCP by observations than to formulating a dynamic theory that might determine H_0, ρ_0, and so on quantitatively.

Fred Hoyle, on the other hand, took the opposite view. He looked for a process – that is, a field theory – that could account for the phenomenon of primary creation of matter. After several attempts he finally adopted the formulation suggested by M. H. L. Pryce. This formulation, known as the C-field theory, was used extensively by Hoyle and the author in the early 1960s. The highlights of the C-field theory are given below.

The action principle

The C-field theory involves adding more terms to the standard Einstein–Hilbert action (see sections 2.8 and 2.9) to represent the phenomenon of the creation of matter. Using Occam's razor, the additional field to be introduced is a scalar field with zero mass and zero charge. We denote this field by C and its derivative with respect to the spacetime coordinate x^i by C_i. The action is then given by

$$\mathscr{A} = \frac{c^3}{16\pi G} \int R(-g)^{1/2} \, \mathrm{d}^4x - \sum_a m_a c \int \mathrm{d}s_a$$

$$- \frac{1}{2c} f \int C_i C^i (-g)^{1/2} \, \mathrm{d}^4x + \sum_a \int C_i \, \mathrm{d}a^i. \qquad (8.6)$$

Instead of the electromagnetic terms (which might be present if we had charged particles), we have in (8.6) the C-field terms. To appreciate the difference between the two interactions, note that the last term of (8.6) is path-independent. If we consider the world line of particle a between the end points A_1 and A_2, we have

$$\int_{A_1}^{A_2} C_i \, \mathrm{d}a^i = C(A_2) - C(A_1).$$

Normally, such path-independent terms do not contribute to any physics derivable from the action principle. So why include such a term? The answer to this question lies in the notion of 'broken' world lines. A theory that discusses creation (or annihilation) of matter *per se* must have world lines with finite beginnings or ends (or both). The C-field interaction term picks out precisely these end points of particle world lines. If we vary the world line of a and consider the change in the action $\mathscr{A}$ in a volume containing the point A_1 where the world line begins (see Figure 8.1) then we get A_1 (which is now varied)

$$m_a c \frac{\mathrm{d}a^i}{\mathrm{d}s_a} g_{ik} - C_k = 0. \qquad (8.7)$$

This relation tells us that *overall energy and momentum are conserved at the creation point*. The 4-momentum of the created particle is compensated by the 4-momentum of the C-field. Clearly, to achieve this balance the C-field must have negative energy. We will return to this point later. We also note that since the interaction term is path-independent, the equation of motion of a is still that of a geodesic:

$$m_a \left[\frac{\mathrm{d}^2 a^i}{\mathrm{d}s_a^2} + \Gamma_{kl}^i \frac{\mathrm{d}a^k}{\mathrm{d}s_a} \frac{\mathrm{d}a^l}{\mathrm{d}s_a} \right] = 0. \qquad (8.8)$$

The constant f in the action (8.6) is a coupling constant. The variation of C gives the source equation in the form

$$C_{;k}^k = cf^{-1}\bar{n}, \qquad (8.9)$$

where $\bar{n}$ = number of net creation events per unit proper 4-volume. In calculating $\bar{n}$ we attach a $+$ sign to the points like A_1 where a world line begins and a $-$ sign to the points like A_2 where a world line ends. Again,

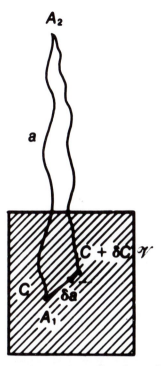

Fig. 8.1 The world line of a begins at A_1 and ends at A_2. If we consider variations in the shaded region, the point A_1 shifts by δa^i. This shift produces a change in the C-field interaction term by an amount $-\delta C = C_i \delta a^i$. The change in the inertial part of the action similarly makes a contribution at A_i of $p_i^{(a)} \delta a^i$, where $p_i^{(a)}$ is the 4-momentum of the particle a. The result (9.7) follows by equating the net contribution of $\delta \mathcal{A}$ at A_i to zero.

we see in (8.9) the relationship between the C-field and the creation/annihilation events.

Finally, the variation of g_{ik} leads to the modified Einstein field equations

$$R^{ik} - \tfrac{1}{2} g^{ik} R = -\frac{8\pi G}{c^4} \left(\underset{(m)}{T}{}^{ik} + \underset{(C)}{T}{}^{ik} \right), \tag{8.10}$$

where $\underset{(m)}{T}{}^{ik}$ is the matter tensor as in the earlier chapters while

$$\underset{(C)}{T}{}^{ik} = -f(C^i C^k - \tfrac{1}{2} g^{ik} C^l C_l). \tag{8.11}$$

Again we note that $\underset{(C)}{T}{}^{00} < 0$ for $f > 0$. Thus the C-field has negative energy density that produces a repulsive gravitational effect. It is this repulsive force that drives the expansion of the universe. The above effect may resolve one difficulty usually associated with the quantum theory of negative energy fields. Because such fields have no lowest energy state, they normally do not form stable systems. A cascading into lower and lower energy states would inevitably occur if we perturb the field in a given state of negative energy. However, this conclusion is altered if we include the feedback of (8.11) on spacetime geometry. This feedback results in the expansion of space and in the lowering of the magnitude of field energy. Both these effects tend to work in opposite directions and help to stabilize the system.

Cosmological equations

Using the Robertson–Walker line element and the assumption that a typical particle created by the C-field has mass m, we get the following equations out of (8.7)–(8.11):

$$\dot{C} = mc^2; \tag{8.12}$$

$$mf\left(\ddot{C} + 3\frac{\dot{S}}{S}\dot{C}\right) = \left(\dot{\rho} + \frac{\dot{S}}{S}\rho\right)c^2; \tag{8.13}$$

$$2\frac{\ddot{S}}{S} + \frac{\dot{S}^2 + kc^2}{S^2} = \frac{4\pi Gf}{c^4}\dot{C}^2, \tag{8.14a}$$

$$3\frac{\dot{S}^2 + kc^2}{S^2} = 8\pi G\left(\rho - \frac{f}{2c^4}\dot{C}^2\right). \tag{8.14b}$$

It is easy to verify that the steady state solution (8.2) follows from these equations for

$$k = 0, \quad S = e^{H_0 t}, \quad \rho = \rho_0 = \frac{3H_0^2}{4\pi G} = fm^2. \tag{8.15}$$

Notice that both H_0 and ρ_0 are given in terms of the elementary creation process; that is, in terms of the coupling constant f and the mass of the particle created. Thus the Hoyle approach gives the quantitative information lacking in the deductive approach of the PCP.

A first-order perturbation of the above equations and of the solution (8.15) also tells us that the solution is stable (see Exercise 20). Indeed, a stability analysis brings out the key role played by (8.7). This tells us that the created particles have their world lines along the normals to the surfaces $C = $ constant. Hoyle has argued that such a result gives a physical justification for the Weyl postulate: it tells us why the world lines of the fundamental observers are orthogonal to a special family of spacelike hypersurfaces. In the C-field cosmology these hypersurfaces are not just abstract notions but have been chosen on a physical basis.

Explosive creation

Although the C-field was introduced primarily to account for the continuous creation of matter, the author showed in 1973 that is also describes explosive matter creation such as is required in the big bang cosmology. We illustrate below how this is achieved for the case $k = 0$.

In equations (8.12) to (8.14), we make use of the idea that all matter is created in an explosive process at $t = 0$. Then the right-hand side of (8.13) is like a delta function $\delta(t)$, leading to the solution

$$\dot{C} = \frac{A}{S^3}, \qquad A = \text{constant}.$$

Notice that this solution is inconsistent with (8.12), except at one epoch $t = 0$. This is hardly surprising, since we have assumed no creation of matter subsequent to $t = 0$. Thus the creation condition (8.9) is not satisfied at $t > 0$.

Substituting for $\dot{C}$ in (8.14a), we can integrate for S and obtain a solution

$$S(t) \propto \left[1 + \frac{(t + t_1)^2}{t_0^2} \right]^{1/3}, \tag{8.16}$$

where t_0 and t_1 are constants related to the initial conditions at $t = 0$ (see Exercise 22).

The scale factor given by (8.16) behaves like that for the standard Einstein–de Sitter model for $t \gg t_0, t_1$. In the C-field model not only is the spacetime singularity at $t = 0$ averted, but we also see the present matter as arising from a primordial explosion *that conserves energy and momentum*.

This conservation of energy and momentum must follow as a general deduction for any C-field model, since the governing equations are derived from an action principle. Hence criticism based on the unexplained origin of new matter, which could be validly applied to the explosive creation of the standard cosmology or to the continuous creation in the Bondi–Gold version of the steady state model, does not apply to the C-field cosmology.

In physical terms, the creation is explained by a process of interchange of energy and momentum between the negative energy C-field and the matter. The divergence of (8.10) gives the mathematical formula for energy conservation:

$$T^{ik}_{(m);k} = f C^i C^k_{;k}. \tag{8.17}$$

It is easy to verify that the idea would not work for a positive energy field (see Exercise 23).

8.3 Observable parameters of the steady state theory

Leaving aside the dynamics of the model, we now come to some of the observable features of the steady state theory. Here we deal essentially with the line element (8.2) and the geometrical properties deducible from it. Indeed, as Bondi and Gold emphasized in their original paper, the steady state model makes precise predictions and is therefore vulnerable to observational disproof, in contrast to the big bang models, which can always be fed with arbitrary parameters. (This comment will become clearer in Chapters 9 and 10, when we discuss observational cosmology.)

Since we have gone through calculations of these observable features at great length in Chapters 3 and 4, we will be brief here and simply state the results.

8.3.1 The redshift

The redshift of a galaxy G_1 at (r_1, θ_1, ϕ_1) emitting light at t_1 that is received by the observer O at $r = 0$ at the present epoch t_0 is given by

$$z_1 = e^{H_0(t_0 - t_1)} - 1 = r_1 \frac{H_0}{c} e^{Ht_0}. \tag{8.18}$$

8.3.2 The luminosity distance

This is given for the above galaxy by

$$D_1 = \frac{c}{H_0} z_1(1 + z_1). \tag{8.19}$$

Equation (8.19) is the *Hubble law* for steady state cosmology. From (8.1) we also see that the deceleration parameter q_0 for this cosmology has the value -1.

8.3.3 Angular size

The angle $\Delta\theta$ ($\ll 1$) subtended at O by an astronomical source of projected linear size d and redshift z is given by

$$\Delta\theta = \frac{H_0}{c} d \cdot \left(\frac{1 + z}{z}\right). \tag{8.20}$$

Thus the angular size tends to a finite minimum as $z \to \infty$.

8.3.4 Flux density

The formula (3.44) becomes in this case

$$\mathcal{F}_{bol} = \frac{L_{bol}}{4\pi \left(\dfrac{c}{H_0}\right)^2 z^2 (1+z)^2} \tag{8.21}$$

For (3.43), we get

$$\widetilde{\mathcal{F}}(\nu_0) = \frac{LJ(\nu_0 \cdot 1 + z)}{4\pi \left(\dfrac{c}{H_0}\right)^2 z^2 (1+z)}. \tag{8.22}$$

8.3.5 Number count

In the notation of section 3.9, the number of sources with redshift less than z is given by

$$N(z) = 4\pi n \left(\frac{c}{H_0}\right)^3 \left[\ln(1+z) - \frac{3z^2 + 2z}{2(1+z)^2}\right]. \tag{8.23}$$

8.3.6 The age distribution of galaxies

New galaxies are always being formed in the steady state universe. Since the universe expands, the galaxies, once formed, move away from each other. Thus the older a population of galaxies, the more sparse its distribution will be. Since the volume bounded by galaxies increases with time as $\exp(3H_0 t)$, we have the following simple result for the age–density relation of galaxies:

$$Q(\tau) \propto e^{-3H_0 \tau}, \tag{8.24}$$

where $Q(\tau)\,d\tau$ is the proper number density of galaxies with ages in the range τ, $\tau + d\tau$. The average age is therefore $(3H_0)^{-1}$. However, caution is necessary in the observational interpretation of (8.24), as we shall see in the following section.

8.4 Physical and astrophysical considerations

This section briefly outlines some of the ideas proposed from time to time in the context of the steady state theory to discuss such problems as the nature of created particles, the formation of galaxies, the origin of the microwave background radiation, and so on. Some of these concepts might still be relevant even if the steady state cosmological picture does not survive.

8.4.1 The hot universe

In 1958 Gold and Hoyle proposed the hypothesis that the created matter
was in the form of neutrons. The creation of neutrons does not violate any
standard conservation laws of particle physics, except the constancy of the
baryon number. Although this was considered an objection in 1958, today
the baryon number is no longer regarded as invariant. Indeed, in Chapter
6 we saw how scenarios based on baryon nonconservation are being
proposed in the context of the early universe to account for the observed
baryon number in the universe.

In the Gold–Hoyle picture the created neutron undergoes a β-decay:

$$n \rightarrow p + e^- + \bar{\nu}. \tag{8.25}$$

The conservation of energy and momentum results in the electron taking
up most of the kinetic energy and thereby acquiring a high kinetic
temperature of $\sim 10^9$ K (see Exercise 29). Gold and Hoyle argued that
such a high temperature produced inhomogeneously would lead to the
working of heat engines between the hot and cold regions, which provide
pressure gradients that result in the formation of condensations of the size
of $\geqslant 50$ Mpc (see Exercise 30). As we have already seen in Chapter 7,
pure gravitational forces are not able to provide a satisfactory picture of
galaxy formation. The temperature gradients set up in the hot universe of
Gold and Hoyle help in this process.

However, the resulting system is not a single galaxy, but a supercluster
of galaxies containing $\sim 10^3$ to 10^4 members. Such large-scale inhomogeneities in the distribution of galaxies were referred to in Chapter 1.
Inhomogeneities on such a large scale $\geqslant 50$ Mpc caution us against
applying the cosmological principle too rigorously. For example, the
formula (8.24) for the age distribution of galaxies will hold over a region
considerably larger than 50 Mpc in such a model. If we are in a particular
supercluster, we expect to see a preponderance of galaxies of age similar
to that of ours in our neighbourhood out to, say, 20 or 30 Mpc. Thus it will
not be surprising if our local sample yields an average age much larger
than the universal average $(3H_0)^{-1} \approx 3 \times 10^9 h_0^{-1}$ years.

Although newly created electrons have a kinetic temperature of
$\sim 10^9$ K, the temperature tends to drop because of expansion. The
average temperature is three-fifths of this value, that is, around 6×10^8 K.
It was suggested by Hoyle in 1963 that such a hot intergalactic medium
would generate the observed X-ray background. However, quantitative
estimates by R. J. Gould soon showed that the expected X-ray background in the hot universe would be considerably higher than what is

actually observed, thus making the hot universe untenable. Although the present background measurements do not rule out such a hot universe for $h_0 \simeq 0.5$, astrophysicists are inclined to look for other explanations for the origin of the X-ray background.

Although now discredited, the hot universe model was the first exercise in linking particle physics (neutron decay) to the formation of large-scale structures in the universe.

8.4.2 The bubble universe

In 1966 Hoyle and the author discussed the effect of raising the coupling constant f by $\sim 10^{20}$. As the formulae (8.15) show, we would then have a steady state universe of very large density $(\rho_0 \simeq 10^{-8} \, \text{g cm}^{-3})$ and very short time scale $(H_0^{-1} \simeq 1 \text{ year!})$. If in such a dense universe creation is switched off in a local region, that is, if we locally have a phase transition from the creative to the noncreative mode,

$$C^i_{;i} = 0, \tag{8.26}$$

then this local region will expand according to (8.16). Being less dense than the surroundings, such a region will simulate an air bubble in water. The reader may look back to Chapter 6 and discover the similarity between this model and the inflationary model that came into fashion fifteen years later.

According to this model, this bubble is all that we see with our surveys of galaxies, quasars, and so on. Hence our observations tell us more about this unsteady perturbation than about the ambient steady state universe. There are, however, observable effects that give indications of the high value of f. For example, these authors showed that particle creation is enhanced near already existing massive objects and that the resulting energy spectrum of the particles would simulate that of high-energy cosmic rays. The actual energy density of cosmic rays requires the high value of f chosen here.

8.4.3 The origin of elements and the microwave background

One of the beneficial influences of the steady state cosmology on astrophysics was that it prompted work on stellar nucleosynthesis. Since the model does not have a high-temperature epoch, it cannot draw on the calculations given in Chapter 5 to explain how nuclei are made from protons and neutrons.

Since the centres of stars provide sites for high temperature and density, astrophysicists looked for nucleosynthesis in such places. The pioneering work of E. M. Burbidge, G. R. Burbidge, W. A. Fowler, and F. Hoyle in 1957 demonstrated in a comprehensive manner how the whole observed range of nuclei can be produced in stellar processes as stars evolve. Thus it became established that the bulk of the nuclei are produced in stars rather than in the early hot universe, as Gamow had envisaged.

In 1964 and 1965 the steady state model received two near-fatal blows. The realization that the observed helium abundance in several parts of the Galaxy is considerably higher than that generated in the stars led astronomers back to Gamow's ideas once again (see Exercise 34). The case for the hot big bang became even stronger with the discovery of the microwave background in 1965. The steady state model never quite recovered from these two blows. Indeed, if today it is to come back as an alternative to the big bang then it must produce an astrophysical interpretation for both the above observations, as well as for the observed abundances of other light nuclei besides helium, like deuterium, Li, Be, and so on.

Energetically, it is realized that increased stellar activity is required to account for the observed helium, and the resulting additional starlight has to be thermalized to produce the microwave background (see Exercise 34). In the final chapter of this book we will describe a possible scenario in which dust grains in the intergalactic space act as thermalizers. In working such scenarios into the steady state model a further constraint has to be placed on any calculations. This is the constraint demanded by the PCP, that is, that the universe in the past was no different from the way it is now.

The main difficulties of such attempts are as follows. Although increased stellar activity can generate sufficient helium, the production of deuterium in stars (or supermassive objects) has not proved so easy to demonstrate, since the deuterium produced is quickly destroyed. Also, the extreme homogeneity of the microwave background (see Chapter 9) places severe limits on any theory that attempts to generate it from discrete sources.

8.5 Mach's principle

There are two ways of measuring the Earth's spin about its polar axis. By observing the rising and setting of stars the astronomer can determine the period of one revolution of the Earth around its axis: the period of

$23^h 56^m 4^s .1$. The second method employs a Foucault pendulum whose plane gradually rotates around a vertical axis as the pendulum swings. Knowing the latitude of the place of the pendulum makes it possible to calculate the Earth's spin period. The two methods give the same answer.

At first sight this does not seem surprising. Closer examination, however, reveals why the result is nontrivial. The first method measures the Earth's spin period against a background of distant stars, while the second employs the standard Newtonian mechanics in a spinning frame of reference. In the latter case, we take note of how Newton's laws of motion get modified when their consequences are measured in a frame of reference spinning relative to the 'absolute space' in which these laws were first stated by Newton.

Thus, implicit in the assumption that equates the two methods is the coincidence of absolute space with the background of distant stars. It was Ernst Mach in the last century who pointed out that this coincidence is nontrivial. He read something deeper in it, arguing that the postulate of absolute space that allows one to write down the laws of motion and arrive at the concept of inertia is somehow intimately related to the background of distant parts of the universe. This argument is known as 'Mach's principle' and we will analyse it further.

When expressed in the framework of the absolute space, Newton's second law of motion takes the familiar form

$$\mathbf{P} = m\mathbf{f}. \tag{8.27}$$

This law states that a body of mass m subjected to an external force $\mathbf{P}$ experiences an acceleration $\mathbf{f}$. Let us denote by S the coordinate system in which $\mathbf{P}$ and $\mathbf{f}$ are measured.

Newton was well aware that his second law has the simple form (8.27) only with respect to S and those frames that are in uniform motion relative to S. If we choose another frame S' that has an acceleration $\mathbf{a}$ relative to S, the law of motion measured in S' becomes

$$\mathbf{P}' \equiv \mathbf{P} - m\mathbf{a} = m\mathbf{f}'. \tag{8.28}$$

Although (8.28) outwardly looks the same as (8.27), with $\mathbf{f}'$ the acceleration of the body in S', something new has entered into the force term. This is the term $m\mathbf{a}$, which has nothing to do with the external force but depends solely on the mass m of the body and the acceleration $\mathbf{a}$ of the reference frame relative to the absolute space. Realizing this aspect of the additional force in (8.28), Newton termed it 'inertial force'. As this name implies, the additional force is proportional to the inertial mass of the body. Newton discusses this force at length in his *Principia*, citing the example of a rotating water-filled bucket (see Figure (8.2)).

According to Mach, the Newtonian discussion was incomplete in the sense that the existence of the absolute space was postulated arbitrarily and in an abstract manner. Why does S have a status in that it does not require the inertial force? How can one physically identify S without recourse to the second law of motion, which is based on it?

To Mach the answers to these questions were contained in the observation of the distant parts of the universe. It is the universe that provides a background reference frame that can be identified with Newton's frame S. Instead of saying that it is an accident that Earth's rotation velocity relative to S agrees with that relative to the distant parts of the universe, Mach took it as proof that the distant parts of the universe somehow enter into the formulation of local laws of mechanics.

One way this could happen is by a direct connection between the property of inertia and the existence of the universal background. To see

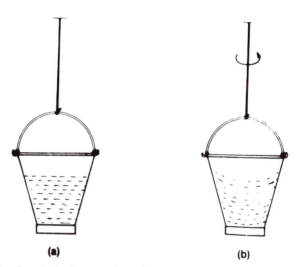

(a)   **(b)**

Fig. 8.2 (a) A bucket full of water hanging by a rope tied to the ceiling. (b) The same bucket turning round and round as a result of the rope unwinding itself from a previously given twist. The water surface in (a) is flat and horizontal, while that in (b) is curved inwards. This curvature of the water surface is due to the centrifugal force that acts on the rotating water mass. This example was discussed by Newton in his *Principia*. Newton argued that in (a) the bucket is at rest relative to the absolute space, while in (b) it is rotating relative to the absolute space and hence an extra force or forces must be postulated to explain the curvature of the water surface. The centrifugal force is the extra force in this example.

this point of view, imagine a single body in an otherwise empty universe. In the absence of any forces (8.27) becomes

$$m\mathbf{f} = \mathbf{0}.$$

What does this equation imply? Following Newton, we would conclude that $\mathbf{f} = \mathbf{0}$, that is, the body moves with uniform velocity. But we now no longer have a background against which to measure velocities. Thus $\mathbf{f} = \mathbf{0}$ has no operational significance. Rather, $\mathbf{f}$ should be completely indeterminate. And it is not difficult to see that such a conclusion follows naturally, provided we argue that

$$m = 0. \tag{8.29}$$

In other words, the measure of inertia depends on the existence of the background in such a way that in the absence of the background the measure vanishes! This aspect introduces a new feature into mechanics not considered by Newton. The Newtonian view that inertia is the property of matter has to be augmented to the statement that inertia is the property of matter as well as of the background provided by the rest of the universe.

Such a Machian viewpoint not only modifies local mechanics, but it also introduces new elements into cosmology. For, except in the universe following the perfect cosmological principle, there is no basis now for assuming that particle masses would necessarily stay fixed in an evolving universe. This is the reason for considering cosmological models anew from the Machian viewpoint. Presented here are some instances of how different physicists have given quantitative expression to Mach's principle and arrived at new cosmological models.

8.6 The Brans–Dicke theory of gravity

In 1961 C. Brans and R. H. Dicke provided an interesting alternative to general relativity based on Mach's principle. To understand the reasons leading to their field equations, we first note that the concept of a variable inertial mass arrived at in section 8.5 itself leads to a problem of interpretation. For, how do we compare masses at two different points in spacetime? Masses are measured in certain units, such as masses of elementary particles, which are themselves subject to change! We need an independent unit of mass against which an increase or decrease of a particle mass can be measured. Such a unit is provided by gravity, the so-called Planck mass encountered earlier:

$$\left(\frac{\hbar c}{G}\right)^{1/2} \approx 2.16 \times 10^{-5} \, \text{g}. \tag{8.30}$$

Thus the dimensionless quantity

$$\chi = m\left(\frac{G}{\hbar c}\right)^{1/2} \tag{8.31}$$

measured at different spacetime points can tell us whether masses are changing. Or alternatively, if we insist on using mass units that are the same everywhere, a change of χ would tell us that G is changing. (We could of course assume that $\hbar$ and c also change. However, by keeping $\hbar$ and c constant we follow the principle of least modification of existing theories. Thus special relativity and quantum theory are unaffected if we keep $\hbar$ and c fixed.) This is the conclusion Brans and Dicke arrived at in their approach to Mach's principle. They looked for a framework in which the gravitational constant G arises from the structure of the universe, so that a changing G could be looked upon as the Machian consequence of a changing universe.

In 1953 D. W. Sciama had given general arguments leading to a relationship between G and the large-scale structure of the universe. We have already come across one example of such a relation in the Friedmann cosmologies:

$$\rho_0 = \frac{3H_0^2}{4\pi G} q_0.$$

If we write $R_0 = c/H_0$ as a characteristic length of the universe and $M_0 = 4\pi\rho_0 R_0^3/3$ as the characteristic mass of the universe, then the above relation becomes

$$\frac{1}{G} = \frac{M_0}{R_0 c^2} q_0^{-1} \sim \frac{M_0}{R_0 c^2} \sim \sum \frac{m}{rc^2}. \tag{8.32}$$

Given a dynamic coupling between the inertia and gravity, a relation of the above type is expected to hold. Brans and Dicke took this relation as one that determines G^{-1} from a linear superposition of inertial contribution m/rc^2 being from a mass m at a distance r from the point where G is measured. Since m/r is a solution of a scalar wave equation with a point source of strength m, Brans and Dicke postulated that G behaves as the reciprocal of a scalar field ϕ:

$$G \sim \phi^{-1}, \tag{8.33}$$

where ϕ is expected to satisfy a scalar wave equation whose source is all the matter in the universe.

8.6.1 The action principle

The intuitive concepts are contained in the Brans–Dicke action principle, which may be written in the form

$$\mathcal{A} = \frac{c^3}{16\pi} \int_{\mathcal{V}} (\phi R + \omega \phi^{-1} \phi^k \phi_k)(-g)^{1/2} \, d^4x + \Lambda. \tag{8.34}$$

Notice first that the coefficient of R is $c^3 \phi / 16\pi$, instead of $c^3 / 16\pi G$ as in the Einstein–Hilbert action. The reason for this lies in the anticipated behaviour of G as given in (8.33). The second term, with $\phi_k \equiv \partial \phi / \partial x^k$, ensures that ϕ will satisfy a wave equation, while the third term includes, through a Lagrangian density L, all the matter and energy present in the spacetime region $\mathcal{V}$. The energy momentum tensor T^{ik} is related to Λ through the relation (2.95). ω is a coupling constant.

The variation of $\mathcal{A}$ for small changes of g^{ik} leads to the field equations

$$R_{ik} - \tfrac{1}{2} g_{ik} R = - \frac{8\pi}{c^4 \phi} T_{ik} - \frac{\omega}{\phi^2} (\phi_i \phi_k - \tfrac{1}{2} g_{ik} \phi^l \phi_l)$$
$$- \frac{1}{\phi} (\phi_{;ik} - g_{ik} \Box \phi). \tag{8.35}$$

Similarly, the variation of ϕ leads to the following equation for ϕ:

$$2\phi \Box \phi - \phi_k \phi^k = \frac{R}{\omega} \phi^2. \tag{8.36}$$

This latter equation can be simplified by substituting for R from the contracted form of (8.35). We finally get

$$\Box \phi = \frac{8\pi}{(2\omega + 3)c^4} T, \tag{8.37}$$

where T is the trace of T^i_k. Thus (8.37) leads to the anticipated scalar wave equation for ϕ with sources in matter, $\Box$ being the wave operator.

Because it contains a scalar field ϕ in addition to the metric tensor g_{ik}, the Brans–Dicke theory is often referred to as the *scalar–tensor theory of gravitation*.

8.6.2 Solar system measurements of ω

It is clear from these field equations that as $\omega \to \infty$ the Brans–Dicke theory tends to general relativity (see Exercise 43). For $\omega = 0(1)$ the theory makes significantly different predictions from general relativity in a number of solar system tests. These tests were briefly reviewed in section 2.10 in the context of general relativity.

The computation of perihelion precession of the planet Mercury gives the prediction of this theory as $(3\omega + 4)/(3\omega + 6)$ times the value given by general relativity. Dicke and his colleagues suggested during the 1970s that if the Sun is oblate, with a quadrupole moment parameter of $\sim 2.5 \times 10^{-5}$, then the resulting change in its gravitational field would lead to a perihelion precession of about 7 per cent of the observed (un-explained) value of ~ 43 arc second per century (see Exercise 44). Had this been the case then the relativistic value of ~ 43 arc second would have been too high, while a Brans–Dicke value for $\omega \simeq 6$ would have correctly accounted for the residual of ~ 40 arc second per century. However, external studies for the Sun's surface do not conform with oblateness even of this order. Hence this test does not give any evidence for ω as small as 6.

The bending angle of a light ray grazing a massive spherical object in the Brans–Dickie theory is $(2\omega + 3)/(2\omega + 4)$ of the relativistic value. Since the accuracy of the radio and microwave measurements of the bending angle is better than ~ 5 per cent and the angle agrees with the relativistic value within this error, the parameter ω has to be as high as ~ 10.

The lunar lasar ranging experiments, however, lead to the conclusion that $\omega \geqslant 29$. Here again, the general relativistic value of the Earth–Moon distance is in excellent agreement with observations, and any departures from it, if they are to be tolerated by the observations, have to be small enough to demand a large value of ω. Radar ranging to probe landers on Mars places an even more severe limit on ω by requiring that $\omega \geqslant 500$.

It therefore follows that at the Solar System level the Brans–Dicke theory has to have a large value of ω in order to survive, thus making it practically indistinguishable from general relativity. However, even for a large ω this theory can produce interesting departures from general relativity at the cosmological level. The following section outlines these differences.

8.7 Cosmological solutions in the Brans–Dicke theory

We will consider only the homogeneous and isotropic cosmological models in the Brans–Dicke theory. Accordingly, we start with the Robertson–Walker line element and the energy tensor for a perfect fluid, as we did in Chapter 4. The scalar-field ϕ is now a function of the cosmic time only. Thus the field equations become

$$\frac{2\ddot{S}}{S} + \frac{\dot{S}^2 + kc^2}{S^2} = -\frac{8\pi p}{\phi c^2} - \frac{2\dot{\phi}\dot{S}}{\phi S} - \frac{\omega\dot{\phi}^2}{2\phi^2} - \frac{\ddot{\phi}}{\phi}, \tag{8.38}$$

$$\frac{\dot{S}^2 + kc^2}{S^2} = \frac{8\pi\varepsilon}{3\phi c^2} - \frac{\dot{\phi}\dot{S}}{\phi S} + \frac{\omega\dot{\phi}^2}{6\phi^2}. \tag{8.39}$$

Compare these equations with the corresponding ones (4.11) and (4.12) of the Friedmann cosmologies. The conservation equation corresponding to (4.15) is the same:

$$\frac{d}{dS}(\varepsilon S^3) + 3pS^2 = 0. \tag{8.40}$$

In addition, we have the field equation for ϕ:

$$\frac{1}{S^3}\frac{d}{dt}(\dot{\phi}S^3) = \frac{8\pi}{(2\omega + 3)c^2}(\varepsilon - 3p). \tag{8.41}$$

We anticipate that big bang solutions will emerge from these equations and set the big bang epoch at $t = 0$. Then the integral of (8.41) is

$$\dot{\phi}S^3 = \frac{8\pi}{(2\omega + 3)c^2}\int_0^t (\varepsilon - 3p)S^3\, dt + C, \tag{8.42}$$

where C is a constant. Two types of solutions are obtained, depending on whether $C = 0$ or $C \neq 0$.

8.7.1 $C = 0$

We will consider a simple example of this type, with $k = 0$, $p = 0$, $\varepsilon = \rho c^2$. This solution is therefore analogous to the Einstein–de Sitter model of general relativity. Write

$$S = S_0\left(\frac{t}{t_0}\right)^A, \qquad \phi = \phi_0\left(\frac{t}{t_0}\right)^B, \tag{8.43}$$

so that $\rho \propto t^{-3A}$ and the field equations give

$$A = \frac{2\omega + 2}{3\omega + 4}, \qquad B = \frac{2}{3\omega + 4} \tag{8.44}$$

and

$$\rho_0 = \frac{(2\omega + 3)B\phi_0}{8\pi t_0^2}. \tag{8.45}$$

The temporal behaviour of S and G ($\propto \phi^{-1}$) is illustrated in Figure 8.3. It can be verified that as $\omega \to \infty$ this solution tends to the Einstein–de Sitter model.

An analogue of the radiation model can be obtained in this theory (see Exercise 47). H. Nariai obtained solutions for $p = n\varepsilon$, with n in the range $0 \leq n \leq \frac{1}{3}$.

8.7.2 $C \neq 0$

In this case the ϕ-terms dominate the dynamics of the universe in the early stages. Thus for small enough t we have

$$\frac{8\pi}{(2\omega + 3)c^2} \int_0^t (\varepsilon - 3p)S^3 \, dt \ll |C|, \tag{8.46}$$

for the cases both of dust and of radiation. For our power law solutions for the case $p = 0$, we have at small enough t

$$3A + B = 1, \qquad t_0 = \frac{S_0^3 \phi_0 B}{C}. \tag{8.47}$$

In the case of a radiation-dominated universe, $p = \frac{1}{3}\varepsilon$ and we can again try a solution of the form (8.43) to get, as $t \to 0$,

$$A^2 = -AB + \frac{\omega B^2}{6}. \tag{8.48}$$

Taking into account (8.47), we can solve (8.48) to get

$$A = \frac{\omega + 1 \pm [(2\omega/3) + 1]^{1/2}}{3\omega + 4}, \qquad B = \frac{1 \pm 3\,[(2\omega/3) + 1]^{1/2}}{3\omega + 4}. \tag{8.49}$$

The upper sign holds when $C > 0$ and the lower sign when $C < 0$. For $C > 0$, $\phi \to 0$ when $S \to 0$, while for $C < 0$, $\phi \to \infty$ for $S \to 0$. These conclusions hold irrespective of the values of k or of the equation of state, since at small values of S the dynamics of the universe are controlled by the ϕ-term.

8.7.3 Production of light nuclei

Dicke and G. S. Greenstein independently investigated the nucleosynthesis problem in the early Brans–Dicke universe. Greenstein followed the same physical approach as was outlined in Chapter 5, for the case $C = 0$. The results obtained by him for $h_0 = 1$ are given in Table 8.1.

For each of three values of the present density of matter ρ_0, Table 8.1 gives three sets of values for the deuterium and helium abundance, corresponding to $\omega = 5$, $\omega = 10$, and $\omega = \infty$. The last case is of course that of general relativity. The differences between the Brans–Dicke theory and general relativity are noticeable for $\omega = 5$ at high values of ρ_0 when more ^{2}H and ^{4}He are formed in the former theory. For $\omega \geqslant 30$, the present observed abundances set an upper limit of $\rho_0 \leqslant 5 \times 10^{-30}$ g cm^{-3} in the Brans–Dicke cosmology.

In the ϕ-dominated models, the constant C can be adjusted to produce

Table 8.1 *Mass fractions of ^2H and ^4He in Brans–Dicke cosmology for matter-dominated models*[a]

ω	ρ_0 (g cm^{-3})		
	10^{-31}	10^{-30}	10^{-29}
5	7.6×10^{-4}	2.6×10^{-5}	3.4×10^{-8}
	0.26	0.33	0.40
10	7.6×10^{-4}	2.1×10^{-5}	$\sim 10^{-9}$
	0.26	0.30	0.35
∞	6.6×10^{-4}	1.3×10^{-5}	$\sim 10^{-11}$
	0.25	0.27	0.29

[a] The deuterium fraction is given above the helium fraction.

any desirable abundances, high or low. For cosmic abundances lower than the above value one has to choose a suitably low value of $|C|$.

However, there is another observational handle on C, which is described briefly below.

8.7.4 The variation of G

Since $G \propto \phi^{-1}$, a time-dependent ϕ will mean a time-dependent gravitational constant. As seen from (8.43), we have for $C = 0$

$$\frac{\dot{G}}{G} = -\frac{2}{3\omega + 4} \cdot \frac{1}{t} = -\frac{H}{\omega + 1}. \tag{8.50}$$

Thus $|\dot{G}|$ is of the order of Hubble's constant unless ω is large and its sign indicates that the gravitational constant should decrease with time (see Figure 8.3).

However, for a large enough $|C|$, the ϕ-dominated solutions differ significantly from the matter-dominated ones, even at the present epochs. In this case, for C large and negative we can have G increasing with time even at relatively recent epochs.

We will review the evidence for or against G-variation in Chapter 10.

8.8 The Hoyle–Narlikar cosmologies

We next consider another gravitation theory that may claim to have given the most direct quantitative expression to Mach's principle. This theory was first proposed in 1964 by Fred Hoyle and the author, and we will refer

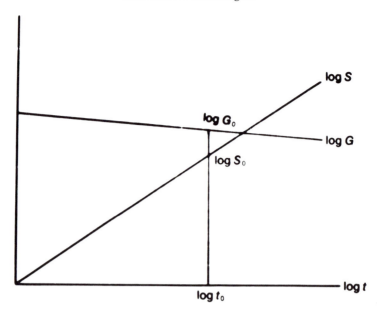

Fig. 8.3 The temporal behaviour of S and G. Both are plotted on a log–log plot for $\omega = 6$. The scales are arbitrary.

to it here as the HN theory and to the cosmological models based on it as HN cosmologies. Throughout this discussion we will set $c = 1$.

Like general relativity and the Brans–Dicke theory, the HN theory is formulated in the Riemannian spacetime. However, there is one important difference between this theory and every other cosmological theory we have discussed so far. The difference lies in the fact that general relativity, the Brans–Dicke theory, and so on are pure field theories, whereas the HN theory is based on the concept of *direct interparticle action*. The difference between the two types of theories is best seen in the description of electromagnetism, to which we will frequently refer in this section and the next for comparison. Until the advent of Maxwell's field theory, it was customary to describe electrical and magnetic interactions as instances of direct action at a distance between particles. The success of Maxwell's theory established the field concept in physics at the expense of the concept of action at a distance (see Figure 8.4).

Since Mach's principle (implying as it does a connection between the local and the distant) suggests action at a distance, even an early convert to it like Einstein later became sceptical as to its validity. However, by the early 1960s it became clear that action at a distance can successfully describe electrodynamics and that it has interesting cosmological implications. Since Hoyle and the author had played an active role in these

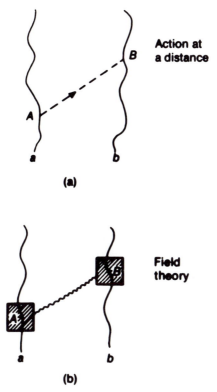

Fig. 8.4 (a) In the action-at-a-distance picture the influence from the point A on the world line of particle a is transmitted directly across spacetime (along the dotted track) to the point B on the world line of particle b. (b) In field theory the field in the neighbourhood of A (shown by the shaded region) is disturbed; the disturbance propagates across spacetime as a wave in the ambient field and reaches the neighbourhood of B. The disturbance then exerts a force on b at B. This is how the influence propagates from a to b.

developments, they naturally adopted an action-at-a-distance approach to Mach's principle.

Accordingly, we use here the somewhat unfamiliar notation of action at a distance. Let us denote by a, b, ... the particles in the universe, m_a, e_a being the mass and charge of the ath particle. As implied by Mach (see section 8.5), the mass m_a is not entirely an intrinsic property of particle a; it also owes its origin to the background provided by the rest of the universe. To express this idea quantitatively, write

$$m_a(A) = \lambda_a \sum_{b \neq a} m^{(b)}(A). \tag{8.51}$$

The above expression means the following. At a typical world point A on the world line of particle a, the mass acquired by a is the net sum of contributions from all other particles $b(\neq a)$ in the universe. The contribution from b at A is given by the scalar function $m^{(b)}(A)$. The coupling constant λ_a is intrinsic to the particle a. However, notice that if a was the only particle in the universe then m_a would be equal to zero and we would have the conclusion arrived at in (8.29).

8.8.1 *A digression into electromagnetic theory*

What are these functions $m^{(b)}(X)$? That they communicate the property of inertia from particles b to any particle placed at the spacetime point X is clear from the context. To arrive at a suitable form for them we take hints from action-at-a-distance electromagnetism, in which it is usual to introduce electromagnetic disturbances that arise specifically from sources, that is, from moving electrical charges. Accordingly, we introduce the 4-potential $A_i^{(b)}(X)$ as denoting the electromagnetic effect at X from the electric charge b. The $A_i^{(b)}(X)$ satisfies the wave equation

$$\Box A_i^{(b)} + R_i^k A_k^{(b)} = 4\pi J_i^{(b)}, \tag{8.52}$$

where $J_i^{(b)}$ is the 4-current generated by the charge b. The solution of (8.52) may be written in the integral form

$$\Box A_i^{(b)}(X) = 4\pi \int e_b G_{ik}(X, B)\, db^k, \tag{8.53}$$

where $G_{ik}(X, B)$ is a Green's function of the wave operator $(g_i^k \Box + R_i^k)$. The well-known Coulomb potential is a special case of (8.53).

The Green's function is not uniquely fixed from the form of the wave operator alone. Boundary conditions must also be specified. The customary boundary condition is that imposed by causality; that is, the influence from B to X must vanish if X lies outside the future light cone of B. The Green's function satisfying this condition is called the *retarded Green's function*. We will denote such a Green's function with a superscript R. Similarly a Green's function confined to the past light cone of B is called the *advanced Green's function* and is denoted with a superscript A (see Figure 8.5).

These Green's functions have played a key role in action-at-a-distance theories. It was originally believed that action at a distance must be instantaneous and hence inconsistent with the framework of special relativity. However, K. Schwarschild, H. Tetrode, and A. D Fokker demonstrated during the first three decades of this century that a

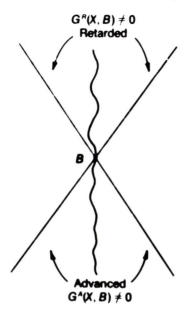

Fig. 8.5 The retarded Green's function of B is nonvanishing only in the future light cone of B, while the advanced Green's function is nonvanishing only in the past light cone.

relativistically consistent action-at-a-distance theory can indeed be formulated. If we consider two spacetime points A and B, with s_{AB}^2 as the invariant square of the relativistic distance between them, then $\delta(s_{AB}^2)$, where δ is the Dirac delta function, is a convenient function for transmitting physical influences between A and B. For, this function acts only when A and B are connectible by a light ray (that is, when $s_{AB}^2 = 0$). This delta function therefore necessarily occurs as the main component in any Green's function in the action-at-a-distance theory. The action principle, which is the basis of the electromagnetic theory in Reimannian spacetime, is described below. We start with the action

$$\mathscr{A} = -\sum_{a}\sum_{<b}4\pi e_a e_b \int\int \bar{G}_{ik}\, \mathrm{d}a^i\, \mathrm{d}b^k, \tag{8.54}$$

where $\bar{G}_{ik}$ is the *symmetric Green's function* given by

$$\bar{G}_{ik}(A,\,B) \equiv \tfrac{1}{2}[G_{ik}^{R}(A,\,B) + G_{ik}^{A}(A,\,B)]. \tag{8.55}$$

Thus $\bar{G}_{ik}(A,\,B) = \bar{G}_{ik}(B,\,A)$ and each term in the action is completely symmetric between each pair of particles. The action (8.54), together with suitable cosmological boundary conditions, reproduces all the electromagnetic effects of the standard Maxwell field theory.

That cosmological boundary conditions are necessary in the action-at-a-distance framework is seen from the following simple illustration. Any retarded signal emitted by particle *a* will get an advanced reaction back from *b*, as shown in Figure 8.6. Thus the theory admits advanced signals and appears to violate causality. Moreover, in Figure 8.4 the signal from *b* arrives at *a* at the same time as the original signal left *a*, no matter how far away *b* is! Thus electromagnetism ceases to be a local theory: any so-called local effect must take account of the response of the universe, which consists of reactions from all such parties *b* other than *a*. A 'correct' response can cancel all the acausal effects. This was pointed out first by J. A. Wheeler and R. P. Feynman in 1945. Later, between 1962 and 1963, J. E. Hogarth, F. Hoyle, and the author showed that this response depends on the model of the universe. In essence, to produce the correct response the universe must be a perfect absorber in the future, i.e., it should be able to absorb all electromagnetic signals directed to the future.

What is the response of the universe? It was shown by Dirac that when an electric charge *a* accelerates, it suffers a force of radiative damping, and that this force can be calculated by evaluating half the difference of the retarded and the advanced fields of the charge *on its worldline*:

$$Q(a) = \tfrac{1}{2}[F^R(a) - F^A(a)]. \tag{8.56}$$

In the Maxwell field theory Dirac's result had remained just a curiosity, without a proper reasoning as to why the radiative reaction must be

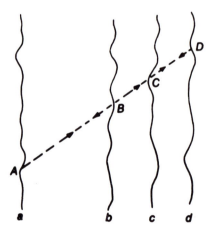

Fig. 8.6 A related signal (shown by dotted line) leaving point *A* on the world line of *a* hits particles *b, c, d, . . .* on points *B, C, D, ...* Their advanced response returns to *A* along the same dotted track, no matter how far these particles are from *a*. Thus even the remote parts of the universe generate instantaneous responses to the retarded disturbance leaving *A*.

determined by the above formula. In the Wheeler–Feynman theory the 'correct' response from the universe to the motion of a is precisely this!

Moreover, if we add (8.56) to the basic time-symmetric direct particle field of a, viz.

$$F(a) = \tfrac{1}{2}[F^{\mathrm{R}}(a) + F^{\mathrm{A}}(a)] \tag{8.57}$$

we get the total effect in the neighbourhood of a to be a pure retarded one. A correct response therefore eliminates all advanced effects except those present in the radiation reaction. It is interesting (and significant) that the steady state model discussed in this chapter generates the correct response, while all Friedmann models fail to do so. Because of the crucial requirement of perfect absorption, this theory is sometimes called the 'absorber theory of radiation'.

8.8.2 Inertia and gravity

Our purpose in the above digression into electromagnetism was to show that a similar approach to inertia leads us to a Machian theory of gravity. In the case of inertia we note that the functions $m^{(b)}(X)$ are scalars and so we have to deal with scalar Green's functions. Thus we write

$$m^{(b)}(X) = \int \lambda_b \widetilde{G}(X, B)\, \mathrm{d}s_b \tag{8.58}$$

and the inertial action as

$$\mathcal{A} = -\sum_a \sum_{<b} \int\!\!\int \lambda_a \lambda_b \widetilde{G}(A, B)\, \mathrm{d}s_a\, \mathrm{d}s_b. \tag{8.59}$$

What is $\widetilde{G}(A, B)$? Again we proceed by analogy with electromagnetism.

From symmetry consideration we need $\widetilde{G}(A, B) = \widetilde{G}(A, B)$. Further, we require $\widetilde{G}$ to be a Green's function of a scalar wave equation. To fix $\widetilde{G}$ completely we use another hitherto undiscussed property of Maxwell's electromagnetic theory known as conformal invariance.

8.8.3 Conformal invariance

Let us consider the transformation

$$\bar{g}_{ik} = \Omega^2 g_{ik}, \tag{8.60}$$

where Ω is a twice-differentiable function of coordinates x^i and lies in the range $0 < \Omega < \infty$. Such a transformation is called *conformal transformation*. Given a spacetime manifold $\mathcal{M}$ with coordinates (x^i) and metric (g_{ik}), we have through (8.60) generated another spacetime manifold $\bar{\mathcal{M}}$ with the same coordinate system (x^i) but with a different metric $(\Omega^2 g_{ik})$. $\mathcal{M}$ and $\bar{\mathcal{M}}$

are said to be *conformal* to each other. If $\mathcal{M}$ is flat $\bar{\mathcal{M}}$ is said to be *conformally flat*.

If we identify the corresponding points (with the same x^i) in $\mathcal{M}$ and $\bar{\mathcal{M}}$, we will find that, in general, distances between the two points are stretched or compressed when we go from $\mathcal{M}$ to $\bar{\mathcal{M}}$. However, the null-cones in both the manifolds are unchanged. This invariance of null cones is distinct from the invariance under coordinate transformations. The coordinate transformations preserve the null directions locally, and they are important in field theories that describe physical interactions locally. The action-at-a-distance theories describe interactions globally and must take account of the global structure of null cones. Hence such theories are expected to preserve their form under conformal transformations as well.

It is easy to verify that the scalar curvature changes under the conformal transformation to

$$\bar{R} = \Omega^{-2}\left(R + 6\frac{\Box\Omega}{\Omega}\right), \tag{8.61}$$

where $\Box$ is evaluated with respect to the metric (g_{ik}). There are, however, certain quantities that do remain the same under a conformal transformation. These are known as *conformally invariant* quantities. It is easy to see, for example, that the action describing Maxwell's field theory is conformally invariant. Consider the changes

$$\bar{A}_i = A_i + \psi_{,i} \qquad (\psi = \text{a scalar function}),$$
$$\bar{F}_{ik} = F_{ik}, \qquad \bar{J} = \Omega^{-4}J_i.$$

These changes leave the form of Maxwell's equations intact.

We now fix the form of $\bar{G}(A, B)$ by demanding that our inertial action (8.59) is conformally invariant. Since under the transformation (8.60)

$$d\bar{s}_a = \Omega(A)\,ds_a, \qquad d\bar{s}_b = \Omega(B)\,ds_b, \tag{8.62}$$

we must have

$$\overline{\bar{G}(A, B)} = \Omega(A)^{-1}\Omega(B)^{-1}\widetilde{G}(A, B). \tag{8.63}$$

The only scalar wave operator that permits (8.63) is then

$$\Box + \tfrac{1}{6}R. \tag{8.64}$$

In other words, $\widetilde{G}(X, B)$ satisfies the wave equation

$$[\Box x + \tfrac{1}{6}R(X)]\widetilde{G}(X, B) = [-g(X)]^{-1/2}\delta_4(X, B). \tag{8.65}$$

$\delta_4(X, B)$ is the four-dimensional Dirac delta function, which vanishes unless $X \equiv B$. Thus we have ensured that the action-at-a-distance theory given by (8.59) does not change under conformal transformations.

8.9 The gravitational equations of HN theory

The action of HN theory is given by (8.59), and with the help of definitions (8.51) and (8.58) we may write as

$$\mathcal{A} = -\sum_a \int m_a \, ds_a. \tag{8.66}$$

Written in this form, this action appears to have only the inertial term of Chapter 2 (see (2.78)). How can such an action yield any gravitational equations?

The answer to this question lies in the fact that the $m'_a s$ in (8.66) are not constants but depend on spacetime coordinates *as well as on spacetime geometry*. For they are defined with the help of Green's functions, which in turn are defined in terms of spacetime geometry. Thus if we make a small variation

$$g_{ik} \to g_{ik} + \delta g_{ik},$$

the wave equation (8.65) will change and so will its solution. Thus we will have

$$\tilde{G}(A, B) \to \tilde{G}(A, B) + \delta\tilde{G}(A, B),$$

and hence $\mathcal{A} \to \mathcal{A} + \delta\mathcal{A}$. We therefore have a nontrivial problem whose solution may be expresed in the following way. To simplify matters we will take all λ_a to be equal to unity. (Later we will relax this assumption.)

Define the following functions:

$$m(X) = \sum_a m^{(a)}(X) = \tfrac{1}{2}[m^R(X) + m^A(X)], \tag{8.67}$$

$$\phi(X) = m^R(X)m^A(X), \qquad m_k \equiv m_{,k}, \ldots, \tag{8.68}$$

$$N(X) = \sum_a \int \delta_4(X, A)[-g(X)]^{-1/2} \, ds_a. \tag{8.69}$$

As in the electromagnetic case, we have chosen the symmetric (half R + half A) Green's function. The gravitational equations then become

$$\begin{aligned} R_{ik} - \tfrac{1}{2}g_{ik}R = {} & - 6\phi^{-1}[T_{ik} - \tfrac{1}{6}(g_{ik}\Box\phi - \phi_{;ik}) \\ & - \tfrac{1}{2}(m_i^R m_k^A + m_k^R m_i^A - g_{ik}g^{pq}m_p^R m_q^A)], \end{aligned} \tag{8.70}$$

together with the 'source' equation for $m(X)$

$$\Box m + \tfrac{1}{6}Rm = N. \tag{8.71}$$

The derivation leading to the final set of equations of the theory may appear somewhat long-winded to anybody unfamiliar with the techniques of direct interparticle action. We have followed here the method used by

Hoyle and the author, who arrived at this theory via their earlier work on electromagnetism. As in the electromagnetic case, the universe responds to a local event. To ensure causality and to eliminate advanced effects, the correct response should be given by

$$\sum_a m^{(a)A}(X) = \sum_a m^{(a)R}(X) = m(X). \tag{8.72}$$

Under these conditions the equations (8.70) further simplify to

$$R_{ik} - \tfrac{1}{2}g_{ik}R = -\frac{6}{m^2}[T_{ik} + \tfrac{1}{6}(g_{ik}\Box m^2 - m^2_{;ik}) + (m_i m_k - \tfrac{1}{2}g_{ik}m^l m_l)]. \tag{8.73}$$

Had we adopted the standard field theoretical approach and introduced a scalar inertia field $m(X)$, we could have arrived at (8.71) and (8.73) from the action given by

$$\mathscr{A} = \int (\tfrac{1}{12}Rm^2 + m^i m_i)(-g)^{1/2}\,d^4x - \sum_a \int m\,ds_a. \tag{8.74}$$

The action-at-a-distance approach, although unfamiliar to a typical theoretical physicist, is useful in that it gives direct expression to Mach's principle. The physical interpretation of the field theoretical term (8.74) is not so easy to see. For this reason, we have discussed the former approach at some length.

Notice that in the former approach our action (8.66) contained only the last term of (8.74), but there m was made up of nonlocal two-point functions. Here m is a straightforward field, with sources in matter whose dynamical properties are defined through the first term in the above action.

Since the property of conformal invariance was used in the formulation of the theory, we expect the final equations (8.71) and (8.73) to exhibit conformal invariance. This expectation is borne out. If (g_{ik}, m) are a solution of these equations, then so are

$$\bar{g}_{ik} = \Omega^2 g_{ik}, \qquad \bar{m} = \Omega^{-1}m. \tag{8.75}$$

Thus apart from coordinate invariance of general relativity, this theory also shows conformal invariance.

We saw in Exercise 33 of Chapter 2 that the coordinate invariance of the action leads to a conservation law for the energy momentum tensor. In this case the conformal invariance of the action leads to a vanishing of trace of the field equations. It may easily be verified that the trace of (8.73) vanishes in view of (8.71). The vanishing of trace represents the fact that the problem is underdetermined. Just as the vanishing of $T^{ik}_{;k}$ in general relativity shows that more solutions can be generated from any given solution by coordinate transformations, so we can generate more solutions through (8.75). All these solutions are physically equivalent, provided we stick to the rule that Ω does not vanish or become infinite.

Suppose we are allowed to choose an Ω in the above range that ensures that

$$\bar{m} = \Omega^{-1} m = \text{constant} = m_0. \tag{8.76}$$

This choice of Ω is possible provided m does not vanish or become infinite. This conformal frame is called the Einstein frame; from (8.76) we get a simplified form for (8.73):

$$R_{ik} - \tfrac{1}{2} g_{ik} R = - \kappa T_{ik}, \tag{8.77}$$

with the constant κ given by

$$\kappa = \frac{6}{m_0^2}. \tag{8.78}$$

Thus we have arrived at Einstein's equations! At first sight we don't seemed to have gained anything. We have no new theory and hence no new predictions, as in the Brans–Dicke theory. Closer examination, however, reveals several ways in which this theory goes beyond relativity.

1. Our starting point was based on Mach's principle. It is only in the many-particle approximation, when the response condition (8.72) is satisfied, that we arrive at the final Einsteinlike field equations. An empty universe in relativity is given by

$$R_{ik} = 0,$$

which can have well-defined spacetimes as solutions. Test particles in such spacetimes will have well-defined trajectories. Such trajectories would not make any sense according to Mach, since we no longer have a material background against which to measure the motion of these particles. These solutions in fact correspond to the $\mathbf{f} = \mathbf{0}$ solutions of Newtonian theory. In the HN theory an empty universe corresponds to

$$m = 0, \quad \text{indeterminate } g_{ik},$$

in accord with the Machian $m = 0$ solution of (8.29).

2. The sign of κ is fixed arbitrarily in general relativity. Neither in the heuristic derivation of Einstein nor in the Hilbert action principle is κ required to be positive. It is only when κ is determined by reference to Newtonian gravity in the weak field approximation (see section 2.9) that we conclude that $\kappa < 0$. In the HN theory (8.78) shows that κ must

necessarily be positive. (This conclusion does not depend on our assumption of $\lambda_a = 1$; the result follows whatever sign the λ_a are given.)

3. In the direct interparticle approach it is not possible to accommodate the λ-term of cosmic repulsion. Thus Occam's razor automatically comes into play. In relativity the λ-term is still possible.

4. The transition from (8.73) to (8.77) is possible provided $0 < \Omega < \infty$. What happens if we break this rule? Suppose in the solution of (8.73) we had a hypersurface on which $m = 0$. If we insist on the transformation (8.76) in a region that contains such a hypersurface, we have to pay the price of $\Omega \rightarrow 0$, which in turn produces spacetime singularities. The work of A. K. Kembhavi in 1979 showed that the well-known cases of spacetime singularities of relativity arise because of the occurrence of zero-mass hypersurfaces in the solution of the equations (8.73). For a simple example of this conclusion let us look at the standard big bang singularity of relativity.

Consider the Minkowski line element (with $c = 1$),

$$ds^2 = d\tau^2 - dx^2 - dy^2 - dz^2, \tag{8.79}$$

as a solution of (8.73). It is easily verified that the mass function satisfying both (8.71) and (8.73) for a uniform number density N of particles is

$$m \propto \tau^2. \tag{8.80}$$

This is the simplest possible cosmological solution in this theory.

If we now insist on going over to a frame with constant mass $\bar{m}$, then from (8.75) we see that the appropriate Ω must be given by

$$\Omega \propto \tau^2. \tag{8.81}$$

However, Ω vanishes on the hypersurface $m = 0$. The transformation to the Einstein conformal frame is therefore 'illegal'. The price paid for insisting that $\bar{m} = $ constant is that the resulting model has a geometrical singularity at $\tau = 0$. In fact it is easily verified that the new model is none other than the singular Einstein–de Sitter model. (Make the time transformation $\tau \propto t^{1/3}$ to demonstrate this result explicitly.)

5. It is instructive to see how the phenomenon of Hubble redshift is explained in the flat spacetime model of (8.79) and (8.80). Clearly, a light photon travelling in Minkowski spacetime does not undergo redshift. Consider, however, what happens to a light photon arriving at the observer at the present epoch τ_0 from a galaxy at a distance r. This photon originated in an atomic (or molecular) transition at time $\tau_0 - r$.

From atomic physics, the wavelength of a photon so transmitted varies inversely as the mass of the electron (making the atomic transition). From (8.80) we see that if λ is the wavelength of this photon and λ_0 the wavelength of a photon emitted in a similar transition at τ_0 at the observer, then

$$1 + z \equiv \frac{\lambda}{\lambda_0} = \frac{m(\tau_0)}{m(\tau_0 - r)} = \frac{\tau_0^2}{(\tau_0 - r)^2}. \qquad (8.82)$$

Thus the redshift in the above HN cosmology arises from the variation of particle masses.

6. A variable gravitational constant arises in the HN cosmologies if we relax the assumption that λ_a are constants. If λ_a change with time then it is possible to generate cosmological models in which G changes with time. We will not discuss such models in detail. The result may be stated in the form

$$\frac{\dot{G}}{G} = -\alpha H, \qquad (8.83)$$

where H is the Hubble constant of the epoch of measurement and α is a constant of order unity.

It was shown by Hoyle and the author in 1972 that λ_a increasing with time may be interpreted as creation of new particles in the universe. They did not give a dynamic theory of matter creation (like the C-field theory of section 8.2), but instead fixed the time dependence of λ_a by an appeal to the Large Numbers Hypothesis. We next describe this hypothesis and its implications for cosmology.

8.10 The Large Numbers Hypothesis

Physics is riddled with units of various kinds and with experimentally determined quantities of various magnitudes. From this vast collection certain constants emerge as having special significance in the framing of basic physical laws; for example, the constant of gravitation G, the charge of the electron e, and so on. The numbers expressing the magnitudes of G, e, and so on depend on the units used. For example

$$e = 4.80325 \times 10^{-10} \text{ electrostatic units}$$

$$= 1.60207 \times 10^{-20} \text{ electromagnetic units.}$$

Clearly these numbers by themselves cannot have absolute significance.

However, certain combinations of these physical constants have no units at all. For example, the combination $\hbar$, c, and e,

$$\frac{\hbar c}{e^2} = 137.036\,02, \qquad (8.84)$$

does not depend on the units used. It must therefore express some physical fact of absolute significance. Indeed, its reciprocal, $e^2/\hbar c$, known commonly as the fine-structure constant, expresses the strength of the electromagnetic interaction, which we believe to be an intrinsic property of nature. A future more complete theory may well give a reason why this constant has this particular value.

Given e, G, and the masses of proton and electron m_p and m_e, we can construct another dimensionless constant (that is, a constant with no units):

$$\frac{e^2}{Gm_p m_e} = 2.3 \times 10^{39} \sim 10^{40}. \tag{8.85}$$

This constant measures the relative strength of the electrical and the gravitational forces between the electron and the proton. Like (8.84), this constant reflects an intrinsic property of nature. However, unlike (8.84), the constant in (8.85) is enormously large! Why such a large number?

Perhaps the appearance of a large dimensionless constant might be dismissed as some quirk on the part of nature. The mystery deepens, however, if we consider another dimensionless number. This is the ratio of the length scale associated with the universe, c/H_0, and the length associated with the electron, $e^2/m_e c^2$. This ratio is

$$\frac{m_e c^3}{e^2 H_0} = 3.7 \times 10^{40} h_0^{-1} \sim 10^{40}. \tag{8.86}$$

Not only do we have another large dimensionless number in (8.86), but it is of the same order as in (8.85).

We can generate another large number of special significance out of particle physics and cosmology. Assuming the closure density ρ_c, let us calculate the number of particles in a Euclidean sphere of radius c/H_0, the mass of each particle being m_p. The answer is

$$N = \frac{4\pi}{3m_p}\left(\frac{c}{H_0}\right)^3 \cdot \frac{3H_0^2}{8\pi G} = \frac{c^3}{2m_p G H_0}$$
$$= 4 \times 10^{79} h_0^{-1}$$
$$\sim 10^{80}. \tag{8.87}$$

Thus, taking N as a standard we see that the large dimensionless numbers of (8.85) and (8.86) are both of the order of $N^{1/2}$.

Reactions among physicists have varied as to the significance of all these numbers. Some dismiss it as a coincidence with the rejoinder: 'so what?' Others have read deep significance in these relationships. The latter class includes such distinguished physicists as A. S. Eddington and P. A. M. Dirac.

Dirac pointed out in 1937 that the relationships (8.86) and (8.87) contain the Hubble constant H_0, and therefore the magnitudes computed in these formulae vary with the epoch in the standard Friedmann model. If so, the near equality of (8.85) and (8.86) has to be a coincidence of the present epoch in the universe, unless the constant (8.85) also varies in such a way as to maintain the state of near equality with (8.86) at all epochs. With this proviso, the equality of (8.85) and (8.86) is not coincidental, but is characteristic of the universe *at all epochs*. The proviso also implies that at least one of the so-called constants involved in (8.85), e, m_p, m_e, and G must vary with the epoch.

This proviso has been generalized by Dirac to what he calls the Large Numbers Hypothesis (LNH). To understand this hypothesis we rewrite the ratio (8.86) as that between the time scale associated with the universe, $T_0 = H_0^{-1}$, and the time taken by light to travel a distance in order of the classical electron radius, $t_e = e^2/m_e c^3$. The LNH then states that any large number that at the present epoch is expressible in the form

$$\left(\frac{T_0}{t_e}\right)^k,$$

where k is of order unity, varies with the epoch t as $(t/t_e)^k$ with a constant of proportionality of order unity.

Applied to (8.85), therefore, the LNH implies that the ratio $e^2/Gm_p m_e$ must vary as $(t/t_e)^{-1}$. Dirac made the distinction between e, m_e, m_p on one side and G on the other in the sense that the former are atomic (microscopic quantities) while G has macroscopic significance. In the Machian cosmologies, G was in fact related to the large-scale structure of the universe. Dirac therefore assumed that if we use 'atomic units' that always maintain fixed values for atomic quantities, then t_e will be constant and $G \propto t^{-1}$. That is, in terms of atomic time units the gravitational constant must vary with the epoch t, with $|\dot{G}/G| \sim H$.

We will now explore the implications of LNH for cosmology.

8.11 The two metrics

Clearly the variation of G predicted by the LNH goes against Einstein's theory of gravitation, which demands a constant G. As in the Brans–Dicke theory, we are forced to modify the relativistic framework to accommodate a varying G. Dirac approached this problem in the following way.

First he took note of the many solar system tests that are in favour of general relativity (see Chapter 2) and argued that the theory should not be abandoned altogether. Instead, Dirac proposed two scales of measurement, one holding in atomic physics and the other in gravitation physics. If we choose the atomic system, we will be able to describe atomic physics in the usual way, that is, with constant values for the atomic constants like e, $\hbar$, m_e, m_p, and so on. However, in this system G will be variable, since Dirac considered it a constant belonging to gravitation physics. If on the other hand we use gravitational units, then according to Dirac G will be constant and atomic quantities will be found variable. And in these latter units the gravitational phenomena can be described by the Einstein equations (2.98).

These two units can be specified in Dirac's framework by having two different spacetime metrics. We will denote these by ds_A^2 and ds_E^2 respectively for the atomic and gravitational systems (the subscript E in the latter case committing us to Einstein's equations of gravity.) We will use these subscripts in general on any physical quantity to indicate what system of measurement is being used. Thus, according to Dirac,

$$G_E, \qquad (m_e)_A, \qquad (m_p)_A$$

are constants, while

$$G_A, \qquad (m_e)_E, \qquad (m_p)_E$$

are variable.

Returning to the astronomical tests of general relativity, we note that the mass of the gravitating body (for example, the Sun) occurs in the Schwarzschild solution. Clearly this mass, which is the gravitational mass, must be a constant in the gravitational units. We denote this mass by M_E. Any measurements made on the Earth however, use atomic systems (such as spectrometers and atomic clocks), and before we interpret any experimental result we must make sure that all observable quantities are transformed to atomic units.

This argument tells us how necessary it is to know the ratio

$$\beta = \frac{ds_E}{ds_A} \tag{8.88}$$

and how the transformation is to be made of any physical quantity from one system of units to another. Here we need a quantitative theory to guide us, a theory that goes further than the above qualitative arguments have so far taken us.

We also note another outcome of our solar system example. If we assume that our astronomical body has N_E nucleons, each of mass m_E, then we may write

$$M_E = m_E N_E = m_E N, \tag{8.89}$$

where we have dropped the suffix E on N because it is a pure number. Whatever metric we use, we will count the same number of particles in the gravitating body. In (8.89) we have $M_E = $ constant, $m_E \neq $ constant, since the latter is an atomic quantity. Thus $N \neq $ constant. In other words, we are forced to conclude that the number of nucleons in the body must change with time. Again we need a quantitive theory to tell us how N changes; but creation (or destruction) of nucleons in a macroscopic object is demanded by Dirac's argument.

So far we have not used the LNH, which started us on the two-metric theory. Let us now see that it helps us in deciding how the nonconservation of nucleon number in the body is regulated.

8.11.1 The creation of particles

If we go back to (8.87) and apply the LNH to N, we easily find that $k = 2$, that is,

$$N(t) \sim \left(\frac{t}{t_e}\right)^2 \propto t^2. \tag{8.90}$$

In other words, the number of particles in the universe in the sense defined in section 8.10 increases with t. Dirac has taken this result to imply that particles are being continually created in the universe.

The creation can occur, according to Dirac, in two possible ways. In *additive creation* the particles are created uniformly throughout space, while in *multiplicative creation* the new particles occur preferentially where matter already exists. Thus in the former mode creation occurs mostly in intergalatic space, while in the latter mode creation occurs mostly in the vicinity of existing astronomical objects.

Using these ideas we return to (8.89). In additive creation the astronomical body will not acquire any significant number of new particles and thus $N = $ constant, giving

$$m_E = \text{constant (additive creation).} \tag{8.91}$$

In multiplicative creation N must increase as t^2 and hence

$$m_E \propto t^{-2} \text{ (multiplicative creation)} \tag{8.92}$$

8.11.2 The determination of β

The connection between ds_A and ds_E can be fixed by considering the motion of a planet (such as the Earth) around a star (the Sun). The dynamic equation in the Newtonian approximation is

$$GM = v^2 r, \tag{8.93}$$

where M = mass of the star, v = speed of the planet, and r = radius of the orbit. The above relation is expected to hold in either of the two systems of units, since $GM/v^2 r$ is a dimensionless quantity. Also, with $c = 1$ the speed v is dimensionless. Thus v = constant in either units. Next, in gravitational units M_E = constant, G_E = constant, hence r_E = constant.

If (8.85) is used with atomic units, we have

$$G_A \sim t^{-1}. \tag{8.94}$$

Also, in multiplicative creation $M_A \propto t^2$ while for additive creation M_A = constant. Hence in these respective units

$$r_A \sim t \quad \text{(multiplicative creation)}, \tag{8.95}$$
$$r_A \sim t^{-1} \quad \text{(additive creation)}, \tag{8.96}$$

thus we have

$$\frac{r_A}{r_E} \sim t \quad \text{(multiplicative creation)}, \tag{8.97}$$

and

$$\frac{r_A}{r_E} \sim t^{-1} \quad \text{(additive creation)}. \tag{8.98}$$

In other words, measured in atomic units, the distance of the planet from the star *increases* with t if the universe has multiplicative creation of matter, and the distance *decreases* with t (as t^{-1}) for additive creation.

From (8.97) and (8.98) we get the behaviour of $β$ defined in (8.88). This ratio of ds_E to ds_A behaves as t^{-1} or t, depending on whether we have multiplicative creation or additive creation in the universe.

8.12 Cosmological models based on the LNH

Using the LNH, Dirac constructed cosmological models in both the circumstances discussed above, namely for multiplicative and additive creation. As in the case of standard cosmologies, the assumptions of homogeneity and isotropy lead to the Robertson–Walker line element in atomic units:

$$ds_A^2 = c^2\,dt^2 - S^2(t)\left[\frac{dr^2}{1 - kr^2} + r^2(d\theta^2 + \sin^2\theta\,d\phi^2)\right]. \quad (8.99)$$

How does the LNH determine k and $S(t)$? We reproduce below the argument given by Dirac.

First we note that the metric proper distance at time t between a galaxy G at $r = 0$ and a galaxy at $r = r_1$ is given by

$$d = S(t)\int_0^{r_1} \frac{dr}{(1 - kr^2)^{1/2}} \equiv S(t)f(r_1). \quad (8.100)$$

According to the LNH, for large t (that is, for $t \gg t_e$) the expression for $S(t)$ should be $\sim (t/t_e)^n$ or $\sim \ln(t/t_e)$. The (metric) recession velocity corresponding to (8.100) will therefore be given by

$$\dot{d} \sim nt_e^{-n}f(r_1)t^{n-1} \quad \text{or} \quad \dot{d} \sim t^{-1}f(r_1). \quad (8.101)$$

The constants multiplying $(t/t_e)^n$ or $\ln(t/t_e)$ in $S(t)$ must be on the order of unity, and hence the constants implied in the ($\sim$) relation above are also on the order of unity. It is then easy to verify that except for $n = 1$, there exists an epoch either in the past (for $n < 1$ or for $S \sim \ln t$) or in the future (for $n > 1$) when $\dot{d} = c$ for any galaxy with $r_1 > 0$. For example, for $n = \frac{1}{2}$ we find that for a galaxy that at present has $\dot{d} \sim 10^{-3}c$, the condition $\dot{d} = c$ occurred in the past epoch given by

$$t_p = \left(\frac{T_0}{t_e}\right) \cdot 10^{-6} t_e \sim 10^{34} t_e.$$

That is, t_p/t_e is a large number. However, by the LNH, t_p is a constant epoch when a significant event took place for galaxy G_1: its recession speed became equal to c. Hence such a constant epoch should not generate a large number. Therefore only the case

$$S(t) \sim (t/t_e) \quad (8.102)$$

is permitted by the LNH.

The arguments given above could be criticized on the following grounds. The epoch when $\dot{d} = c$ is not unique to the model as a whole; it depends on $f(r_1)$ and hence on the galaxy chosen. So it is not necessary that LNH should apply to this epoch. Nor is it clear why $\dot{d} = c$ should be considered significant. Nothing special happens to the galaxy in question when its metric velocity of recession becomes equal to c for the observer at $r = 0$. No global property like the event horizon or the particle horizon enters the argument.

Nevertheless, if we follow the argument further then we can write the cosmological line element as

$$ds_A^2 = c^2 dt^2 - (At)^2 \left[\frac{dr^2}{1 - kr^2} + r^2(d\theta^2 + \sin^2 \theta\, d\phi^2) \right], \quad (8.103)$$

where A is a constant. We next consider multiplicative creation. Since in this case, from section 8.11,

$$ds_E \equiv t^{-1}\, ds_A, \quad (8.104)$$

it is easy to see that a transformation

$$dt_E = \frac{dt}{t} \quad (8.105)$$

gives us

$$ds_E^2 = c^2 dt_E^2 - A^2 \left[\frac{dr^2}{1 - kr^2} + r^2(d\theta^2 + \sin^2 \theta\, d\phi^2) \right]. \quad (8.106)$$

Now we recall that the above line element must be a solution of Einstein's equations. In Chapter 3 we did obtain such a static solution for homogeneous and isotropic dust with the use of the λ-term (see section 3.2), namely, the Einstein universe with $k = +1$. With a suitable scaling of the r-coordinate we can express (8.106) in the form (3.7). Notice, however, that unlike the Einstein universe this Dirac universe does show the phenomenon of redshift of galaxies. For, redshift measurements involve comparisons of the rates at which atomic clocks run at the emitting and receiving galaxies; and for such comparisons the line element (8.103) instead of (8.106) must be used.

For additive creation the situation is more complicated. In the multiplicative creation case the gravitational mass of an astronomical object was held constant in the gravitational units in spite of creation of new particles, by letting the particle masses decrease with time. In the additive creation case the particle masses remain constant even though their number increases (see (8.91)). Dirac was therefore faced with apparent nonconservation of energy. To conserve energy Dirac proposed that along with positive mass particles an equal number of negative mass particles is also created. The negative mass distribution is homogeneous and remains undetectable by standard astronomical observations. In a completely homogeneous situation, the positive and negative mass distributions compensate gravitationally to produce flat Minkowski spacetime. The formation of stars and galaxies by the accumulation of positive mass particles in the actual universe is a result of small departures from this completely homogeneous situation.

It is worth pointing out that when Dirac first proposed a cosmological model based on the LNH between 1937 and 1938 he assumed no matter creation. In this model the number of particles per unit coordinate volume is constant, as in standard cosmologies. Hence the number of particles per unit proper volume goes as S^{-3}, and since the proper volume of the universe goes as $(c/H)^3$, the number of particles in the universe denoted earlier by N goes as

$$S^{-3} \left(\frac{c}{H} \right)^3 \propto (\dot{S})^{-3}.$$

However, by the LNH we know that

$$N \propto t^2. \tag{8.107}$$

Therefore we have

$$t^2 \dot{S}^3 = \text{constant},$$

that is,

$$S \propto t^{1/3}. \tag{8.108}$$

Thus for no particle creation S increases much more slowly with t. (Of course, this solution is ruled out if we apply the LNH to the function S, as we did in the beginning of this section.)

8.12.1 HN cosmology revisited

Some of the ideas of Dirac are found in a version of HN cosmology proposed by its authors between 1971 and 1972. In the HN cosmology we considered the cases where λ_a, λ_b, ..., the constants that denote the strength of the inertial interaction are true constants. However, if these constants vary with time then new cosmological models emerge. In these models the following properties hold: (1) there is particle creation at all epochs in such a way that the LNH is satisfied, (2) in atomic units G varies, while (3) in the gravitational units G is constant and particle masses vary. Thus, the model is like the multiplicative creation model later proposed by Dirac, although its motivation and quantitative details were different. We briefly illustrate how this model works.

Consider a homogeneous and isotropic Minkowski universe given by

$$ds_M^2 = d\tau^2 - dr^2 - r^2 (d\theta^2 + \sin^2 \theta \, d\phi^2), \tag{8.109}$$

where we have $c = 1$ for convenience. Let $n(\tau)$ be the particle number density and $\lambda(\tau)$ the time-varying inertial coupling constant of (8.58). The functions $n(\tau)$ and $\lambda(\tau)$ vary in such a way as to compensate each other's effect; that is, to maintain

$$\lambda n = \text{constant}. \tag{8.110}$$

Thus the mass function $m(\tau)$ is the same as if we had a universe of uniform particle number density $n = $ constant and fixed λ. As in (8.80) we then get

$$m(\tau) \propto \tau^2. \tag{8.111}$$

Since $Gm^2 = $ constant, we get the gravitational constant in the Minkowski framework as

$$G_M \propto \tau^{-4}. \tag{8.112}$$

However, the mass of a typical particle is not $m(\tau)$ but $\lambda m(\tau)$. To determine it we need to know $\lambda(\tau)$. Hoyle and the author determined $\lambda(\tau)$ from a requirement that the universe is opaque to electromagnetic radiation along the future light cone. This requirement comes from the absorber theory of radiation discussed in section 8.8. We omit the details and quote the result.

This requirement fixes $\lambda(\tau) \propto \tau^{-1}$ and $n(\tau) \propto \tau$. It is then verified that the LNH is incorporated by the fact that the dimensionless number

$$\lambda^2(\tau^3 n)^{1/2} \equiv \text{constant} = 0(1). \tag{8.113}$$

A conformal transformation,

$$ds_E = \Omega_E \, ds_M, \qquad \Omega_E \propto \tau^2, \tag{8.114}$$

then takes us to the gravitational framework in which $G_E = $ constant. Also, the gravitational mass of an astronomical body remains constant. Thus, as in Dirac's multiplicative creation theory, the local solar system tests give the same answer as in relativity.

To transform to atomic framework we need another conformal transformation:

$$ds_A = \Omega_A \, ds_M, \qquad \Omega_A \propto \tau, \tag{8.115}$$

By writing $t \propto \tau^2$ the line element now becomes

$$ds_A^2 = dt^2 - 2H_0 t[dr^2 + r^2(d\theta^2 + \sin^2 \theta \, d\phi^2)]. \tag{8.116}$$

In this framework the gravitational constant varies as

$$G_A \propto t^{-1}. \tag{8.117}$$

There is therefore considerable similarity between this theory and the model proposed by Dirac a few years later.

8.13 Conclusion

This brings us to the end of our brief excursion through some of the better-known parts of alternative cosmologies. Our survey is by no means exhaustive. We have not discussed such important models as the matter–antimatter symmetric cosmology of Alfven and Klein, the Einstein–Cartan cosmologies, or Milne's kinematic relativity; nor have we discussed such unusual ideas as Segal's chronometric cosmology or McCrea's notion of cosmological uncertainty.

Our purpose here was to summarize a few nonstandard cosmologies, in order to show that 'respectable' cosmology has not been confined to the standard Friedmann models, to the ideas outlined in the earlier chapters. To what extent do the theoretical ideas (standard or nonstandard) presented so far in this book stand up to observations available today? We consider this question next.

Exercises

1   Discuss the considerations that led to the formulation of the steady state cosmology. Are any of these considerations valid today?

2   What is the perfect cosmological principle? What shortcoming of the ordinary cosmological principle is it designed to remove?

3   By considering various astronomical objects, arrive at a length scale over which you would expect the PCP to apply. What are the corresponding time scales over which you would expect the universe to obey this principle?

4   Compute the scalar curvature R for the Robertson–Walker model. Use the PCP to demand that R is constant and show that this leads to S as specific functions of t for $k = 0, \pm 1$. Use the constancy of Hubble's constant to deduce that $k = 0$.

5   Show that the deceleration parameter for the steady state universe is equal to -1 at all epochs.

6   Deduce from the PCP and the local observation of a departure from thermodynamic equilibrium that the steady state universe must expand.

7   Show that if the steady state universe has a proper number density n of sources each radiating with luminosity L, then the total intensity in a solid angle $d\Omega$ of the sky is given by $F\,d\Omega$, where

$$F = \frac{1}{16\pi} Ln\left(\frac{c}{H_0}\right).$$

Estimate F by substituting characteristic values of L, n for galaxies and deduce that the night sky is quite dark.

8   Discuss the validity of the following statement: 'Of the various

ways of resolving Olbers's paradox, the only way open to the steady state model is that of the expansion of the universe.'

9   Show that according to the C-field cosmology the rate of creation of matter needed to sustain the steady state model is given by

$$Q = 4 \times 10^{-46} h_{08}^3 g \; \text{cm}^{-3} \, \text{s}^{-1}.$$

10   Using the result of Exercise 9, express the creation rate in terms of solar masses per year per cubic megaparsec.

11   Discuss the merits and limitations of the C-field cosmology.

12   It is claimed that the steady state theory is readily testable and therefore more prone to an observational disproof than the standard big bang cosmology. Give examples to justify this claim.

13   Compare and contrast the C-field and the electromagnetic field.

14   Show with the example of the C-field that path-independent terms in the action can lead to nontrivial results.

15   Show that in the absence of any other forces, a created particle in the C-field cosmology follows a geodesic.

16   Show that in the C-field theory overall energy and momentum are conserved when a particle is created.

17   Derive the form of energy momentum tensor of the C-field by the variation of g_{ik} and setting the first-order variations of the action equal to zero. Show that

$$T^{ik}_{(C);k} = -fC^i C^k_{;k}.$$

Evaluate this relation near the spacetime point where a particle is created and deduce the law of conservation of the 4-momentum.

18   Discuss the physical implications of the negativity of the C-field energy.

19   Obtain the cosmological equations (8.12) through (8.14). Derive the general solution of these equations for the case $k = 0$.

20   Consider a perturbation of the steady state line element of the following form:

$$ds^2 = g_{ik} \, dx^i \, dx^k, \qquad x^0 = ct,$$
$$g_{00} = (1 + h_{00}), \qquad g_{0\mu} = 0, \qquad g_{\mu\nu} = -(\delta_{\mu\nu} + h_{\mu\nu})e^{2H_0 t},$$

where h_{ik} are general functions of spacetime coordinates. Further, take the density and the flow vector as

$$\rho = \frac{3H_0^2}{4\pi G} + \rho_1, \qquad u^i = (1, 0, 0, 0) + u_1^i.$$

Treating ρ_1, u_1^i, $h_{\mu\nu}$ as small quantities of the first order, show that they decay with time as

$$\rho_1 = Ae^{-3H_0t} + Be^{-5H_0t}, \qquad u_1^i = \bar{u}^ie^{-5H_0t},$$

$$h_{00} = 0, \qquad h_{\mu\nu} = \alpha_{\mu\nu} + \beta_{\mu\nu}e^{-2H_0t} + \gamma_{\mu\nu}e^{-3H_0t} + \varepsilon_{\mu\nu}e^{-5H_0t}.$$

The functions A, B, $\bar{u}^i$, $\alpha_{\mu\nu}$, $\ldots$, $\varepsilon_{\mu\nu}$ depend on x^μ. Prove that even the inhomogeneity corresponding to $\alpha_{\mu\nu}$ becomes less and less important as the universe expands.

21 Deduce that the flow vector of created matter has zero spin. What implication does this result have for Mach's principle?

22 Deduce the solution (8.16) for the explosive creation process. Relate the parameters of your solution to the amount of matter created in the universe. In particular, show that

$$t_1 = t_0,$$

and that the maximum density occurs at $t = 0$ and is given by fm^2.

23 Consider a reservoir of energy ξ in a volume V that expands. Show that if $\xi > 0$, expansion as well as conversion of energy to matter will reduce ξ to zero in a finite time. Show further that if $\xi < 0$ this conclusion is drastically altered.

24 Show that in steady state cosmology the redshift of a galaxy is proportional to its radial proper distance from us.

25 Show that a steady state cosmology does not have a particle horizon, but that it does have an event horizon of proper radius c/H_0. That is, show that a galaxy whose radial proper distance from us exceeds c/H_0 cannot ever communicate with us.

26 Estimate the difference in apparent bolometric magnitudes of a galaxy of redshift $z = 1$ computed according to the steady state model and the Friedmann model with $q_0 = 1$.

27 A family of radio sources with the same luminosity and with energy spectrum given by $\sim \nu^{-1}$ as a function of frequency ν are being counted in the steady state universe. Show that the flux density S varies with the source redshift z as

$$S \propto z^{-2}(1 + z)^{-2}.$$

Calculate the slope $d\log N/d\log S$ as a function of z where N is given by (8.23). Tabulate this function for $z = 10^{-n}$, $n = 4, 3, 2, 1, 0$. What do you conclude from this table?

28 In a universe with the line element

$$ds^2 = c^2\,dt^2 - S^2(t)[dr^2 + r^2(d\theta^2 + \sin^2\theta\,d\phi^2)],$$

$Q(t, \tau)\,d\tau$ denotes the proper number density of galaxies at epoch

t with ages between τ and $\tau + d\tau$. Suppose that $\eta(t)$ denotes the rate (per unit proper volume) at which new galaxies are being injected into the universe. Show that $Q(t, \tau)$ satisfies the differential equation

$$\frac{\partial Q}{\partial t} + \frac{\partial Q}{\partial \tau} + 3 \frac{\dot{S}}{S} Q = \eta(t)\delta(\tau).$$

Deduce from this equation the age distribution of galaxies in the steady state universe.

29   Look up the rest mass energies of the neutron and the proton in the Table of Constants at the end of this book. Assuming that ~20 per cent of this energy difference is acquired by the electron in the β-decay of the neutron, estimate the velocity and the kinetic temperature of the electron.

30   For a density of hydrogen atoms of 2×10^{-5} cm^{-3} and a kinetic temperature of 10^9 K, estimate the velocity of sound. Equate this to the expansion velocity $H_0 D$ according to Hubble's law and estimate the distance D of the irregularity that would develop in the Gold–Hoyle hot universe where thermal pressures are pitted against the force of universal expansion.

31   Compare the bubble universe with the inflationary models.

32   Discuss the bubble universe. Do you see any similarity between the way the hot universe generates spatial inhomogeneities and the way the bubble universe generates temporal unsteadiness? What happens to the PCP in these models?

33   An expanding bubble may be considered as a cloud of gas moving radially outwards. In a uniform spherical bubble with mass $M(r)$ within radius r the expansion is given by

$$\dot{r}^2 = \frac{2GM(r)}{r}.$$

Suppose next that a supermassive object of mass μ appears at the origin when $r = r_0$. Show that the cloud now expands to a maximum radius given by

$$r_{max} = \left(1 + \frac{M}{\mu}\right)r_0.$$

In the bubble universe theory this idea serves as a basis for forming an elliptical galaxy as a cloud of gas out to r_{max} which is gravitationally controlled by a supermassive object at the galactic nucleus.

34 Assuming that our Galaxy has been radiating at the rate of 4×10^{43} erg/s for a time 3×10^{17} s and that this energy is derived from a conversion of hydrogen to helium, estimate how much helium is formed in this way. (Energy of 6×10^{18} erg g^{-1} is released when hydrogen is converted to helium.) Comment on this answer in relation to the primordial mass fraction of helium obtained in Chapter 5.

35 Describe the observations that still need to be explained if the steady state cosmology is to stage a comeback as a viable cosmological model.

36 Discuss how inertial forces arise in Newtonian dynamics. A stone tied to a string is whirled around in a circle. How can the motion of the stone be understood in terms of inertial forces?

37 What observation led Mach to formulate his famous principle?

38 Why is it unsatisfactory to conclude from $mf = 0$ for a particle in an otherwise empty universe that $f = 0$? Interpret any other conclusion that could be drawn from the above equation.

39 Set up the problem corresponding to that described in Exercise 38 in general relativity. Does this theory provide a satisfactory solution to the problem?

40 Construct a mass unit from the fundamental constants c, $\hbar$, and G that could be used as a standard to decide whether particle masses change with epoch. Under what circumstances can we assert that G is changing with epoch?

41 Give the qualitative argument of Brans and Dicke leading to the conclusion that G^{-1} satisfies a scalar wave equation with sources in matter.

42 Derive the field equations of Brans–Dicke theory from an action principle. Why is the theory called a scalar–tensor theory?

43 Show that in the approximation $\omega \gg 1$, the wave equation satisfied by ϕ gives a solution

$$\phi = \text{constant} + O\left(\frac{1}{\omega}\right).$$

Interpreting the constant as proportional to G^{-1}, show that the Brans–Dicke field equations take the form

$$R_{ik} - \tfrac{1}{2}g_{ik}R = -\frac{8\pi G}{c^4}T_{ik} + O\left(\frac{1}{\omega}\right).$$

44 In Newtonian gravity an oblate Sun will generate a gravitational potential

$$\phi = \frac{GM_\odot}{r}\left[1 - J\left(\frac{R_\odot}{r}\right)^2 P_2(\cos\theta)\right],$$

where J is the quadrupole moment parameter and P_2 is the second Legendre polynomial. Show that the orbit of a planet precesses because of the above gravitational effect at the rate $3\pi R_\odot^2 J/l^2$, while l is the semi latus rectum of the orbit. Estimate the precession rate for Mercury for $J = 2.5 \times 10^{-5}$. What significance does this calculation have for the Brans–Dicke theory?

45   Discuss the Solar System tests of the Brans–Dicke theory.

46   Calculate the age of a Brans–Dicke universe for the simplest case $C = 0$, $p = 0$, $k = 0$. Does this model have a greater or a smaller age than the corresponding relativistic model?

47   Show that for a radiation universe in Brans–Dicke cosmology with $C = 0$, we have $S \propto t^{1/2}$ and $\phi = $ constant. Comment on why this case gives exactly the same answer as relativistic cosmology.

48   Show that the inequality (8.46) is satisfied for a dust universe as well as for radiation universe in Brans–Dicke cosmology with $C \neq 0$.

49   Derive the behaviour of S and ϕ as functions of t in the early ϕ-dominated Brans–Dicke universe.

50   Discuss primordial nucleosynthesis in the Brans–Dicke cosmology.

51   The Brans–Dicke theory can be re-expressed as a theory in which $G = $ constant but the particle masses change with epoch. Show that this is achieved by a conformal transformation

$$\bar{g}_{ik} = \frac{\phi}{\bar{\phi}}\, g_{ik}, \qquad \bar{\phi} = \text{constant}.$$

The field equations then become (in the new metric)

$$\bar{R}_{ik} - \tfrac{1}{2}\bar{g}_{ik}R = -\kappa\bar{T}_{ik},$$

where κ is constant. Although these look like Einsteins's equations, the $\bar{T}_{ik}$ contain ϕ and its derivatives. Show from the new field equations that

$$\bar{\square} \ln\phi = \frac{8\pi G}{(2\omega + 3)c^4}\,\bar{T}$$

with $G = $ constant. The form of the theory was obtained by Dicke in 1962. The particle masses in this version vary as

$$\bar{m} = m(\bar{\phi}/\phi)^{1/2}, \qquad m = \text{constant}.$$

52   Show that in a ϕ-dominated Brans–Dicke cosmology it is possible to have an increasing gravitational constant at an epoch t, provided

$$\int_0^t (\varepsilon - 3p)S^3 \, dt < -c^2 \left(\frac{2\omega + 3}{8\pi} \right) C.$$

53 Illustrate qualitatively the difference between a field theory and an action-at-a-distance theory by an example from electrodynamics.

54 Verify that in Minkowski spacetime the electromagnetic Green's function has the simple form

$$G_{ik} = \frac{1}{4\pi} \delta(s^2) \eta_{ik},$$

where s^2 is the invariant square of the distance between the two world points at which G_{ik} is defined.

55 Use the Green's function of Exercise 54 to drive the potential for a static electric charge.

56 Show how the definition of mass in the HN theory satisfies Mach's principle.

57 Show by a time transformation that Robertson–Walker spacetime with $k = 0$ is conformal to flat (Minkowski) spacetime.

58 Show with the help of the following series of transformations that the $k = +1$ Robertson–Walker spacetime is conformally flat:

$$r = \sin R, \qquad T = \int^t \frac{du}{S(u)}, \qquad c = 1,$$

$$\xi = \tfrac{1}{2}(T + R), \qquad \eta = \tfrac{1}{2}(T - R),$$

$$\tau = \tfrac{1}{2}(\tan \xi + \tan \eta), \qquad \rho = \tfrac{1}{2}(\tan \xi - \tan \eta).$$

What are the corresponding series of transformations to show that the $k = -1$ Robertson–Walker models are also conformally flat?

59 Show that the following tensor is conformally invariant:

$$C^h_{ijk} = R^h_{ijk} + \tfrac{1}{2}(g^h_j R_{ik} - g^h_k R_{ij} + g_{ik} R^h_j - g_{ij} R^h_k)$$
$$+ \tfrac{1}{6} R(g^h_k g_{ij} - g^h_j g_{ik}).$$

This tensor is known as the *Weyl conformal curvature tensor*.

60 Show that a null geodesic is invariant under conformal transformations.

61 Explain why conformal invariance should play an important role in action-at-a-distance theories.

62 Show that Maxwell's equations remain unchanged under a conformal transformation provided the potential and the field transform as

$$\bar{A}_i = A_i + \psi_{;i}, \qquad \psi \text{ a suitable scalar}, \qquad \bar{F}_{ik} = F_{ik}.$$

63   Verify by direct substitution that $\widetilde{G}(A, B)$ defined by (8.63) does satisfy the conformal transform of (8.65).

64   Use the conformal flatness of the Einstein–de Sitter model to calculate the explicit form of $\widetilde{G}(A, B)$ in that universe.

65   Suppose a symmetric Green's function $G(A, B)$ satisfies the wave equation

$$\Box_X G(X, B) = [-g(X)]^{-1/2}\delta_4(X, B).$$

Show that a small variation of the metric tensor in a region $\mathcal{V}$ produces a small variation of $G(A, B)$ given by

$$\delta G(A, B) = \int_{\mathcal{V}} \delta(-g^{1/2}g^{ik})G^R(A, X),_i G^A(X, B),_k \, d^4x.$$

(Note that A and B need not lie in $\mathcal{V}$.)

66   Show that the action (8.74) leads to the field equation (8.73).

67   Compare the degree of underdeterminacy of the gravitational equations of the HN theory with that of general relativity.

68   Show that any conformally invariant action leads to an energy tensor of vanishing trace.

69   Discuss the aspects in which the HN theory of gravity differs from general relativity.

70   Construct dimensionless constants from

$$\text{(a) } e, \hbar, c; \quad \text{(b) } G, m_p, \hbar, c; \quad \text{(c) } G, m_p, c, H_0.$$

71   Which of the dimensionless constants of Exercise 70 are very large or very small?

72   Compute N exactly for the closed Friedmann model with $h_0 = 1$, $q_0 = 1$. Show that N is constant at all epochs. Can this result be reconciled with LNH?

73   Find the relation connecting the three large numbers in (8.85), (8.86), and (8.87).

74   Deduce from the LNH that the gravitational constant must decrease with epoch at a rate (of fractional decrease) of the order of Hubble's constant.

75   Give the arguments that led Dirac to postulate particle creation in the universe.

76   Show that in gravitational units multiplicative creation demands the particle masses decrease with time t as t^{-2}.

77   In what way does the difference between additive and multiplicative creation show up in the long-term evolution of planetary orbits? How are the orbital angular speeds of the planets affected by the variation of G?

78   Give the arguments based on LNH that lead to the conclusion that the scale factor of the expanding universe can be proportional only to cosmic time. Comment on the plausibility of these arguments and compare them with Bondi and Gold's derivation of the steady state line element based on the perfect cosmological principle.

79   Derive the formula for redshift in the Dirac universe with multiplicative creation. Explain how this redshift arises even though the gravitational metric is static.

80   Plot the atomic time t_A against the gravitational time t_E for the Dirac universe with multiplicative creation. Show that although the Einsteinlike universe in the gravitational metric has t_E going to $-\infty$, the atomic time goes only as far back as $t_A = 0$.

81   Compare and contrast Dirac's cosmological ideas on creation of negative as well as positive mass with (a) Dirac's ideas about vacuum as a sea of undetectable negative-energy electrons and (b) the cosmological reservoir of negative energy C-field in the steady state cosmology.

82   Show that in Dirac's cosmology with no particle creation the gravitational constant decreases as

$$\frac{\dot{G}}{G} = -3H.$$

Estimate this rate in terms of the present estimate of the Hubble constant. How is this rate modified in the Dirac models with particle creation?

9

Local observations of cosmological significance

9.1 Introduction

In this chapter we will review those astronomical observations that attempt to determine the large-scale structure of the universe from relatively local surveys. These tests do not tell us about the geometrical structure of the universe, since they do not extend far enough. Nevertheless, we shall see how even local measurements place restrictions on what can be said about the distant parts of the universe. This may sound paradoxical, but it is a consequence of the symmetry assumptions made by most models of the universe, in particular the cosmological principle. These models provide the background against which to assess the observed data. Should an inconsistency develop in a particular model then either the model is wrong or the data are imperfect (or both!). We will encounter examples of both kinds of difficulties in the tests to be discussed here.

Briefly these tests are as follows:

1. The measurement of Hubble's constant.
2. The anisotropy of large-scale velocity field in our local neighbourhood.
3. The distribution and density of matter in our local neighbourhood.
4. The age of the universe and of the various objects in it.
5. The abundance of light nuclei.
6. The evidence for antimatter in the universe.
7. The microwave radiation background in our local neighbourhood.

Since our main purpose is to compare theoretical predictions with observations, we will refer to earlier chapters in the following discussion. We will refer not only to standard cosmology but also to the various nonstandard cosmologies discussed in Chapter 8.

A survey of observational cosmology today reveals a number of issues on which there are disagreements among different observers and theoreticians. Sometimes the more important points of physical significance get buried under heaps of numerical data. In some cases new data have replaced old data, so that fresh interpretation becomes necessary. The approach adopted in this text emphasizes the significant issues that the observations are supposed to reveal rather than the many controversial numerical details. While every attempt is made to present 'up to date' data, newer observations than those discussed here are bound to arise in the course of time.

9.2 The measurement of Hubble's constant

Modern cosmology began with Hubble's observations, which were referred to in Chapter 1. Hubble obtained a value of $h_0 \sim 5.3$ from his original observations, whereas present-day observations suggest that h_0 lies in the range $0.5 \leqslant h_0 \leqslant 1$. The reader may wonder not only at such a drastic change in h_0 over the last six decades, but also at the fact that even today considerable uncertainty exists about the true value of this important parameter of modern cosmology. This section attempts to clarify the situation.

To begin with, let us recall that the Hubble constant H_0 relates the redshift z of a nearby galaxy to its distance D from us:

$$cz = H_0 D. \tag{9.1}$$

Therefore if we measure z and D for a number of galaxies (as Hubble did), we should be able to estimate H_0. The observations measure z fairly accurately. The difficulties arise in estimating D. The large value obtained by Hubble was due to the fact that he grossly underestimated the distance of the galaxies in his survey.

Figure 9.1 shows, for example the original relation of Hubble alongside the plot of the same extragalactic objects with modern revised distance estimates. Readers may draw their own conclusions as to whether Hubble would have got a linear relation with the revised data.

How does an astronomer measure distances of galaxies? We will outline below the methods available to him, all of which follow the philosophy outline by van den Bergh in 1975: 'All determinations of the extragalactic distance scale are ultimately based on the assumption that recognizable types of distant objects are similar to nearby objects of the same type.' We will see how this philosophy operates in practice.

Before we begin it is useful to introduce the concept of a *distance*

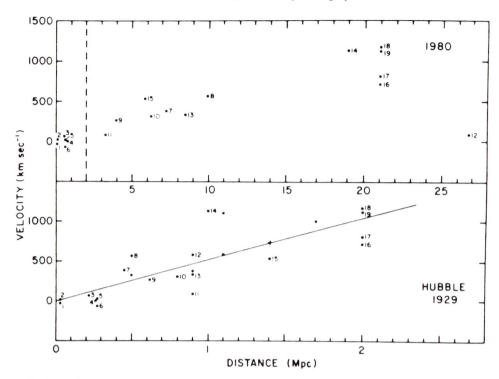

Fig. 9.1 The Hubble plot by Hubble side by side with the modern plot with revised distances of the same objects. (By courtesy of A. Hewitt).

modulus, which is familiar to the stellar astronomer. Recall that for an object of luminosity L at a distance D from us, the apparent and absolute magnitudes are defined by the formulae

$$m = -2.5 \log \frac{L}{4\pi D^2} + \text{constant}, \tag{9.2}$$

$$M = -2.5 \log L + \text{constant}. \tag{9.3}$$

The constant in (9.2) is fixed by assigning a given magnitude $m = 0$ to an object with $L/4\pi D^2 = 2.52 \times 10^{-5} \, \text{erg cm}^{-2} \, \text{s}^{-1}$. The constant in (9.3) is fixed by defining M as the apparent magnitude of an object if it was viewed from a distance of 10 pc. Hence if D is measured in parsecs, (9.2) and (9.3) give

$$m - M = 5 \log D_{\text{pc}} - 5. \tag{9.4}$$

The stellar astronomer usually measures distances in parsecs. Hence the above relationship is convenient to him. The cosmologist, on the other hand, measures distances in megaparsecs. For him the convenient form of (9.4) is therefore

$$m - M = 5 \log D_{\text{Mpc}} + 25 = \mu. \tag{9.5}$$

μ is called the distance modulus.

If we substitute the Hubble relation (9.1) in (9.5) with $H_0 = 100 \, h_0 \, \text{km s}^{-1} \, \text{Mpc}^{-1}$ and also subsitute the value of $c = 2.997\,929 \times 10^5 \, \text{km s}^{-1}$, we arrive at the following relation for the Hubble law:

$$5 \log h_0 = 42.38 + (M - m) + 5 \log z$$
$$= 42.38 + \mu + 5 \log z. \tag{9.6}$$

It is therefore necessary to determine μ and z for a galaxy in order to estimate h_0.

9.2.1 Galatic extinction

The above definitions do not take into account an important correction arising from the fact that we are looking at any other galaxy through our own. Thus the flux of light from outside our Galaxy is liable to be partially reduced by absorption and scattering within our Galaxy. The extinction suffered by this light will depend on the column density; that is, on the distance travelled by the light through our Galaxy and the density of absorbing and scattering agents on the way. How much allowance should be made for this effect? Clearly, the true luminosity of the observed galaxy must be higher and its true absolute magnitude lower than the corresponding values estimated without taking this correction into account. Accordingly, if we wish to use the above formulae then the estimate of M must be reduced by an extinction function A.

Alternatively, if we know the true value of M for a distant galaxy, then before calculating its distance modulus we must reduce its measured apparent magnitude by A.

Observers are not unanimous on the value of A. A. Sandage and G. Tammann use the following extinction law for blue magnitudes:

$$A = 0 \text{ for } |b| > 50°,$$
$$A = 0.13(|\text{cosec } b| - 1) \text{ for } |b| \leq 40°, \tag{9.7}$$

while G. de Vaucouleurs uses a uniform cosecant law

$$a = 0.20(|\text{cosec } b| - 1) \tag{9.8}$$

for all galactic latitudes b. Figure 9.2 illustrates the galactic models underlying these formulae. Already, it is clear that different corrections for extinction are liable to lead to different answers for h_0.

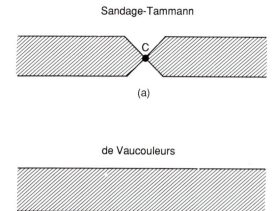

Fig. 9.2 The Galaxy models assumed by (a) Sandage and Tammann and (b) de Vaucouleurs to compute the extinction of visual light from outside our Galaxy. In the former the absence of a shaded region at high latitude describes the assumption that there is no extinction for $|b| > 50°$.

9.2.2 Measurements of extragalactic distance

The distances of planets and satellites within the Solar System are accurately measured with the help of trigonometry and Kepler's laws. The distances of stars up to ~ 25 to $50\,\mathrm{pc}$ can be measured with the help of trigonometric parallax. Going still further, a more reliable method is that based on the Hyades main sequence. A comparison of the main sequence of the Hyades cluster with the main sequences of more remote clusters in our Galaxy enables us to measure distances of stars in these clusters. These methods, however, do not work beyond our Galaxy. New techniques are needed for measurements of extragalactic distances. We discuss some more of them below.

Cepheid variables

Cepheid variables are a group of stars whose luminosity varies by about 10 per cent, but with a great deal of regularity. One can associate a period P for one cycle of variation of each Cepheid variable. Though the first of these variable stars, the star known as δ Cephei, was discovered as early as 1784 by John Goodricke, the crucial property that made the Cepheids so useful for extragalactic distance measurement was discovered in 1912 by Henrietta Leavitt. This property is a unique relationship between P and the luminosity L of the star. Figure 9.3 illustrates this relationship.

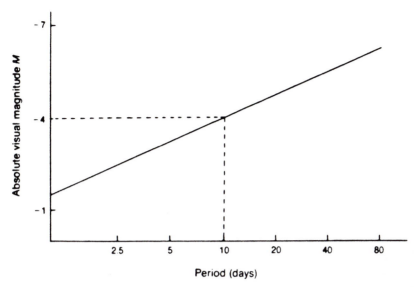

Fig. 9.3 Luminosity plotted on a logarithmic scale against period for a number of Cepheid variables. The straight line illustrates the fact that their luminosity increases with period, which enables us to calculate the luminosity of a distant Cepheid by measuring its period. Thus in the above figure a Cepheid with period of 10 days will have absolute magnitude $M = -4$. (Based on H. Arp, Southern hemisphere photometry VIII: Cepheids in the small Magellanic Cloud, *A. J.*, **65**, 404.)

Because Cepheids are bright and variable, they can be detected in nearby galaxies with relative ease. Thus if we detect a Cepheid in a galaxy and measure its period, we can accurately estimate its luminosity L and hence its absolute magnitude M. Then (9.5) gives its distance modulus and hence the distance modulus of the galaxy in which it is located.

It was with the help of Cepheids that Hubble established the fact that galaxies exist outside our own. His early work leading to the discovery of the expansion of the universe was also based on Cepheids.

This method takes us to distances $\sim 10\,\text{Mpc}$; that is, to galaxies in our local neighbourhood.

Brightest star

This method of measuring distant galaxies makes use of the assumption that in similar spiral Sc galaxies of comparable luminosities, the brightest stars also have comparable luminosities. Since, as Hubble found in galaxies M31 and M33, the brightest stars are significantly brighter than the brightest Cepheids, this method takes us as far as the Virgo cluster of galaxies; that is, to distances of ~ 10 to $15\,\text{Mpc}$.

H II regions

H II regions are large domains of ionized hydrogen. These are found not only in our Galaxy but also in others. The linear diameter of the largest H II region, or better still the mean linear size of the three largest H II regions, shows a strong variation with the luminosity and the luminosity class (see section 1.3) of the parent galaxy. For dwarf galaxies this mean size is as low as 75 pc, while for supergiant galaxies the size goes up to 460 pc. By comparing the angular sizes of such H II regions in remote and nearby galaxies of similar type, we can estimate the ratio of their distances. Then, if the distance of the nearby galaxy is known, the distance of the remote galaxy can be estimated. Note that this method, unlike the others so far mentioned, relies on the size rather than the luminosity of the distance indicator.

Supernovae

A new technique based on supernovae has recently shown promise of reliablility and does not require many arbitrary assumptions. Basically, this method involves determining the actual flux of light at different frequencies leaving the photosphere of the exploding star, and it does not depend on other step-by-step methods of distance determination. The method consists of measurements of the rate of expansion of the photosphere of the supernova, and it makes use of a variant of a method used by W. Baade in 1926 for variable stars.

If we approximate the photosphere (see Figure 9.4) by a blackbody of temperature T and radius R, and suppose that its distance from us is D, then its angular size θ and the flux density $f(v)$ at frequency v are given by

$$\theta = \frac{R}{D}, \tag{9.9}$$

$$f(v) = \frac{R^2}{D^2} \cdot \frac{2\pi h v^3}{c^2(e^{hv/kT} - 1)}. \tag{9.10}$$

(Here the redshift has been ignored.)

Thus we get from (9.9) and (9.10)

$$\theta = \left[\frac{f(v)c^2(e^{hv/kT} - 1)}{2\pi h v^3} \right]^{1/2}. \tag{9.11}$$

Hence if we measure $f(v)$ and T we can get θ. Further, if we measure R, we get D from (9.9). Spectral scans of the continuum spectrum of the

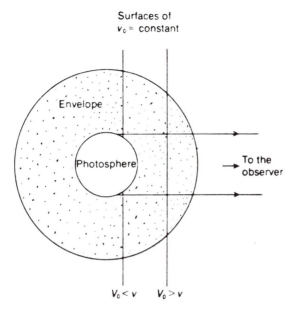

Surfaces of
V_0 = constant

Envelope

Photosphere

To the
observer

$V_0 < v$ $V_0 > v$

Fig. 9.4 The shaded region is the expanding envelope of scattering particles around the photosphere. The surfaces of constant velocity V_0 of the particles in the envelope relative to the observer are shown (in idealized condition) as planes. In the plane on the left V_0 is less than the photospheric velocity v, while in the plane on the right $V_0 > v$. The switch over from $V_0 < v$ to $V_0 > v$ can be related to the extent of scattering produced by the particles in the envelope and is seen in the line profiles of the supernova in the form of varying depletion. A study of the line profiles enables the astronomer to fix the value of v.

supernova give a good estimate of T. To measure R, R. P. Kirshner and J. Kwan suggested the following method.

In this method we approximate the rate of expansion of the photosphere by a constant value v, so that if the expansion started at $t = t_0$ when the radius was $R = R_0$ then the radius at subsequent times is given by

$$R = v(t - t_0) + R_0 \qquad (9.12)$$

(constancy of v is justified by the fact that pressure in the interstellar medium is negligible and the expansion is nearly free). The photosphere is surrounded by a tenous atmosphere whose atoms scatter the photospheric radiation. As explained in Figure 9.4, observation of the line profiles enable us to measure v, the photospheric expansion velocity. This is because in the expanding atmosphere some scattering atoms are moving faster towards the observer and some slower than the rate at which the photosphere is moving towards him. Thus there is a small Doppler effect in the scattering process, and this affects the absorption line profiles.

The photospheric velocity v corresponds to that velocity at which the depletion of the continuum is maximum, if it is sharp and well defined. If it is not sharp but has a flat trough, then the red edge of the depletion trough represents the photospheric velocity.

This process is claimed to be relatively unambiguous and free from the uncertainties surounding other methods that broadly require 'similar looking' objects to have 'equal' physical parameters such as luminosity and size. However, the method requires a good series of supernova observations, which may not be readily available in the galaxy whose distance we want to measure.

H I line profiles of spirals

R. B. Tully and J. R. Fisher found a good correlation between the luminosity of a spiral galaxy and its 21-cm line width, a correlation that does not depend on galaxy type. Thus in principle, if we determine the 21-cm line profile of a remote spiral we can estimate its luminosity. However, the line width is best determined for spirals viewed edge-on, but for these the internal absorption in the galaxy is large. Thus the observer is forced to use those spirals that are viewed at an inclined angle and yet give a reasonably reliable line width. The distances of the M81 and M101 groups have been estimated this way after using data on the nearer galaxies M31 and M33 for calibration.

Brightest galaxy

If we consider the thousand-odd galaxies in the Virgo Cluster, one galaxy, M87, stands out as being significantly brighter, more massive, larger than the rest. It is an elliptical galaxy. A. Sandage noted that other, more distant clusters of galaxies also contain similar dominating elliptical galaxies. On the assumption (supported by observations of nearby clusters) that such ellipticals have comparable luminosity, we can estimate M and hence the distance modulus of clusters as remote as 1000 Mpc.

9.2.3 The Hubble constant

The above methods are some of the many used in obtaining extragalatic distance estimates. These and some others have been visually summarized in an 'Eiffel Tower' constructed by de Vaucouleurs, shown in Figure 9.5. Notice that the distances are determined in progression from one stage to the next. At each stage there is scope for errors of calibration. For example, even a revision of the stellar distance scale, such as that of the

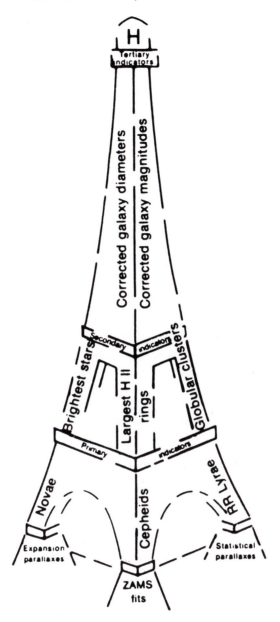

Fig. 9.5 The Eiffel Tower shown here describes how cosmological distances are measured in stages. The different levels used for calibration are shown starting from the nearest at the first level and leading to the furthest at the top. ZAMS stands for 'zero age main sequence', which refers to the method of measuring distances using the Hyades main sequence mentioned in the text. (Based in C. Balkowski & B. E. Westerlund, eds, *Proceedings of the IAU–CNRS Colloquium*, held in Paris, 6–9 September, 1976 (Paris: CNRS, 1977).)

Hyades main sequence in our Galaxy, will lead to revision of all subsequent scales. Such systematic errors were present in Hubble's original method, and when pooled together they gave a value of $h_0 \geqslant 5$. For example, the Cepheid period–luminosity relation available to Hubble was incorrect. He also used too faint an absolute magnitude for the brightest star in other galaxies. From (9.6) we see that a value of M will lead to a high value of h_0.

Another major source of uncertainty comes in distinguishing the 'true' Hubble flow from the peculiar motions caused by other relatively local inhomogeneities. In the following section we highlight this problem.

For these reasons, it would be wise on the part of the cosmologist to be cautious about the exact value of Hubble's constant. Sandage and Tammann prefer a value of $h_0 \sim 0.5$, while de Vaucouleurs prefers $h_0 \sim 1.0$. In view of the prevailing uncertainties of various distance indicators, it is customary nowadays to say that h_0 lies between these two limits. Unless a dramatic breakthrough in extragalactic obsevations (such as that promised by the Hubble Space Telescope) occurs, we have to live with the conclusion that the best guess of h_0 is in the range

$$0.5 \leqslant h_0 \leqslant 1.0. \tag{9.13}$$

9.3 The anisotropy of local large-scale velocity fields

The simple picture of a homogeneous and isotropic universe based on the Robertson–Walker line element is now beginning to look oversimplified, especially with the discovery of large-scale velocities that appear superimposed on the Hubble flow. Since by definition the Hubble flow is small in our 'local' neighbourhood, it tends to get swamped by these other velocities. It is an immensely complicated problem to untangle the two and to pinpoint the causes of the non-Hubble velocities. Here we will give the bare outline of the situation on theoretical and observational fronts, beginning with the latter.

It was in the mid-seventies that work by V. C. Rubin, W. K. Ford, and others gave a glimpse of the problem. The so called Rubin–Ford effect showed that the Hubble constant is *not* isotropic when measuring the radical velocities of 184 Sc I and Sc II galaxies (I and II are the van den Burgh luminosity classes for spiral Sc galaxies, with Sc I being the brightest class of galaxies.) The anisotropy was of the dipole type and could be accounted for by the assumption that our Galaxy is moving with a substantial speed against the background of the galaxies. The speed was

(454 ± 125) km s^{-1} towards $l = 163° \pm 15°$, $b = -11° \pm 14°$

relative to the distant part of the sample and

$(474 \pm 164$ km s^{-1} towards $l = 167° \pm 20°$, $b = 5° \pm 20°$

relative to the nearer part.

This was the first indication that the Galaxy is not in the cosmological rest frame.

9.3.1 The local distribution

The *Nearby Galaxies Atlas* published by Tully and Fischer contains detailed maps of the distributions and speeds of galaxies in the relatively local region. These maps are helpful in constructing the topography of the nearby region. Figure 9.6 gives a schematic plot of the distribution over a cubical region around our Galaxy, with each side of the cube measuring approximately a speed differential of 10 000 km s^{-1}. (That is, if H is the Hubble constant, the linear size is approximately H^{-1} times this value. We may find it convenient to use speeds as distances in this way.)

The shaded region of the cube is the galactic zone of avoidance which is perpendicular to the supergalactic plane (see Chapter 1). One may consider the motions of these objects as made up of several components:

1. The flow towards the Great Attractor located at a distance of ~ 4200 km s^{-1} from the local group. The GA is located approximately at $l = 309°$ and $b = +18°$ (galactic coordinates). The two Centaurus clusters are, for example, falling into the GA with speeds ~ 1000 km s^{-1} (away from us).
2. The infall of matter towards the Virgo cluster.
3. The 'Local Anomaly' which appears to require a bulk velocity correction of 360 km s^{-1} for a region extending from the Local Group out to distances of 700 km s^{-1}.
4. The Hubble flow.

This multicomponent model had several parameters which can be determined by the least-square technique, by using the Tully–Fisher relation for spirals to measure their distances (see section 9.2) and the redshifts for radial velocities. The model determines velocities which are compared with the observed values and the differences are minimized by the least-square method. This technique was first used in 1988 by the 'Seven Samurai', the authors D. Lynden–Bell, S. M. Faber, D. Burnstein, R. L. Davies, A. Dressler, R. Terlevich, and G. Wegner. The broad conclusions are as follows.

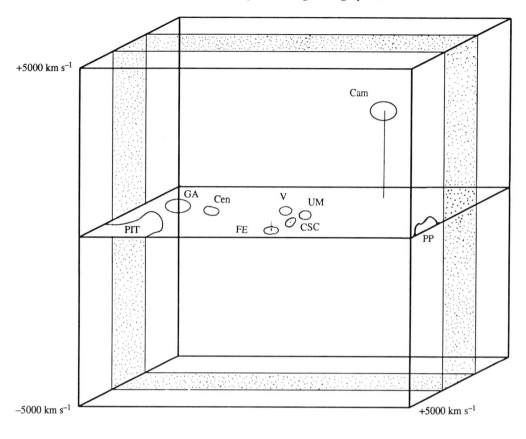

+5000 km s⁻¹

−5000 km s⁻¹

+5000 km s⁻¹

Zone of avoidance due to galactic plane

Fig. 9.6 A cubical volume containing some significant large-scale structures in our neighbourhood. Shown in the cube are GA: Great Attractor; V: Virgo Cluster; CSC: Coma–Sculptor Cloud containing the Local Group; UM: Ursa Major Cluster; Cen: Centaurus; FE: Fornax–Eridanus; Cam: Camelopardalis; PP: Perseus–Pisces; PIT: Pavo–Induiis–Telescopium. (After S. M. Faber and David Burstein, Motions of galaxies in the neighbourhood of the Local Group. In V. C. Rubin & G. V. Coyne, SJ, eds, Proceedings of the Vatican Study Week on 'Large-Scale Motions in the Universe', Princeton, 1988, p. 118.)

The GA is a large mass attracting matter towards it, causing large-scale streaming motion in its direction. On a smaller and nearer scale, the Virgo Cluster has neighbouring galaxies falling towards its centre, including the Local Group. However, the Local Group has a further anomalous motion relative to the Virgocentric flow. It is perhaps too early to take all the numerical estimates of speeds and direction as very accurate. More observations in the future will certainly help in making these estimates more reliable.

There is, however, considerable discussion (at the time of writing this book) as to whether the GA exists at all. For example D. A. Mathewson, V. L. Ford and M. Buchhorn have measured the peculiar velocities of 1355 spiral galaxies in the southern sky and used the Tully–Fischer relation to estimate their distances. They find no backside infall into the GA region, rather they find a bulk flow of about 400 km s^{-1} on scales of $100h_0^{-1}$ Mpc. Thus there is considerably doubt about the existence of an attracting mass there.

An independant piece of information that we will consider in section 9.8 is the motion of our Galaxy with respect to the rest frame of the cosmic microwave background.

9.3.2 The Hubble constant revisited.

We return to the question as to why the controversy over the value of H_0 (i.e., whether $h_0 \approx 1$ or $h_0 \approx 0.5$) persists. R. B. Tully has argued that the local velocity anomaly is the culprit confusing the issue. The argument may be illustrated by a simplified example.

Imagine a local mass concentration M superimposed on a Hubble flow. At a distance R from the mass, the radially outward velocity V may be given by

$$V = - \left(\frac{2GM}{R}\right)^{1/2} + HR \equiv H_{\text{eff}} R. \tag{9.14}$$

The first term is an inward velocity corresponding to a zero value at infinity, while the second term is the Hubble flow with the Hubble constant H. We may look upon (9.14) as a Hubble flow only, with an 'effective' Hubble constant

$$H_{\text{eff}} = H - \left(\frac{2GM}{R^3}\right)^{1/2}. \tag{9.15}$$

Thus the effective Hubble constant is *smaller* than the 'true' Hubble constant, *closer* to the mass concentration. As we go away from M, the effective Hubble constant approaches the true value.

Hence the possible presence of a mass concentration in the Coma–Sculptor Cloud that causes a local velocity anomaly couped with the Virgocentric flow manages to reduce the value of Hubble's constant for relatively nearby galaxies and makes h_0 closer to 0.5. However, more remote samples of galaxies tend to give $h_0 \approx 1$ which therefore corresponds to the true value of the Hubble constant.

Tully estimates that the local anomaly may be caused by a mass of the order $10^{14} M_\odot$ as compared with the $\sim 10^{15} M_\odot$ in the Virgo Cluster. By contrast, the Great Attractor mass may be as high as $5 \times 10^{16} M_\odot$. These values illustrate how important it is to chalk out the topography of the universe both in mass and in velocities before drawing firm conclusions about the values of the cosmological parameters.

We next consider the attempts to determine the mean density of matter in the universe, a parameter that has far-reaching consequences for the cosmological theories we have considered so far.

9.4 The distribution and density of matter in our neighbourhood

In Chapter 4 we introduced the density parameter Ω_0 through the relation

$$\rho_0 = \frac{3H_0^2}{8\pi G} \Omega_0 \equiv \rho_c \Omega_0, \tag{9.16}$$

where ρ_c is the present closure density in Friedmann cosmology. In numerical terms, (9.16) implies

$$\rho = 2 \times 10^{-29} (h_0^2 \Omega_0) \, \text{g cm}^{-3}. \tag{9.17}$$

Thus a direct measurement of ρ_0 is of interest, since it places limits on the parameters h_0 and Ω_0.

The present approach to the problem involves setting limits on the density of matter in the form of galaxies, clusters of galaxies, and so on; that is, matter in the standard luminous form. This is done as follows. Suppose we know the average mass/light ratio for galaxies, which is conventionally expressed in solar units:

$$\left\langle \frac{M_G}{L_G} \right\rangle = \eta \frac{M_\odot}{L_\odot}. \tag{9.18}$$

Next we determine the mean luminosity density l_G of galaxies. The best value of l_G comes from the Revised Shapley Ames Catalogue and is given by

$$l_{GS} \approx 4.4 \times 10^7 L_\odot h_0 \, \text{Mpc}^{-3} \quad \text{for spiral galaxies,}$$

$$l_{GE} \approx 17.4 \times 10^7 L_\odot h_0 \, \text{Mpc}^{-3} \quad \text{for E/SO galaxies.}$$

The total luminosity density is therefore of the order

$$l_G \approx 2.2 \times 10^8 L_\odot h_0 \, \text{Mpc}^{-3}. \tag{9.19}$$

From (9.18) and (9.19) we then get the mean cosmological density in the form of galaxies in our neighbourhood as

$$\rho_G \approx 2.2 \times 10^8 \eta M_\odot h_0 \text{ Mpc}^{-3}$$
$$\approx 1.5 \times 10^{-32} \eta h_0 \text{ g cm}^{-3} \qquad (9.20)$$

What is the estimate for η? The main difficulty in estimating η lies in the measurement of the galactic masses. By comparison, the measurement of luminosities is easy, the only uncertainty in the process arising from the lack of precision in H_0. We will briefly review the methods employed in the measurement of η for various types of objects before summarizing the results in Table 9.1.

9.4.1 Mass/light ratios

Methods of measuring η for individual galaxies and for clusters of galaxies are summarized below.

Spiral galaxies

The best handle on the mass contained in a typical spiral is given by its rotation curve. Figure 9.7 illustrates the principle by means of a flat, disc-shaped object representing a circular distribution of stars moving round a common centre C. The rotation velocity v of a star S at a distance

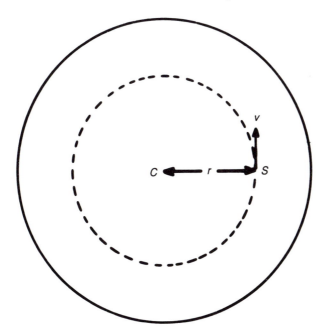

Fig. 9.7 Galactic disc approximated as a system of stars S moving in circular orbits round a common centre C. The velocity v of S is governed by Newton's laws of gravitation and motion.

r from C is related (in an equilibrium distribution) to the gravitational force F_r acting on S towards the centre:

$$m \frac{v^2}{r} = F_r \tag{9.21}$$

Therefore if we have v as a function of r, we get F_r as a function of r. Then, by Newton's law of gravitation (which is applicable here because the gravitational fields are weak), we can determine the mass distribution. For example, if most of the mass was concentrated in the nuclear region around C then we would have $F_r \propto r^{-2}$ and $v \propto r^{-1/2}$. The light distribution across a spiral galaxy does suggest the above to be a good approximation. However, in actual fact the rotation curve – the function $v(r)$ – is flat for most galaxies. That is, after rising sharply outside the nuclear region, v remains constant $= v_0$ (say). Moreover, this relation extends well beyond the visible disc.

The implications of this result are either that there is more mass in the outer parts of the galaxy than is indicated by its luminosity distribution, or that Newton's law of motion and the inverse square law of gravity might not be valid over the galactic distance range ($\sim$ few kpc.) Taking the former (and less radical) view, astronomers have estimated the masses of spirals. S. M. Faber and J. S. Gallagher have listed the rotation velocities and masses contained within the Holmberg radius (where the surface brightness drops to $\sim 26.5\, m_{pg}$ (arc second)$^{-2}$) for 39 spirals. Since the luminosities are also known, we can estimate the mean value of η for this sample. The result is

$$\eta \approx (9 \pm 1) h_0.$$

Elliptical galaxies

These galaxies show hardly any rotation, hence the rotation curve technique employed for spirals fails here. Instead, the mass estimates are based on the variation of the velocity dispersion σ across the galaxy.

In the spherical mass approximation the star distribution function in statistical equilibrium attains the form

$$f \propto \exp - \left[\frac{v^2/2 + \phi(r)}{\sigma^2} \right], \tag{9.22}$$

where $\phi(r)$ is the gravitational potential and $\langle v^2 \rangle = 3\sigma^2$. Assuming that the number density of stars varies as $r^{-\varepsilon}$, the above relation and the Poisson equation give the mass interior to radius r as

$$M\,(< r) = \frac{\varepsilon \sigma^2 r}{G}.$$ (9.23)

If the luminosity density $j(r)$ varies as the number density, the mass/luminosity ratio varies as $r^{\varepsilon-2}$. It is not yet possible to make precise statements based on observations about the value of ε. Near the centre of the elliptical galaxy, however, the observations of σ are more precise. I. King proposed a model of a galaxy in which

$$j(r) \propto \left(1 + \frac{r^2}{a^2}\right)^{-3/2}.$$ (9.24)

This model works well in most cases. A notable exception is the giant galaxy M87, for which it was argued by two sets of observers in 1978 that the rapid increase of $j(r)$, as well as a rapid increase of σ towards the centre, indicates a concentration of mass in the centre over and above that given by the King model.

The mean mass/light ratio in the central region of large ellipticals is found to lie in the range

$$\eta = (10 \pm 2)h_0.$$

Statistics of groups of galaxies

A catalogue of galaxies lists them by their coordinates on the celestial sphere, two galaxies with nearly the same coordinates being seen near each other. However, can we be certain that groups of apparently nearby galaxies are indeed close to one another and part of one physical system? The answer is being sought along two different lines, both statistical in nature and both leading to estimates of η.

S. J. Aarseth, J. R. Gott, and E. L. Turner adopted the approach of N-body simulations in which galaxies move under each other's gravitational pulls and tend to cluster together in small or large groups. A comparison of such distributions with real galaxy catalogues helps in identifying groups of galaxies and hence in estimating η.

The other approach, pioneered by E.L. Scott and J. Neyman and used extensively by P. J. E. Peebles and others, involves galaxy–galaxy correlation functions. We referred to it earlier in Chapter 7. In this approach the probability of finding a galaxy in a small volume δV at a distance r from a typical galaxy is defined as

$$\delta P = n\,\delta V[1 + \xi(r)]$$ (9.25)

where n is the number density of galaxies on the average. In a uniform distribution $\xi(r) = 0$. A positive $\xi(r)$ indicates enhancement of galaxy density near the typical galaxy, hence $\xi(r)$ is called the two-point correlation function.

In actual measurements the position vector r from the typical galaxy has two components with respect to the observer. The radial component π can be measured from the observed difference in the redshifts of the two galaxies using Hubble's law. The transverse component σ is measured by noting the angular separation of the two galaxies and multiplying it by their mean Hubble distance. However, apart from the universal velocity, the galaxies also have peculiar (random) velocities relative to their local cosmological rest frames. Such velocities tend to distort the radial component, with the result that if a plot is made of the two components of r on a Cartesian coordinate system, then the distribution of points tends to cluster round the axis corresponding to the radial component.

Using such plots for NGC and IC galaxies, Peebles concluded that a reasonably good estimate of $\xi(r)$ is given by

$$\xi(r) = \left(\frac{r_0}{r}\right)^{\gamma}, \qquad \gamma = 1.77, \qquad r_0 = 4.2h_0^{-1}\,\text{Mpc}. \qquad (9.26)$$

Aarseth and his colleagues arrived at similar results from their computer simulations. The peculiar velocities of galaxies can be estimated from the above-mentioned concentration effect, and the velocity dispersion comes out as

$$\langle v^2 \rangle^{1/2} \approx (600 \pm 250)\,\text{km s}^{-1}. \qquad (9.27)$$

From this result we can estimate η as follows. The mean number of neighbours within the characteristic distance $r_0 \sim R = 5h_0^{-1}\,\text{Mpc}$ is given by

$$N = n \int_0^R [1 + \xi(r)]\,\text{d}^3 r$$

$$= 42 \qquad (9.28)$$

for $n = 0.03h_0^3\,\text{Mpc}^{-3}$ (estimated for bright galaxies). The peculiar velocity v_i of ith galaxy having N_i neighbours of mass M at distance R is expected to be of the order

$$v_i^2 \sim \frac{GMN_i}{R}.$$

The result follows from the so-called virial theorem, which essentially states that in the equilibrium N-body distribution an equipartition exists between the kinetic and potential energy. From these results and from estimates of $\langle N_i^2 \rangle$ we get

$$M \sim \frac{R\langle v_i^2\rangle\langle N_i\rangle}{G\langle N_i^2\rangle} \sim 5 \times 10^{12} h_0^{-1} M_\odot \qquad (9.29)$$

A detailed calculation using the data on luminosities then gives η in the following range:

$$\eta \sim (500 \pm 200) h_0.$$

Clusters of galaxies

Similar correlation-function analysis has been applied to Abell clusters up to redshifts $z \leqslant 0.2$. The value of η comes out close to that for nearby groups of galaxies:

$$\eta \sim (500 \pm 100) h_0.$$

As early as 1933 F. Zwicky pointed out what has now become well known as the *missing mass problem* in clusters. The problem can be briefly stated as follows. If we estimate the mass of galaxies moving in one another's gravitational field in a cluster, then the virial theorem gives the mass of the cluster in terms of the velocity dispersion and the effective mean radius:

$$M = \langle v^2 \rangle \frac{R}{G}. \qquad (9.30)$$

From observations of the velocity dispersion $\langle v^2 \rangle^{1/2}$ we can therefore estimate the total mass M in the cluster. This value comes out considerably higher than that estimated on the basis of mass/light ratios η_G of individual galaxies. That is, if we see n galaxies in the cluster and if the total luminosity in the cluster is L, then the mass in the cluster is $L\eta_G$. Zwicky was the first to point out that

$$L\eta_G \ll M. \qquad (9.31)$$

For the Coma Cluster, for example, $M|L\eta_G \sim 30$ (see Exercise 20).

We will return to the speculations about the implications of the above inequality towards the end of this section.

The local supercluster

It was pointed out by G. de Vaucouleurs that we are situated in a region that seems to be on the outskirts of a galaxy concentration centred on the Virgo Cluster of galaxies located at a distance

$$D = 11 h_0^{-1} \text{ Mpc}.$$

Estimates of the average mass/light ratio per galaxy in the supercluster are still tentative, but are believed to be in the range

Table 9.1. *Average mass/light ratio per galaxy*

Object	ηh_0^{-1}
Our Galaxy (inner part)	6 ± 2
Our Galaxy (outer part)	40 ± 30
Spiral galaxies	9 ± 1
Elliptical galaxies	10 ± 2
Galaxy pairs	80 ± 20
Local Group	160 ± 80
Statistics of clustering	500 ± 200
Abell clusters	500 ± 200
Local superclusters	80 ± 30

$$\eta \sim (80 \pm 30)h_0.$$

Table 9.1 summarizes the above results, as well as some others not discussed here.

9.4.2 Dark matter

Returning to (9.20), we now see that the density parameter can be determined, at least within broad limits, from the values given in Table 9.1. Since the estimate is based on galaxy data we will denote the estimate of Ω_0 by Ω_G.

According to Table 9.1, ηh_0^{-1} ranges between values of 4 and 700. This gives the value of Ω_G in the extreme range

$$0.003 \leqslant \Omega_G \leqslant 0.53. \tag{9.32}$$

Note that this range does not depend on h_0.

In the inflationary cosmology we require $\Omega_0 = 1$. Clearly Ω_G falls short of this value. Can we therefore conclude that the universe is open? Is inflation ruled out? The answers are not so simple, however.

It is already noticeable that a considerable part of the matter in the universe might be nonluminous. We have seen that if we stick to the Newtonian inverse-square law of gravitation, the flat rotation curves of spiral galaxies imply more mass in the outer regions in these galaxies than is observed in the form of stars. In clusters of galaxies, the virial theorem (which again is based on the Newtonian law of gravitation) demands higher mass than observed.

In Chapter 5 we found that there are stringent limits on the baryonic density of the universe, limits imposed by the observations of primordial deuterium. We will review the deuterium evidence in section 9.6, but will now take note of its implications for Ω_0. The large M/L ratios in Table 9.1 imply that even within the above limits there is a lot of baryonic nonluminous matter. This could be in the following forms.

1. *Low-luminosity stars and stellar remnants.* One possibility is of 'brown dwarfs', that is, stars with masses too low ($\leq 0.08 M_\odot$) for them to be able to shine through nuclear hydrogen fusion. Such stars may form during the star formation process but are very difficult to detect unless they are part of binaries. At the other end of stellar evolution, high-mass stars may have reached their final states of white dwarfs/neutron stars/black holes. But, as calculated by B. Carr and others, the density in such remnants cannot account for more then $\Omega_0 \approx 0.03$; otherwise their integrated light intensity would be unacceptably high.

2. *Small solid bodies like comets, asteroids, dust grains, etc.* But there is a limit to how much these can contribute to Ω_0, because they are mostly made of heavy elements whose abundances together do not exceed ~ 0.01 of the hydrogen abundance.

3. *Neutral and ionized gas in the form of hydrogen.* This, however, is too small in amount to account for dark matter. For example, the X-ray halo of M87 shows a gas content of only 3 per cent of the total mass of the galaxy.

   There are also stringent limits on the intergalactic neutral hydrogen. In the spectrum of a high red-shift ($z > 2$) quasar, the blue side of the Lyman α ($\lambda = 1215$ Å) line should show a significant dip in the continuum as a result of *en-route* absorption by neutral hydrogen. In 1965 J. Gunn and B. A. Peterson looked for this effect in the quasar 3C-9 and placed an upper limit (on the basis of no detectable effect within the limits of sensitivity of observations) of $\Omega_{HI} \leq 4 \times 10^{-7} h_0^{-1}$. Molecular hydrogen can also be ruled out as a possible contender for dark matter on similar tests. The Lyman α absorption line systems found in the quasars may be due to discrete clouds of neutral hydrogen. However, most of the hydrogen in these intergalactic clouds may have been photoionized by the quasar radiation, and hence these data can be used to put an estimate of $\Omega_{HII} \approx 10^{-3}$. Further, the condition that the clouds have not been overheated by conduction sets a limit on the density of ionized intergalactic medium of $\Omega_{HII} \leq 0.02 h_0^{-2}$.

4. *Massive black holes.* These, with masses exceeding a few hundred solar masses, might also be candidates for dark matter. Such black holes form from the collapse of massive stars which do not explode as supernovae and so do not eject heavy elements into the surrounding medium. (Smaller-mass black holes cannot number too many, since they are formed by supernova explosions and hence pollute the interstellar medium with the heavy elements.) Based on the maximum such effects seen, B. Carr and others have argued that the contribution to Ω_0 from such black holes is not more than $\sim 10^{-4}$. For the massive ones, however, another restriction applies.

C. Canizares has argued that too many such massive black holes would exaggerate the gravitational lensing effect on quasars. The absence of any significant lensing distortion makes such massive black holes also negligible.

And so we are led to nonbaryonic alternatives, which are certainly required if the inflationary cosmology with $\Omega_0 = 1$ is to be believed. We have considered various forms of the nonbaryonic dark matter in Chapter 7. At present none seem satisfactory for any scenario of structure formation. None have been confirmed as 'existing' in laboratory experiments. The nearest to experimental credibility are massive neutrinos, for which there are conflicting claims of nonzero rest mass.

Can 'inflation' survive as an idea if astronomers see no direct evidence for $\Omega_0 = 1$? It can – by resurrecting the λ term (see Exercise 22)! This, of course, leaves the problem of the fine tuning of the λ-term unsolved (see Chapter 6).

9.4.3 q_0 and the deceleration of nearby galaxies

In Chapter 10 we will describe the attempts to estimate q_0 from the Hubble diagrams of distant objects in the universe. As an illustration we briefly mention an attempt initiated by A. Sandage, G. Tammann, and A. Yahil in 1976 that made use of velocity measurements of nearby galaxies.

From an examination of the nearby galaxies in the Revised Shapley Ames Catalogue, these authors noted that our local group of galaxies has a Hubble radial velocity of $\sim 1000\,\mathrm{km\,s^{-1}}$ relative to the Virgo cluster of galaxies and that the Virgo cluster is surrounded by galaxies in all directions in an extended spherical region that they called the Virgo complex. Our local group is near the periphery of this complex. The mean density excess in this complex is found to be

$$\left\langle \frac{\delta\rho}{\rho_0} \right\rangle \simeq 3.$$

Now this density excess will produce a deceleration of galaxies in the complex. The amount of deceleration depends not only on the excess but also on the overall mean density ρ_0. Working only within the Friedmann framework of cosmology, these authors related ρ_0 to q_0 through (4.42) or (4.58).

From the actual observation of the peculiar velocity field, it was then possible to derive ρ_0 (or q_0) with the requirement that the decelerations produced were compatible with the density excess of $\langle \delta\rho/\rho_0 \rangle = 3$. If, for example, q_0 were as high as 0.5, the expected peculiar velocity of the Local Group towards the Virgo Cluster would have been $27\,000\ \mathrm{km\,s^{-1}}$, which is too high. Clearly $q_0 \ll 0.5$, and the range of q_0 was claimed by these authors to be $\approx 0.06 \pm 0.015$.

This method needs to be reviewed periodically as the data on the large scale structure gets updated.

9.5 The age of the universe

The formulae (4.37), (4.51) and (4.64) give the age of the universe according to the various Friedmann models. Since these formulae depend on two parameters, H_0 and q_0 (or Ω_0), both of which have been discussed above, we are now in a position to take a look at the problem of whether the Friedmann age estimates are consistent with the various astrophysical estimates of the age of the universe. Table 9.2 gives a few characteristic values of the estimates of the Friedmann models for purposes of comparison.

At present there are two different ways of estimating the ages of galaxies, both of which have been applied to our Galaxy. A primary requirement of consistency is of course that the age of a Friedmann model (as given for example in the last column of Table 9.2) must exceed the age of any object in it.

9.5.1 Stellar evolution

This method, applied to globular clusters in our Galaxy, is based on the principle that stars become redder and brighter when they leave the main sequence to become red giants. Since the red giant phase in the star's life lasts a comparatively short time, say up to about 10 per cent of the time the star spends on the main sequence, the turning point from the main sequence to the giant branch provides the cluster age with 10 per cent uncertainty.

Let the *cluster age*, the time when the stars turn off from the main sequence, be denoted by $t_c \times 10^9$ years, and let Y and Z be the helium and metal abundances in the star at this stage. The calculations of stellar evolution then show that

$$\log t_c = 1.035 + 2.085\,(0.3 - Y) - 0.03(\log Z + 3). \qquad (9.33)$$

Thus the age depends critically on the helium abundance Y. Y can be estimated from a comparison of the lifetime a star spends on the

Table 9.2. *Ages of a few characteristic Friedmann models*

Model	h_0	$\Omega_0 = 2q_0$	Age in 10^9 yr
Open ($k = -1$)	$\frac{1}{2}$	0	19.6
Open ($k = 1$)	1	0	9.8
Einstein–de Sitter ($k = 0$)	$\frac{1}{2}$	1	13.0
Einstein–de Sitter ($k = 0$)	1	1	6.5
Closed ($k = +1$)	$\frac{1}{2}$	2	11.2
Closed ($k = +1$)	1	2	5.6

horizontal branch to the time it spends on the red giant branch. If this ratio is R, then calculations show that

$$Y = 0.3 - 0.39 \log \frac{f}{R} \tag{9.34}$$

where $f = 2$ if the stellar model takes account of semiconvection and certain other effects, while $f = 1$ if these effects are not taken into account. R can be estimated from the observed ratio of horizontal branch stars and red giant stars in the cluster.

Cluster ages deduced by this method fall in the range from 13×10^9 to 18×10^9 years.

9.5.2 Nuclear cosmochronology

In 1960 F. Hoyle and W. A. Fowler first demonstrated how the relative abundances of radioactive nuclei of long lifetimes can lead to estimates of the age of our Galaxy. The method was already used for estimating the age of the Solar System. For example, current observations of the ratios of $^{87}Sr/^{86}Sr$ plotted against $^{87}Rb/^{86}Sr$ in different solar system materials (such as meteorites) give the age accurately as $t_S \simeq 4.54 \times 10^9$ years. (See Exercise 24.)

As illustrated in Figure 9.8, the method of nuclear cosmochronology attempts to estimate the time elapsed before the Solar System was formed. According to this method, we start our nuclear clock at $t = 0$ with the birth of the Galaxy. The stars evolve and the more massive ones become supernovae, which manufacture long-lived radioactive nuclei in the so-called *r-process* (the rapid absorption of neutrons by heavy nuclei). The rate at which this process goes on is denoted by a function $p(t)$, which declines to negligible value at $t = T$. Between the epoch and the formation of the Solar System there occurs a short time gap Δ, known as

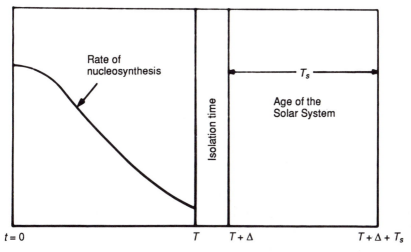

Fig. 9.8 Time chart showing how the age of the Galaxy is estimated. The details are explained in the text. (Based on J. Audouze, Ages of the universe. In R. Balian, J. Audouze, & D. N. Schramm, eds, *Physical Cosmology*, Les Houches Lectures Session XXXII, p. 195 (Amsterdam: North Holland, 1979).)

the *isolation time*, during which we may ignore nucleosynthesis, in particular the *r*-process. Thus the total nuclear age of the Galaxy is

$$t_G = T + \Delta + t_S. \tag{9.35}$$

In brief, T and Δ can be estimated as follows. The formalism, a variation on the earlier work of Hoyle and Fowler, is due to D. N. Schramm and G. J. Wasserburg. We consider a series of nuclei i $(i = 1, 2, \ldots)$ with decay constants λ_i and production rates $P_i p(t)$. We also assume that the abundance N_i of nucleus i is reduced exponentially at the rate ω owing to dilution of stellar matter with external gas and the cycling of matter back into stars. Thus N_i satisfies the following differential equation:

$$\frac{dN_i}{dt} = -\lambda_i N_i - \omega N_i + P_i p(t). \tag{9.36}$$

It is assumed that the relative production rate P_i/P_j of two nuclei i and j is constant.

Equation (9.36) can be integrated from 0 to T to give

$$N_i(T) = P_i e^{-(\lambda_i + \omega)T} \int_0^T p(t) e^{(\lambda_i + \omega)t} \, dt. \tag{9.37}$$

Between T and $T + \Delta$ we may ignore the ω and the $P_i p$ terms of (9.36) and deduce

$$N_i(T + \Delta) = N_i(T) e^{-\lambda_i \Delta}. \tag{9.38}$$

For long-lived nuclei $\lambda_i T \gg 1$ and certain approximations can be made. Define

$$R_{ij} = \frac{P_i N_j(T + \Delta)}{P_j N_i(T + \Delta)} \tag{9.39}$$

$$\langle \tau \rangle = \frac{\int_0^T t p(t)\, dt}{\int_0^T p(t)\, dt}. \tag{9.40}$$

It is easy to see that for $p(t) = \text{constant}$, $\langle \tau \rangle = T/2$, while for $p(t) \propto \delta(t)$, $\langle \tau \rangle = 0$. The value of $\langle \tau \rangle$ will in general lie between these two extreme limits.

Simple algebra and calculus then gives, from (9.37) and (9.38),

$$T = \langle \tau \rangle + \Delta_{ij} - \Delta \tag{9.41}$$

where

$$\Delta_{ij} = \frac{\ln R_{ij}}{\lambda_i - \lambda_j}. \tag{9.42}$$

Radioactive isotopes of thorium (^{232}Th) and uranium (^{238}U) and more recently the osmium (^{187}Os)-rhenium (^{187}Re) pair have been used to estimate Δ_{ij} and hence T and t_G. The decay constants λ_i and λ_j and the quantity R_{ij} are required. The ratio $N_i(T + \Delta)/N_j(T + \Delta)$ in R_{ij} is that prevailing at the time of formation of the Solar System, and it can be estimated from the present ratio in meteorites and so on and from the knowledge of t_S. The ratio P_i/P_j is taken from theories of nucleosynthesis.

Short-lived isotopes ($\lambda_i T \ll 1$) are used to estimate Δ. We have from (9.37) and (9.38)

$$N_i(T + \Delta) = \frac{P_i}{\lambda_i} p(T) \exp(-\lambda_i \Delta). \tag{9.43}$$

Hence

$$\Delta = \frac{1}{\lambda_i - \lambda_j} \ln\left(R_{ij} \frac{\lambda_j}{\lambda_i}\right). \tag{9.44}$$

From the short-lived isotopes of iodine (^{129}I) and plutonium (^{244}Pu) one finds that Δ lies in the range between 1 and 2×10^8 years.

The nuclear age so estimated lies in the range between 6 and 20 billion years, the width of this range indicating the span of uncertainties in the various quantities used for determining the time intervals Δ_{ij} and $\langle \tau \rangle$.

Nevertheless, it is clear when these age estimates and the estimates from globular clusters are compared with those of Table 9.2, that models with $h_0 = 1$ and $\Omega_0 \geqslant 1$ will find it very difficult to accommodate the above astrophysical estimates of the age of our Galaxy. In particular, the inflationary model is ruled out because it predicts $\Omega_0 = 1$ unequivocally.

To make the problem easier for the conventional point of view, attempts are being made to see whether the stellar and radioactive ages can be brought down significantly. For example, if significant mass loss occurs during the main sequence stage of stellar evolution then the time spent by the star on the main sequence is reduced. (For it started with higher mass and evolved faster.) Arguing in this way, L. A. Willson, G. H. Bowen and C. Struck-Marcell claim that it may be possible to reduce the ages of globular clusters to values as low as $7\text{--}10 \times 10^9$ years. Likewise, W. A. Fowler and C. C. Meisl have recalculated the nuclear age of the Galaxy using a time–dependent model for nucleosynthesis in which an early 'spike' is followed by a uniform synthesis. They claim that the age then comes down to $11 \pm 1.6\,(1\sigma)$ billion years. Even these exercises, however, do not help the inflationary model if $h_0 \approx 1$.

It is worth pointing out that the steady state model discussed in Chapter 8 predicts the average age of a galaxy as $\frac{1}{3}H_0^{-1}$. For $h_0 = 1$, this average will be ~ 3 billion years – very much lower than the above astrophysical age estimates. It is nevertheless possible to accommodate older-than-average objects in the steady state model. For example, condensation in the hot universe, which requires galaxies to form in groups at a time, can explain why our Galaxy and its neighbours are older than the average age. The age problem is therefore not so acute as in the steady model as it is in the big bang models.

9.6 The abundance of light nuclei

It is generally recognized that nuclei with atomic weights $A \geqslant 12$ are synthesized in stars through various processes discussed in theories of stellar evolution. The nuclei ^6Li, ^9Be, ^{10}B, and possibly ^{11}B could be produced in galactic cosmic rays by the breakup of heavy nuclei as they travel through the interstellar gas. It is the lighter nuclei, in particular ^2H, ^3He, ^4He, and ^7Li, that appear to pose difficulties of production in stars in the amounts observed. Further, their abundances are such that they could have been produced in the big bang nucleosynthesis. We will therefore discuss what constraints their observations place on standard cosmology, as well as on other cosmologies discussed in Chapter 8.

9.6.1 ^{4}He

The observed helium abundances (always denoted by mass fraction Y) in the universe are quoted as lying in the broad range

$$0.13 \leqslant Y \leqslant 0.34. \tag{9.45}$$

The scatter is wide because of the uncertainties of various observational estimates. Further, the estimate of primordial helium in the Sun at the time the Solar System formed $\sim 4.54 \times 10^9$ years ago depends on the model and hence cannot be uniquely fixed. M. Peimbert, S. Torres Peimbert, and J. F. Rayo have suggested that the breakup of Y at any location is as follows:

$$Y = Y_0 + \Delta Y,$$
$$Y_0 = 0.23 \pm 0.02, \tag{9.46}$$
$$\Delta Y \approx (2.5 \pm 0.5)Z,$$

where $Y_0 = $ primordial helium abundance, $\Delta Y = $ stellar helium abundance, and $Z = $ abundance of heavy elements made by stars. Since $Z \leqslant 0.02$, it follows that $\Delta Y \leqslant 0.06$. Table 9.3 gives some indication of the spread in Y and Z in various galaxies.

By way of comparison, the Orion Nebula in our Galaxy has $Y = 0.280$ and $Z = 0.016$. Further, F. Caputo, V. Castellani, and A. Martini have reported that on the basis of the observed ratio of horizontal branch stars to red giants in several globular clusters in the Galaxy (see (9.34)), the Y-fraction is as low as 0.10.

More recently, E. Terlevich, R. Terlevich, E. Skillman, J. Stepanian, and V. Lipovetskii have looked for helium content in extremely metal-poor galaxies, since it would be closer to the primordial value. In the first sample of such galaxies they find that the galaxy SB5 0335-052 has $y = 0.215 \pm 0.01$.

Clearly the survival of the theory of big bang nucleosynthesis depends on such low values of Y becoming exceptions rather than the rule. Since ^{4}He once produced and ejected in interstellar medium is difficult to get rid of, low-Y objects have to be explained as arising from inhomogeneities in the primordial setup. What is the tolerable range for the standard big bang nucleosynthesis?

It is helpful to go back to Chapter 5 and recall Figure 5.2, reproduced here as Figure 9.9. We note that in the primordial picture Y_0 is relatively insensitive to h_0 and Ω_0. However, the introduction of new leptons would push up the neutron/proton ratio, and hence the value of Y_0. The

Table 9.3. *Some determinations of Y and Z*

Galaxy	Y	$Z(A \geqslant 12)$
1 Zw 18	0.233	0.0004
II Zw 70	0.250	0.0039
IC 10 1	0.244	0.0039
II Zw 40	0.227	0.0041
NGC 6822 V	0.243	0.0058
NGC 6822 X	0.250	0.0069
IC 10 2	0.236	0.0075
NGC 4449	0.251	0.0091

Source: J. Lequeux, M. Peimbert, J. F. Rayo, A. Serrano, and S. Torres-Peimbert, 1979, 'Chemical composition and evolution of irregular and blue compact galaxies', *Astronomy and Astrophysics*, **80**, 155.

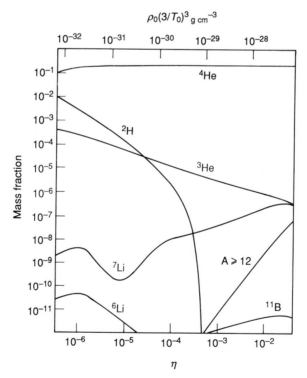

Fig. 9.9 Primordial abundances of light nuclei as functions of the present density of matter in the universe. The relation between ρ_0 and η is given by (5.52). (After R. V. Wagoner, The early universe. In R. Balian, J. Audouze, & D. N. Schramm, eds, *Physical Cosmology*, Les Houches Lectures Session XXXII, p. 395 (Amsterdam: North Holland, 1979).)

following formula, due to R. V. Wagoner, summarizes this result for the fraction η defined in (5.52) exceeding $\sim 10^{-5}$:

$$Y_0 = 0.333 + 0.0195 \log \eta + 0.380 \log \xi. \qquad (9.47)$$

Here the fraction $\xi = 1$ if no new particles except those considered in Chapter 5 are assumed to be present in the early universe. In terms of our notation of Chapter 6, this implies $g = 9$. If there are more particles, $g \to g + \Delta g$ where $\Delta g = \Delta g_b + \frac{7}{8}\Delta g_f$, and

$$\xi^2 = 1 + \frac{\Delta g}{g}. \qquad (9.48)$$

For $Y_0 \leqslant 0.25$ and $\Omega_0 = 0.01$, only one new lepton is allowed, the so-called τ-lepton with its τ-neutrino. It is interesting that the accelerator experiments in high–energy particle interactions independantly corroborate this conclusion. If, however, Y_0 were as high as 0.28, then up to four new leptons would be permitted by (9.48), whereas a value as low as 0.21 would land the standard model in real trouble. We state these limits without comment, since at present it is hard to say what the 'true' value of Y_0 is.

We have already commented on the implications for Y_0 in the Brans–Dicke cosmologies (Chapter 8). These models were consistent within the present range of uncertainties for an acceptable range of parameters.

The cosmology that is truly in trouble is of course the steady state cosmology. To produce the observed values of Y, this model must invoke either increased stellar activity in earlier epochs (thus departing from the strict application of the PCP) or the existence of supermassive stars of mass $\geqslant 10^8 M_\odot$ in whose interior big-bang-like conditions can prevail for sufficient time to generate the observed Y. Such a recourse might still be needed even in the big bang cosmology if it turns out from the scatter in the values of observed Y that there is no true Y_0.

Brans–Dicke cosmology is the only G-varying cosmology to work out Y_0 in detail. Similar calculations will have to be done in other G-varying cosmologies, since the fraction Y_0 depends sensitively on the rate of expansion of the early universe (see Chapter 5), which in these cosmologies differs from the canonical big bang value considerably.

9.6.2 ^2H

The deuterium abundance, which we will denote here by $X(^2H)$, was measured mainly from the Lyman series absorption lines in the ultraviolet spectra of the bright stars observed with the Copernicus satellite. It is found that

$$9 \times 10^{-6} \leqslant X(^2H) \leqslant 3.5 \times 10^{-5}.$$

Although a mean interstellar value of $X(^2H) \simeq 2 \times 10^{-5}$ is often quoted, there is considerable variation in its value from cloud to cloud. It is not clear whether these variations are due to partial destruction of the primordial deuterium through various processes. It has to be destruction, since so far no satisfactory stellar scenario for production of deuterium is known. Thus the primordial value would correspond to the upper end of the range of observations. At least we expect it to exceed $\sim 2 \times 10^{-5}$. (Contrast this situation with that for ^4He, for which there is no destruction mechanism but for which processes of production exist in stars.)

Referring back to Figure 9.9, we see that by primordial abundance $X(^2H) \geqslant 2 \times 10^{-5}$ implies that the baryonic density at present cannot exceed 7×10^{-31} g cm^{-3}, which in turn sets an upper limit on the present baryon density parameter $(\Omega_B)_0$:

$$h_0^2(\Omega_B)_0 \leqslant 0.0375. \tag{9.49}$$

Thus, if matter in the universe is predominantly in baryons, the universe must be open. Notice that since black holes are expected to be made of baryons, the hypothesis that most of the deficit between Ω_B and 1 is made of unseen matter in the form of black holes is not tenable. The missing mass or the unseen mass could be nonbaryonic, as discussed earlier.

9.6.3 ^7Li and ^3He

The ^7Li-abundance curve has a plateau with a dip, touching a minimum value of Li/H of 10^{-10} for $\eta = 3.2 \times 10^{-10}$. The observed data rule out the plateau value of $\sim 10^{-9}$. Even the minimum value is only marginally consistent with the minimum. The upper limit on Li/H by number was placed at 0.8×10^{-10} by K. C. Sahu, M. Sahu and S. R. Pottasch by observing interstellar absorption in the direction of the Large Magellanic Clouds.

The standard hot big bang nucleosynthesis predictions have the merit of being well defined. If there are discrepancies, what does one do? One way tried is to consider an earlier epoch when the nucleons had not formed: when the matter existed in the form of quark gluon plasma. The introduction of inhomogeneity at this stage can lead to some parts of the universe being 'proton-rich', while some parts become 'neutron rich' compared with the standard neutron/proton ratio. It is then possible to have an additional parameter to get a better agreement between the observed and predicted abundances of light nuclei. This method works partially successfully, but cannot explain away the above ^{7}Li problem.

The ^{3}He nucleus does not provide a powerful check on cosmological models because it could be produced in the observed amounts in stars. Thus, by and large, only ^{2}H and ^{4}He give us the most stringent limits on the parameters of the early universe.

9.7 The evidence for antimatter

In Chapter 6 we discussed recent attempts to account for the predominance of matter over antimatter in the universe, attempts that make use of the Grand Unified Theories (GUTs). How firm is the evidence that the universe is indeed made up only of matter? During the late 1950s and 1960s, H. Alfven and O. Klein produced cosmological models that start off with perfect symmetry between matter and antimatter. In their model, which we will not discuss here in detail, the symmetric components of the plasma that make up the universe are subsequently separated into matter-dominated and antimatter-dominated regions by a hydromagnetic process. Baryon-symmetric big bang models were also discussed by R. Omnes, F. W. Stecker, and others in the late 1960s and 1970s.

In Chapter 6 we found that unless specific symmetry-breaking techniques such as those proposed by the GUTs are employed, the standard big bang universe would end up with a net baryon number zero. GUTs attempt to explain not only why there is a net baryon number in the whole universe but also why the photon/baryon number ratio is of the magnitude implied by (5.57). By contrast, in the baryon symmetric cosmology there is separation between regions of matter and antimatter, while the overall baryon number is zero for the universe.

Theoretical speculations apart, what is the direct evidence for antimatter in the universe?

Space probes in the Solar System appear to rule out the existence of antimatter there. Interaction with the solar wind would have produced strong γ-rays had any of the planets been made of antimatter. Since observations beyond the Solar System are based largely on electromagnetic radiation, which treats matter and antimatter alike, it is hard to obtain a firm answer to the above question for a star or a galaxy. Cosmic rays do bring nuclei from the distant parts of the Galaxy (and even from beyond the Galaxy). However, intensive searches have failed to detect significant antimatter nuclei in cosmic rays. A few antiprotons (1 part in $\sim 10^4$) are found, but these could be produced by the interaction of cosmic rays with interstellar matter. Nevertheless, heavy antinuclei cannot be produced in this way and hence their detection in cosmic rays would confirm the existence of antimatter in the universe. The present evidence is somewhat tentative, though it cannot rule out a substantial antimatter component in extragalactic cosmic rays.

Faraday rotation is one form of indirect evidence. This is the rotation of the plane of polarization of light passing through a medium containing charged particles and a magnetic field. Because they are light, electrons (rather than protons) contribute more to the Faraday rotation. If positrons were also present they would also produce Faraday rotation but in the opposite sense. Since net Faraday rotation is observed in radiation from sources inside and outside the Galaxy, G. Steigman has interpreted this result by showing an imbalance in the abundance of electrons and positrons. However, this conclusion is based on the magnetic field retaining the same sign throughout. If the field changes sign as radiation enters an antimatter region, the Faraday rotation produced by positrons will be of the same sign as that produced by electrons.

Other indirect evidence could come from observations of the γ-ray background. Such a background can arise from various astrophysical causes – such as primordial black holes, blackbody radiation, and the inverse Compton process – in addition to the annihilation of nucleons and antinucleons. Each process, however, carries its own signature and its own limits on the magnitudes of the physical quantities involved. From an analysis of the γ-ray spectrum over the energy range of ~ 1–10^2 MeV, F. W. Stecker has concluded that the interpretation involving matter–antimatter annihilation is the one that fits the data best. Such regions of matter and antimatter would have to be separated from each other. However, Steigman has criticized this claim on the grounds that the fit is based on a number of parameters that could be adjusted to fit any spectrum of γ-rays.

The symmetry between matter and antimatter was also considered by G. R. Burbridge and F. Hoyle in the 1950s in the context of the steady state universe. If newly created particles were also accompanied by newly-created antiparticles, the symmetry in the universe would be preserved. However, it turned out that the γ-ray background resulting from the annihilation of particles and antiparticles would be very strong – far above that observed today.

9.8 The microwave background

We now come to an observation that in its importance to standard cosmology ranks next only to Hubble's discovery of nebular redshifts. This important discovery was first made in an unexpected fashion in 1965 by A. A. Penzias and R. W. Wilson, scientists at the Bell Telephone Laboratory. While looking for radio wave intensities in the plane of the Milky Way with the help of an antenna having a 20-foot horn reflector of low noise, Penzias and Wilson decided to use the wavelength 7.35 cm because at this wavelength the noise from the Galaxy was negligible. After making measurements in various directions and allowing for numerous unknown causes of radiation, they discovered that an unaccounted isotropic noise remained. Was this radiation background genuine? And if so what was its cause? Not knowing the answers to these questions, they hesitated before announcing their discovery.

Penzias and Wilson would not have waited to publish this result had they been aware of the prediction Gamow and his colleagues Alpher and Herman made some fifteen years earlier. This was the prediction that if the universe had a hot phase soon after the big bang, it should now possess a cooled-down relic radiation background. Alpher and Herman had estimated the present background temperature of around 5 K, whereas Gamow had made a guess of ~ 7 K. Penzias and Wilson had assigned a temperature of ~ 3.5 K to the background radiation they observed.

While Penzias and Wilson were puzzling over their discovery, news of it reached Princeton, where P. J. E. Peebles, himself a leading worker in the early universe calculations, grasped its significance. Indeed, the Princeton group including Peebles and R. H. Dicke, P. G. Roll, and D. T. Wilkinson, had already set up an experiment to measure this relic radiation. Although their own measurement of 3.2 cm came in late 1965, it was anticipated by the announcement of the discovery of Penzias and Wilson on 13 May 1965.

9.8.1 Spectrum

The background temperature has since been measured at several wavelengths by ground-based radiometers at frequencies upwards from 0.015 cm^{-1} and by balloon-, rocket-, or satellite-borne instruments at higher frequencies. The results are summarized in Table 9.4, which does not claim to be exhaustive. It is convenient to express the frequencies in units per centimetre by dividing the frequency expressed in Hertz by c.

Thus $3 \text{ cm}^{-1} \equiv 9 \times 10^{10}$ Hz. The observed flux is expressed in the form of a temperature of the blackbody radiation with the corresponding flux in the given frequency range.

The entries against the CN molecule experiment in Table 9.4 were obtained as follows. The ground state of the CN molecule has rotational levels $J = 0, 1, 2, 3, \ldots$. The transition from $J = 0$ to $J = 1$ is effected by incident radiation of frequency 3.79 cm^{-1}, while that from $J = 1$ to $J = 2$ is caused by incident radiation at a frequency of 7.58 cm^{-1}. Observations of CN molecules in interstellar space show that upper levels are partially populated, thus indicating the presence of a radiation field. The ambient radiation temperature can be determined from the degree of excitation of these levels (see Exercise 33). Such observations (first made as long ago as 1941) tell us that the microwave background extends beyond our local neighbourhood. However, it is desirable to ascertain its existence beyond our local neighbourhood by using detectors on spacecraft.

To check the true blackbody character of the radiation it is necessary to have detectors above the Earth's atmosphere. There were several early attempts using balloons and rockets. However, many of these reported departures from the Planckian spectrum that later turned out to be false alarms. Perhaps the most accurate and exhaustive study at the time of writing is that reported at the end of Table 9.4.

The Cosmic Background Explorer satellite (COBE) was lauched in 1989 and gave a beautiful spectrum shown in Figure 9.10. The COBE measurements give a very precise Planckian spectrum with a blackbody temperature of

$$T_0 = 2.735 \pm 0.06 \text{ K}. \tag{9.50}$$

The overall senstivity and accuracy of the experiment made it clear that some of the earlier claims of significant departures from the Planckian spectrum at high frequencies (e.g. by Woody and Richards, Matsumoto *et al.* in Table 9.4) were erroneous.

9.8.2 Anisotropy

If the microwave background is indeed of primordial origin, its anisotropies can tell us a lot about the present and the past history of the universe. The early developments in the post-recombination era imprint their signature on the radiation background, imprints that are expected to survive to this day. Observations of anisotropies are discussed below, taking the small-angle measurements first.

Table 9.4. *Measurements of the microwave background*

Experiment type	Frequency (cm^{-1})	Temperature (K)	Observers	Reference
Ground-based radiometers	0.0136–0.0207	3.7 ± 1.2	T. F. Howell and J. R. Shakeshaft	*Nature*, **216**, 753 (1967)
	0.079	2.70 ± 0.07	N. Mandolesi *et al.*	*Astrophys. J.*, **310**, 561 (1986)
	0.136	3.3 ± 0.33	A. A. Penzias and R. W. Wilson	*Astrophys. J.*, **142**, 419 (1965)
	0.31	2.69 ± $^{0.16}_{0.21}$	R. A. Stokes, R. B. Partridge and D. T. Wilkinson	*Phys. Rev. Lett.*, **19**, 1199 (1967)
	0.313	3.0 ± 0.5	P. G. Roll and D. T. Wilkinson	*Phys. Rev. Lett.*, **16**, 405 (1966)
	0.413	2.783 ± 0.025	D. G. Johnson and D. T. Wilkinson	*Astrophys. J. Lett.*, **313**, L1 (1986)
	0.633	2.78 ± $^{0.12}_{0.17}$	R. A. Stokes, R. B. Partridge and D. T. Wilkinson	*Phys. Rev. Lett.*, **19**, 1199 (1967)
	0.667	2.0 ± 0.4	W. J. Welch, S. Keachie, D. D. Thornton and G. Wrixon	*Phys. Rev. Lett.*, **18**, 1068 (1967)
	1.08	3.16 ± 0.26	M. S. Ewing, B. F. Burke and D. H. Staelin	*Phys. Rev. Lett.*, **19**, 1251 (1967)
	1.17	2.56 ± $^{0.17}_{0.22}$	D. T. Wilkinson	*Phys. Rev. Lett.*, **19**, 1195 (1967)
	1.22	2.9 ± 0.7	V. J. Puzanov, A. E. Salomonovich and K. S. Sankevich	*Soviet Phys.-Astronomy*, **11**, 905 (1968)
	1.89	2.70 ± 0.04	D. M. Meyer and M. Jura	*Astrophys. J.*, **297**, 119 (1985)
	2.79	2.4 ± 0.7	A. G. Kislyakov, V. I. Chernyshev, Yu. V. Lebskii, V. A. Maltsev and N. V. Serov	*Soviet Astronomy-AJ*, **15**, 29 (1971)

Method	Wavelength	Temperature (K)	Authors	Reference
	3.0	2.61 ± 0.25	M. F. Miller, M. McColl, R. J. Pederson and F. L. Vernon Jr	Phys. Rev. Lett., **26**, 919 (1971)
	3.0	$2.46^{+0.40}_{-0.44}$	P. E. Boynton, R. A. Stokes and D. T. Wilkinson	Phys. Rev. Lett., **19**, 462, (1968)
	3.0	$2.46^{+0.40}_{-0.54}$	P. E. Boynton and R. A. Stokes	Nature, **247**, 528 (1974)
CN molecule	3.79	2.93 ± 0.06	P. Thaddeus	Ann. Rev. Astron. and Astrophys., **10**, 305 (1972)
	7.58	$2.9^{+0.4}_{-0.5}$	D. J. Hegyi, W. A. Traub and N. P. Carlton	Astrophys. J., **190**, 543 (1974)
Rocket	1.67–33.3	$3.8^{+0.8}_{-1.9}$	K. D. Williamson, A. G. Blair, L. L. Catlin, R. D. Hiebert, E. G. Lloyd, and H. V. Romero	Nature, Phys. Sci., **241**, 79 (1973)
	7.69–25	$3.4^{+0.7}_{-3.4}$	J. R. Houck, B. T. Soifer, M. Harwit, and J. L. Pipher	Astrophys. J., **178**, L29 (1972)
	4.31	2.795 ± 0.018	T. Matsumoto, S. Hayakawa, H. Murakami, S. Sato, A.E. Lange, and P. L. Richards	Astrophys. J., **329**, 567 (1988)
	7.05	2.963 ± 0.017		
	10.4	3.150 ± 0.026		
	3–16	2.736 ± 0.017		
Balloon	1–11.5	$2.5^{+0.25}_{-0.45}$	H. Gush, M. Haplern and E. Wishnow	Phys. Rev. Lett., **65**, 537 (1990)
			D. Muehlner and R. Weiss	Phys. Rev. Lett., **30**, 757 (1973)
	2.38–13.53	$2.96^{+0.13}_{-0.08}$	D. P. Woody and P. L. Richards	Phys. Rev. Lett., **42**, 925 (1979); and Astrophys. J. **248**, 18 (1981)
Satellite (COBE)	1–20	2.735 ± 0.06	J. C. Mather, et al.	Astrophys. J. Letts., **354**, L37 (1990)

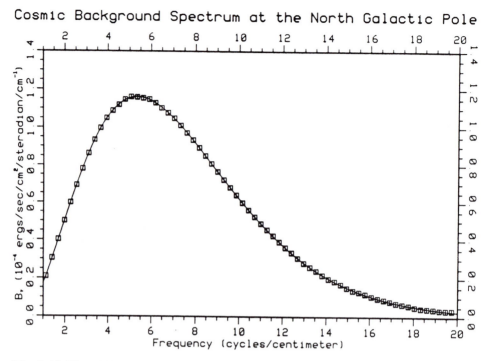

Fig. 9.10 The COBE measurements of the microwave background radiation at different frequencies. The continuous curve represents the best-fit Planckian curve to the data points. (See Table 9.4 for source.)

To look for *small-angle anisotropies* of angle θ, a large antenna with beam width $B \approx \theta$ is pointed at a fixed angle relative to the Earth and swept across the sky by the rotation of the Earth. The antenna temperature T_A records a small fluctuation ΔT_A composed of the intrinsic fluctuation of the background as well as receiver noise. Thus ΔT_A sets an upper bound on the intrinsic fluctuation. If $B \gg \theta$ then we may look upon the beam as covering $\sim (B/\theta)^2$ patches of angular size θ. Detailed calculations then show that the intrinsic fluctuation is less than

$$\left(1 + \frac{B^2}{\theta^2}\right)^{1/2} \Delta T_A. \tag{9.51}$$

Table 9.5 gives the data on small-scale fluctuations. These are all upper limits with no positive detection on any angular scale larger than a few arc minutes.

The 1992 measurements from COBE do, however, reveal small-scale fluctuations of $\Delta T/T \approx 6 \times 10^{-6}$. The COBE radiometer should ultimately be capable of detecting $\Delta T/T$ as low as 2×10^{-6}.

Such a high degree of isotropy also poses difficulties for a theory that

Table 9.5. *Small-angle anisotropy of the microwave background*

Frequency (cm^{-1})	Angular scale (arc min)	$\Delta T/T$	Observers	Reference
0.090	10–20	$<1.5 \times 10^{-4}$	K. C. Stankevich	Sov. Astron. **18**, 126 (1974)
0.278	>2	$<7 \times 10^{-4}$	R. L. Carpenter, S. Gulkis and T. Sato	Astrophys. J., **182**, L61 (1973)
0.0357	10	$<1.8 \times 10^{-3}$	E. K. Conklin and R. N. Bracewell	Nature, **216**, 777 (1967)
0.325	4.5	$<4.5 \times 10^{-5}$	J. M. Uson and D. T. Wilkinson	Astrophys. J. Lett., **277**, L1 (1984)
0.34	480	$<4 \times 10^{-5}$	R. D. Davies et al.	Nature **326**, 462 (1987)
0.357	>5	$<8.0 \times 10^{-5}$	Y. N. Parijskij	IAU Symp. No. **79**, 315 (1978)
0.500	>1.25	$<7.0 \times 10^{-4}$	J. C. Pigg	IAU Symp. No. **79**, 317 (1978)
0.66	2	$<1.7 \times 10^{-5}$	A. C. S. Redhead et al.	Astrophys. J., **346**, 566 (1989)
1.03	7	$<8.0 \times 10^{-5}$	R. B. Partridge	Astrophys. J., **235**, 681 (1980)
3.0	20	$<3.5 \times 10^{-5}$	P. R. Meinhold and P. M. Lubin	Astrophys. J., **370**, L11 (1991)
7.692	30	$<1.2 \times 10^{-4}$	N. Caderni, V. De Cosmo, R. Fabbri, B. Melchiorri, F. Melchiorri and V. Natale	Phys. Rev., **D16**, 2424 (1977)

attempts to explain the microwave background as arising from superpositions of radiation from discrete sources. As discusssed in Exercise 34, the sources would have to be more numerous and more closely spaced than galaxies.

The fact that $\Delta T/T$ is less than $\sim 10^{-5}$ on the scale of a few arc minutes poses severe difficulties for theories of galaxy formation. For, according to our discussion of Chapter 7, temperature fluctuations larger than this should have been observed in the relic background today.

So far as the cold dark matter hypothesis is concerned, the COBE data (at the time of writing) seem to rule out moderate values of the biasing parameter b. As the limits of the sensitivity improve it will be possible to make the CDM scenario more and more constrained. Indeed, most theories of galaxy formation known to date find it difficult to explain this extraordinary smoothness of the microwave background. The present limits imply that the universe was far too homogenous in the past to have initiated galaxy formation.

We next consider the possibility that the microwave background in clusters of galaxies interacts with high-energy electrons in the clusters and is partially scattered into X-rays. This interaction, known as the Zeldovich–Sunyaev effect, is another indication of the existance of a microwave background in remote clusters of galaxies. Although this effect leads to a dip in temperature of the order of $\Delta T/T \sim 10^{-4}$ across the cluster, its detection so far has been only marginal.

Large-angle anisotropies in the microwave background can arise from two sources. One cause is the limitations of the particle horizon discussed in Chapter 6. The particle horizon at decoupling subtends on angle θ_H at the observer today. It can be shown that measurements of radiation in different directions separated by angles large compared with

$$\theta_H \approx 2 \left(\frac{2q_0}{z_R} \right)^{1/2} \sim 5°(q_0)^{1/2}, \tag{9.52}$$

should show differences reflecting the early inhomogeneities on the scale of the particle horizon at $z = z_R$. The second cause of anisotropy is the motion of the Earth relative to the cosmological rest frame.

Measurements over large angles show no evidence for anisotropy on the scale of (9.52); but they do show evidence for the second effect. In early experiments E. S. Cheng, P. R. Saulson, D. T. Wilkinson, and B. E. Corey observed the anisotropy described by a temperature variation with direction of the following kind:

$$T = T_0 + T_1 \cos \theta, \tag{9.53}$$

with $T_1 \sim 3 \times 10^{-3}$ K. Such a variation can be explained by the assumption that the centre of the Galaxy has a velocity of ~ 540 km s^{-1} in the direction $l = 280°$, $b = 30°$. Another set of measurements by G. F. Smoot, M. V. Gorenstein, and R. A. Muller indicates a similar effect, but the corresponding Galaxy velocity is ~ 630 km s^{-1} in the direction $l = 261°$, $b = 33°$. The first group used balloons for measurement, while the second group used a U-2 aircraft. The COBE measurements also tend to agree with this conclusion. These result in a velocity of the Galaxy of 547 ± 17 km s^{-1} towards $l = 260°$, $b = 29° \pm 2°$. Note that this motion is different from the large-scale streaming motions discussed in section 9.3.

9.8.3 Interaction with cosmic rays

An interesting effect of the microwave background is to deplete cosmic rays of very-high-energy protons. At low energies in the centre-of-mass frame of the proton and the photon, the cross-section is of a second order in the fine-structure constant α. At higher energies, however, pions are produced and the cross-section becomes of first order in α. Thus, for a nucleon (that is, a neutron or a proton) N colliding with a microwave background photon γ the result is the reaction

$$\gamma + N \rightarrow \pi + N,$$

with the scattered nucleon having smaller energy than the incident nucleon. Let us estimate the energetics of the problem.

Let the 4-momentum of the microwave photon be given by

$$(q, 0, 0, q),$$

as measured by an observer at rest in the universal rest frame (see Figure 9.11). Let a nucleon of rest mass m and of momentum p strike it at angle θ in the XY-plane. The 4-momentum of the nucleon is then

$$(p \cos \theta, p \sin \theta, 0, (p^2 + m^2 c^2)^{1/2}).$$

What is the minimum proton energy E_p required to produce a pion and a nucleon? To calculate E_p it is convenient to work in the centre-of-mass frame of the system. For the total energy in the centre-of-mass frame is simply given by E_0, where

$$\left(\frac{E_0}{c}\right)^2 = [q + (p^2 + m^2 c^2)^{1/2}]^2 - (q + p \cos \theta)^2 - p^2 \sin^2 \theta$$

$$= m^2 c^2 + 2q [(p^2 + m^2 c^2)^{1/2} - p \cos \theta]. \tag{9.54}$$

The energy E_0 must exceed $(m_\pi + m)c^2$ where m_π is the mass of the pion. Since $m_\pi \ll m$, we get finally

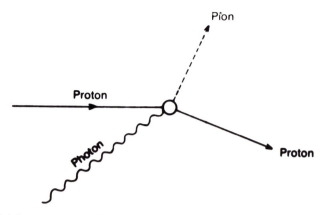

Fig. 9.11 A high-energy cosmic ray proton that collides with a microwave photon can produce a pion and a proton of less energy. (A charged pion and a neutron can also be produced.)

$$(p^2 + m^2c^2)^{1/2} - p\cos\theta \geqslant \frac{mm_\pi}{qc},$$

that is,

$$E_p \geqslant \frac{mm_\pi}{qc}. \tag{9.55}$$

For $qc = kT_\gamma, T_\gamma \sim 3$ K, we get

$$E_P \geqslant 3 \times 10^{20} \text{ eV}. \tag{9.56}$$

In other words, the pion production at energies exceeding E_p will lead to a sharp cutoff in the number of cosmic ray protons with energies exceeding E_p. A confirmation of this effect would establish the existence of the microwave background at distances beyond our Galaxy, since cosmic rays of such high energies are believed to be of extragalactic origin.

The cosmic ray spectrum in the range 10^{20} to 10^{21} eV, however, does not steepen as required by the above effect. Instead it flattens, thus indicating a relative increase in the number of high-energy nucleons. Supporters of the standard relic interpretation of the microwave background argue that it is still possible to prevent depletion of high-energy nucleons from the cosmic ray spectrum, provided they do not travel distances in excess of $\sim 10^{26}$ cm. That is, they should not come from beyond the local supercluster (see Exercise 36).

This concludes our discussion of local tests of cosmological importance. We will postpone a survey of the overall observational situation until we have looked at the surveys and tests relating to the large-scale structure of the universe in the following chapter.

Exercises

1 A galaxy has an apparent magnitude of 18 and an absolute magnitude of -17. Show that its distance from us is 100 Mpc.

2 Define the distance modulus suitable for cosmological distances. Show that an uncertainty of 1.5 magnitudes in the distance modulus can lead to an uncertainty of factor 2 in the estimate of the Hubble constant.

3 Comment on the way galactic extinction affects the measurement of extragalactic distances. If the effect is ignored, will the estimate of Hubble's constant be higher or lower than the true value?

4 Discuss the galactic extinction models currently in force. Show that de Vaucouleurs's model always leads to a higher value for the extinction parameter A than the Sandage–Tamman model. Estimate this difference for galactic latitudes $b = 30°$ and $b = 60°$.

5 The distance of a nearby galaxy at $b = 30°$ is being estimated by observing Cepheids in it and using the period–luminosity relation. Show that its estimated distance using de Vaucouleurs's extinction model will be smaller than that using the Sandage–Tamman model, after correcting for galactic extinction. What will be the corresponding ratio of the Hubble constants measured in the two models?

6 The period P (days) and the absolute visual magnitude M of galactic Cepheids are related by

$$M = -1.18 - 2.90 \log P \ (3 < P < 50).$$

A Cepheid in a nearby galaxy has a period of 10 days and an apparent magnitude (corrected for galactic extinction) of 20. Estimate the distance of the galaxy from these data.

7 In the supernova expansion method of determining distance the estimates of v, the photospheric velocity, are v_1 and v_2 at times t_1 and t_2. If the angular radii at t_1 and t_2 are θ_1 and θ_2, show that an estimate of the supernova distance D is given by

$$\frac{v^2(t_2 - t_1) + R_0(1 - v_2/v_1)}{\theta_2 - \theta_1(v_2/v_1)},$$

where the radius follows the law $R = v(t - t_0) + R_0$.

8 A supernova in NGC 1058 had photospheric velocity of $8.6 \times 10^8 \ \mathrm{cm \, s^{-1}}$ on Julian Date (JD) 2 400 568, while on JD 2 440 589 its photospheric velcity was $6 \times 10^8 \ \mathrm{cm \, s^{-1}}$. The angular radii of the supernova on these dates were 0.039 and 0.15 ($\times 10^{15} \ \mathrm{cm \, Mpc^{-1}}$), respectively. Show with the help of Exercise 7, and ignoring R_0,

that the distance of the supernova is about 12 Mpc and that its outward expansion started on JD 2 440 558. (Julian Date is counted from 1 January 4713 BC.)

9   Outline the observational difficulties that stand in the way of a precise determination of Hubble's constant.

10   In the Newtonian framework applicable to our local neighbourhood, the isotropic Hubble law may be expressed as the velocity distance relation

$$\mathbf{V}(\mathbf{r}) = H_0\mathbf{r},$$

r being the position vector of a galaxy relative to the origin. If the observer at the origin has a peculiar velocity **w**, he observes an anisotropic velocity distance relation given by

$$\mathbf{V}'(\mathbf{r}) = \mathbf{V}(\mathbf{r}) - \mathbf{w} = H_0\mathbf{r} - \mathbf{w}.$$

Show that effective Hubble constant $H(\theta)$ in a direction making an angle θ with the direction of the observer's peculiar velocity is given by

$$H(\theta) = H_0 - \frac{w\cos\theta}{r}.$$

Thus $H(\theta)$ is maximum at antapex $(\theta = \pi)$ and minimum at apex $(\theta = 0)$.

11   Imagine that the GA exists at a distance r_0. This will pull galaxies in its local neighbourhood towards itself. Show that the velocity distance curve as observed from our Galaxy would have an 'S' shape as a result of this perturbation.

12   Comment on the fact that although the redshift of a nearby extragalactic source is measurable very accurately, its interpretation as the velocity to be used in Hubble's velocity–distance relation is likely to contain errors.

13   Using the information of section 9.2 on extragalactic distance scales, deduce that the luminosity density of galaxies scales as h_0. Show also that Ω_G determined from the mass/light ratio of luminous objects is independent of h_0.

14   Let $\sigma(r)$ denote the surface mass density at a point P located at distance r from the centre of a thin disc-shaped galaxy. Show that the gravitational force F_r at P is directed towards the centre of the galaxy and is given by

$$F_r = G\int_0^\infty \sigma(rx)x\,\mathrm{d}x\int_0^{2\pi} \frac{(1 - x\cos\theta)\mathrm{d}\theta}{(1 - 2x\cos\theta + x^2)^{3/2}}.$$

15 Show that the integral in Exercise 14 can be evaluated for $\sigma(r) \propto r^{-1}$ and that it gives flat rotation curves

$$v^2 = 2\pi G r \sigma(r) = \text{constant}.$$

16 Discuss the implications of the flat rotation curves of elliptical galaxies. If there is no unseen mass involved, but Newton's laws are modified, how is the gravitational force expected to behave with distance?

17 In a spherical mass distribution in an SO galaxy, the star distribution function is given by (9.22). Assuming that all stars have equal mass and that their number density varies as $r^{-\varepsilon}$ ($\varepsilon < 0$), show that the mass contained in a sphere of radius r concentric with the galaxy is given by (9.23).

18 Discuss qualitatively how peculiar velocities of galaxies in a cluster distort the distribution of points on a two-dimensional plot for galaxies in a group, a plot that gives the radial separation of galaxies from a typical member against their transverse separation.

19 Let σ and π denote the components of the separation vector of a typical galaxy G from a fixed galaxy G_0, as seen by a remote observer perpendicular and parallel to his line of sight. The redshift difference between G and G_0, $\pi H_0/c$, is made of the cosmological component and the Doppler component due to a peculiar velocity w. If w has a distribution function $f(w)$, then show that the two-point correlation function $\xi(\sigma, \pi)$ is related to the spatial correlation function $\xi(r)$ by the relation

$$\xi(\sigma, \pi) = \int_{-\infty}^{\infty} f(w)\xi\left[\sigma^2 + \left(\pi - \frac{w}{H_0}\right)^2\right]^{1/2} dw.$$

20 In the Coma cluster of galaxies the observed velocity dispersion is $\sim 861 \text{ km s}^{-1}$, while the radius of the cluster is $\sim 4.6 h_0^{-1}$ Mpc. Show that the cluster mass given by the virial theorem is $\sim 2.3 \times 10^{15} h_0^{-1} M_\odot$. The total luminosity of the cluster is estimated at $\sim 75 \times 10^{12} h_0^{-2} L_\odot$. Show that the mass/light ratio parameter η for the cluster is $\sim 300 h_0$.

21 Discuss the missing mass problem in clusters and galaxies.

22 Show that if $\Omega_0 < 1$ then a closed universe requires a λ-term exceeding the value

$$\frac{3H_0^2}{c^2}(1 - \Omega_0).$$

23 In a globular cluster the metal content $Z \sim 10^{-3}$ and the ratio of

horizontal branch stars to red giants is 0.9. Show that in the $f = 1$ model the age of the globular cluster is around 11.9×10^9 years, while in the $f = 2$ model it is increased to around 2.0×10^{10} years.

24   The nucleus ^{87}Rb decays to ^{87}Sr with a half-life of $\tau = 4.7 \times 10^{10}$ years. Let $X(t)$ and $Y(t)$ denote the numbers of these nuclei in a meteorite at any time t, so that the quantity $X(t) + Y(t)$ is conserved. Let t_0 denote the epoch when the Solar System was formed. Show that a plot of relative abundances $X(t)/Z$ against $Y(t)/Z$, where Z = number of ^{86}Sr nuclei (which are unchanged) leads to a straight line whose slope is given by

$$\exp{(\lambda t_0)} - 1,$$

where $\lambda = \tau^{-1} \ln 2$.

25   Deduce (9.41) from (9.36)–(9.40).

26   Comment on the statement that very low values of ^{4}He abundance (for example, $Y \leqslant 0.15$) are embarrassing to the standard picture of the big bang nucleosynthesis. Contrast this situation with that of the abundance of deuterium.

27   Discuss how many leptons are permitted in the primordial era by the observed abundance of ^{4}He.

28   What limits does the present deuterium abundance place on the baryon density of the universe? Suppose we are given that the universe is closed. What modifications can you suggest in the standard big bang picture to reconcile the two results?

29   Determine from (9.49) the minimum value of λ necessary to reconcile the present deuterium abundance with a closed universe.

30   Discuss with the help of Figure 9.9 whether the limits on the density of baryons needed for producing deuterium and lithium in the observed amount are consistent with each other.

31   Using the theory describing the passage of a plane-polarized electromagnetic wave of frequency v travelling through a medium containing n charged particles of mass m and charge e and a magnetic field, show that if $H_\parallel$ is the component of the magnetic field along the direction of propagation of the wave then the plane of polarization turns through an angle

$$\Delta \theta = \frac{ne^3}{2\pi m^2 c^2} H_\parallel \Delta l v^{-2}$$

as the wave traverses a distance Δl. Comment on how this result (known as Faraday rotation) has been used to argue about the possible existence of antimatter in the universe.

32 Discuss the evidence for or against the presence of antimatter in the universe.

33 The ratio of occupied levels for $J = 1$ and $J = 0$ states for the CN molecule in the star ζ Ophiuchi is 0.55 ± 0.05 and in the star ζ Persei it is 0.48 ± 0.15. The energy difference between the two levels is equal to kT, $T = 5.47\,\text{K}$ and occupation weights are $g_1/g_0 = 3$. Deduce that the temperatures of the incident radiation lie in the respective ranges 3.22 ± 0.15 K and 3.00 ± 0.6 K.

34 Let n be the number density of sources generating a cosmic radiation background. Construct a cone of angle 2θ at the observer with the requirement that if such a cone is extended to cosmological distances it contains typically one source. Then θ denotes the typical angle over which the generated background would show patchiness. Show that $\theta \sim (H_0/c)^{3/2} n^{-1/2}$. Apply this result to the small-angle anisotropy of the microwave background.

35 Using the formulae of the Lorentz transformation of flux density from one inertial frame to another, show that if the observer is moving with a uniform velocity $\mathbf{V}$ relative to the cosmological rest frame, then in the approximation $|\mathbf{V}| \ll c$ he will measure a temperature deviation in the direction of the unit vector $\mathbf{k}$ of magnitude

$$\frac{\Delta T}{T} \approx \frac{\mathbf{V} \cdot \mathbf{k}}{c}.$$

36 It is given that a high-energy proton loses 13 per cent of its energy in a single collision with a microwave photon that produces a pion. In the microwave background there are ~ 550 photons per cm^3 and the mean cross-section of the above interaction is $\sim 2 \times 10^{28}\,\text{cm}^2$. Show that the proton loses most of its energy over a distance of $\sim 4 \times 10^{25}$ cm. Discuss the implications of this result for high-energy cosmic rays of extragalactic origin.

10

Observations of distant parts of the universe

10.1 The past light cone

Technically speaking, all our observations of the universe are confined to the past light cone. Nevertheless it is possible to make a distinction between the observational tests of cosmology described in Chapter 9 and those to be described here. This distinction is illustrated with the help of Figure 10.1. This diagram describes schematically the past light cone of a present observer in terms of the cosmological redshift. The observations described in Chapter 9 fall in the topmost region I with $z \leq 0.1$. In this region it is usually possible to go over to the locally inertial frame and use Newtonian physics and special relativity (see section 2.4). Although most cosmological models sink their geometrical differences close to the observer, it is still possible to test their physical differences in this region. For example, we can attempt to measure q_0 and Ω_0 from observations of galaxies (see section 9.4).

Region II, $0.1 \leq z \leq 1.0$, has been a hotbed of cosmological controversies. In this region, observations of galaxies and radio sources have been used to determine the geometrical nature of the universe. We will examine this region more closely in sections 10.2–10.4.

Region III, $1 \leq z \leq 5.0$, consists almost entirely of observations of quasars. Whatever geometrical differences exist between the various cosmological models in region II are magnified in region III. For this region quasars were expected to be useful probes of cosmology. There are, however, reservations still held by a few astronomers regarding the cosmological origin of quasar redshifts. We will discuss these reservations in the final chapter, while taking the more conventional view in this one.

Region IV, which extends from $z = 5.0$ to $z \approx 1000$, has so far proved inaccessible to cosmological observations, while region V, which takes us

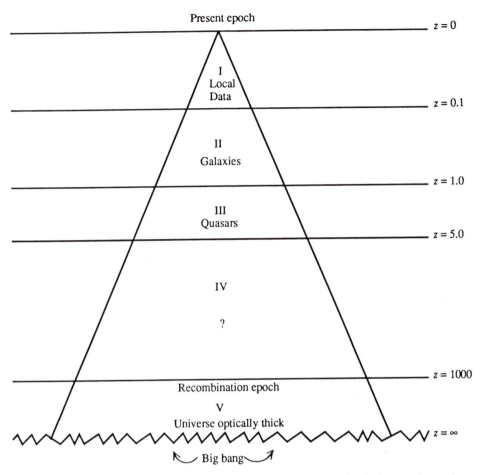

Fig. 10.1 A schematic description of the status of cosmological observations along the past light cone (shown by thick lines). Regions I to V, described in the text, are marked out by the epochs of redshifts $z = 0, 0.1, 1.0, 5.0, 1000, \infty$.

right back to the big bang ($z = \infty$), is in principle unobservable, since it is supposed to be optically thick (see Chapters 5–7).

Of course, the classification shown in Figure 10.1 has been guided largely by the standard hot big bang picture. In some alternative models regions IV and V do not have any theoretical significance; and if quasars turn out to be local objects, even region III is eliminated from cosmological observations.

With this background we will first describe four important tests for region II:

1. The redshift-magnitude relation.

2. The number counts of extragalactic objects.
3. The variation of angular size with distance.
4. The variation of surface brightness with redshift.

10.2 The redshift–magnitude relation

Basically, this is an extension of Hubble's relation to region II. In Chapter 9 we saw that in the nearby region the relation is described by (9.6), which is reproduced below:

$$m - M = 42.38 - 5 \log h_0 + 5 \log z. \tag{10.1}$$

What is the form of this relation in general, when the redshifts are not small?

The answer to this question is provided by the relation (3.44) between the flux density and luminosity for the Robertson–Walker models. The useful quantity in these relations is the luminosity distance D, which for Friedmann models is a function of H_0, q_0, and z. From (3.45) of Chapter 3 and (4.70) of Chapter 4, we get the following relation:

$$m - M = 5 \log D - 5$$

$$= 5 \log \left(\frac{c}{H_0 q_0^2} \right) - 5 + 5 \log \{ q_0 z + (q_0 - 1)[(1 + 2 q_0 z)^{1/2} - 1] \}. \tag{10.2}$$

For any Friedmann model we can express m as a function of z. If we are interested in regions of small redshifts where a first-order Taylor expansion is valid, we can reduce (10.2) to the form

$$m - M = 42.38 - 5 \log h_0 + 5 \log z + 1.086(1 - q_0)z + O(z^2). \tag{10.3}$$

This form helps in understanding how the different q_0 curves behave as z increases. Starting from the same Hubble relation (10.1), the curves gradually fan out, with the curves for high q_0 to the left and low q_0 to the right, as shown in Figure 10.2. This figure also shows the curve for the steady state model, which has the formal value of $q_0 = -1$. Thus it appears that if we make measurements in region II of Figure 10.1, we should be able to tell which q_0 model is best represented by the data.

A. R. Sandage and his colleagues spent a number of years on this cosmological test with the hope that the correct geometry of the universe would be revealed. Although in the 1960s Sandage often quoted a value of $q_0 \approx 1$, it gradually became clear that a number of uncertainties combine to make this test ineffective, at least at present. The various issues that

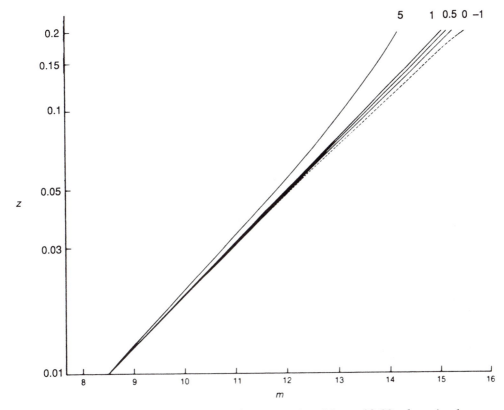

Fig. 10.2 A set of theoretical (m, z) curves for $M = -23.88$, $h_0 = 1$, drawn according to the approximate relation (10.3) for the cases $q_0 = 0$, 0.5, 1.5. The dotted curve represents the steady state model with $q_0 = -1$.

arise in practical applications of this test are discussed below. Some of the issues have been understood and partially resolved; other continue to be difficult to settle.

10.2.1 Observational uncertainties

Local motions

Small corrections are necessary for the fact that our observational frame is not the cosmological rest frame. Thus the Sun moves in the Galaxy and, as we saw in section 9.3, the Galaxy has a peculiar velocity. Fortunately, however, these corrections matter less and less for observations of more and more remote galaxies (see for example Exercise 10 at the end of Chapter 9).

The uncertainty of h_0

The relations (10.2) and (10.3) show that for relative comparison to two model curves with different q_0, the uncertainty in h_0 is eliminated.

The aperture correction

The importance of this effect was only gradually realized. It arises from the fact that galaxies are not objects with sharp boundaries; they tend to fade gradually into the background light of the sky. Therefore the amount of light received from the galaxy in relation to the background depends on the aperture of the telescope.

J. E. Gunn and J. B. Oke have suggested the following recipe for aperture correction.

Suppose we are measuring magnitudes in a wavelength band centred in λ_0. Then (3.42) gives the flux density at λ_0 as

$$\mathcal{F}(\lambda_0) = \frac{LI(\lambda_0/1 + z)}{4\pi(1 + z)D^2(q_0, z)}, \tag{10.4}$$

where $D(q_0, z)$ is the luminosity distance in the appropriate Friedmann model. Suppose the luminosity L is distributed across the galaxy according to a power law with respect to the projected radius ρ from the centre of the galaxy:

$$L(\leq \rho) = L_0\left(\frac{\rho}{\rho_0}\right)^\alpha, \qquad \rho = \text{constant}, \tag{10.5}$$

where α is a number of order unity; $\alpha = 0.7$ is a good approximation.

Now if we lived in the Einstein–de Sitter universe, the projected radius ρ_0 would subtend an angle at our location that is given by (see (4.76))

$$\gamma = \frac{H_0\rho_0(1 + z)^2}{cD(\frac{1}{2}, z)}. \tag{10.6}$$

In a general Friedmann model, the same angle will be subtended at the observer by a radius ρ given by

$$\gamma = \frac{H_0\rho(1 + z)^2}{cD(q_0, z)}. \tag{10.7}$$

A comparison of (10.6) and (10.7) shows that if the astronomer decides to measure the apparent magnitude of the galaxy within a given angular radius γ, he collects light from within different radii (ρ_0 and ρ) of the galaxy, depending on which Friedmann model is being used. If we standardize with respect to the Einstein–de Sitter model, we must correct for the luminosity according to (10.5). Thus instead of a fixed L, we must have

$$L = L_0 \left[\frac{D(q_0, z)}{D(\frac{1}{2}, z)} \right]^\alpha, \qquad (10.8)$$

so that (10.4) is changed to

$$\mathscr{F}(\lambda_0) = \frac{L_0 I(\lambda_0/1 + z)}{4\pi(1 + z)D(q_0, z)^{2-\alpha}D(\frac{1}{2}, z)^\alpha}. \qquad (10.9)$$

To fix ideas Gunn and Oke used $\rho_0 = 16$ kpc for $h_0 = 0.6$.

The formula (10.9) corrects for the fact that by fixing γ we make allowance for light coming from a smaller $(q_0 > \frac{1}{2})$ or larger $(\rho_0 < \frac{1}{2})$ region of the galaxy than that with $\rho = \rho_0$ for $q_0 = \frac{1}{2}$.

The K-correction

This effect was briefly hinted at towards the end of section 3.6 through the relation (3.42). It arises from the fact that when an astronomer measures the magnitude of a galaxy of large redshift z at a wavelength λ_0, he is receiving light from the galaxy at the emission wavelength of $\lambda_0(1 + z)^{-1}$. Hence, for a comparison of $m(\lambda_0)$ of two galaxies of different redshifts we must allow for the fact that their absolute magnitudes are being observed at different redshifts.

Taking the logarithm of (10.4) and converting to magnitudes, we get

$$m(\lambda_0) - M_{\text{bol}} = -2.5 \log I(\lambda_0/1 + z) + 2.5 \log (1 + z) + 5 \log D - 5.$$
$$(10.10)$$

If we apply the standard bolometric correction appropriate for wavelength λ_0 we get

$$m_{\text{bol}} = m(\lambda_0) + \Delta m(\lambda_0).$$

However, from (10.10) we see that if we wish to use (10.2) we must make a further correction and write instead

$$m_{\text{bol}} = m(\lambda_0) + \Delta m(\lambda_0) - K(\lambda_0), \qquad (10.11)$$

where

$$K(\lambda_0) = 2.5 \log (1 + z) - 2.5 \log I(\lambda_0/1 + z). \qquad (10.12)$$

Thus it is necessary to know the intensity distribution function $I(\lambda)$ in the source galaxy. Oke and Sandage estimated this effect in 1968 in a quantative way. More recently, ultraviolet (UV) astronomy is giving more information about $I(\lambda)$ for galaxies at UV wavelengths. For large z, these wavelengths get redshifted to the observed optical wavelengths (see Exercise 6).

Oke and Sandage also pointed out that the correct estimate of the K-term eliminates the so-called *Stebbins–Whitford effect* observed in the

1950s. This effect was based on the observation that galaxies appear to be redder and redder as their redshift increases. If it was genuine, such an effect implied that the more distant galaxies – that is, those of an earlier epoch – were systematically redder than the galaxies of the present epoch, and hence the universe must be evolving. This was therefore an argument against the steady state universe, an argument that has now been shown to be void.

The Malmquist bias

If we use some average galaxy luminosity in a distant cluster as a standard candle, this bias creeps in. For, as we examine more and more remote clusters of galaxies, we would tend to miss out larger and larger fractions of the intrinsically faint ones. So, in a magnitude-limited sample the luminosity distribution gets truncated at the lower end, the effect becoming more and more severe in more and more remote clusters.

The standard candle

Sandage found that as far as remote clusters were concerned, a good standard candle was provided by the brightest and most massive elliptical in the cluster. The luminosity variation of such elliptical galaxies from cluster to cluster is found to be remarkably small in nearby clusters. This is important, since we might notice spurious effects simply by a systematic variation of the standard candle.

The Scott effect

First pointed out by E. Scott, this effect is of the type against which caution was expressed above. Since the brightness distribution of galaxies in clusters has no sharp upper limit, we would tend to pick out more and more intrinsically bright galaxies as we look farther and farther away. This effect leads to an overestimate of q_0.

Intergalactic absorption

In 1976 S. M. Chitre and the author estimated the effect of absorption by intergalactic dust on the measurements of q_0. Since dust overestimates magnitudes, this effect leads to an underestimate of q_0. The effect may be considerable for even a miniscule proportion of intergalactic dust (see Exercise 8).

Luminosity evolution

One of the most serious difficulties in the redshift magnitude test arises from the uncertain corrections to the observed luminosities of galaxies to take account of luminosity evolution. If galaxies all formed at an epoch t_G ($< t_0$), and as they grew older their luminosity $L(t)$ changed as a function of $t - t_G$, then the present luminosity $L(t_0)$ may not give a reliable estimate of $L(t)$. As we know, for any redshift z the epoch of emission is given by

$$1 + z = \frac{S(t_0)}{S(t)}. \tag{10.13}$$

The interval $t_0 - t$ is called the *look-back time* of the galaxy. For very small z ($\ll 1$), $t \approx t_0$ and $L(t) \approx L(t_0)$. However, for z in region II, $L(t)$) may differ substantially from $L(t_0)$ and hence our extrapolations based on nearby clusters (which give $L(t_0)$ may not be good for remote clusters of galaxies. B. Tinsley was the first to appreciate the importance of this effect and to work it out quantitatively. Basically, the evolution of $L(t)$ comes from the evolution of stars in the galaxy. We will not go into the details of Tinsley's arguments here but simply state the empirical rule that seems to emerge from them.

The stellar population in giant ellipticals of the type used by Sandage is predominantly very old and metal-rich like the Population II stars in the disc of our Galaxy. To estimate their integrated luminosity it is necessary to know the 'initial mass function' (IMF). The IMF essentially specifies the relative number distribution of stars (in a cluster) within different mass ranges at the time of formation. Since the rate of evolution of a star depends on its mass, the IMF is important in determining the future composition of the cluster. In the visual region the integrated luminosity of such a population satisfies the law

$$L(t) \propto t^{-1.3+0.3x}, \tag{10.14}$$

where x is the slope of the IMF for stars in the mass range $0.8M_\odot \leqslant M \leqslant 1.2M_\odot$. (Since $t_G \ll t$ for region II, (10.14) shows L as a function of t rather than $t - t_G$.) The Salpeter IMF in the solar neighbourhood has $x \approx 1.35$. Thus the t-dependence of $L(t)$ is close to t^{-1} in (10.14).

10.2.2 The Hubble diagram

All these effects taken together pose a formidable array of problems from which it is very difficult (and currently impossible!) to extract the 'true' value of q_0. For example, taking the aperture effect and the luminosity evolution effect together, the first order $m-z$ relation becomes, instead of (10.3), the following:

$$m - M = 42.38 - 5 \log h_0 + 5 \log z$$

$$+ 1.086z \left(\frac{2 - \alpha}{2} \right) \left[q_0 - \frac{2}{2 - \alpha} \frac{1}{H_0} \frac{1}{L_0} \left(\frac{dL}{dt} \right)_{t_0} \right] + 1.086 \frac{\alpha}{4} z$$

$$+ 0(z^2). \tag{10.15}$$

Here we have made a Taylor expansion to first order to estimate $L(t)$. From (10.15) it easy to see that for a luminosity evolution given by (10.14), a first-order correction of q_0 is

$$\Delta q_0 = (2 - 0.5x)(H_0 t_0)^{-1}. \tag{10.16}$$

The product $H_0 t_0$ depends on q_0 (and λ if λ-cosmology is used). It is clear, however, that the 'true' q_0 is less than the observed q_0 by an amount Δq_0.

Figure 10.3 illustrates what a Hubble diagram looks like for galaxies in region II after making corrections for various effects (with the exception

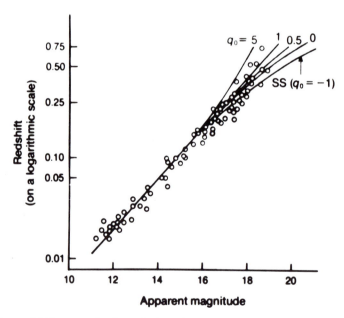

Fig. 10.3 The redshift magnitude relation for the brightest cluster members. A number of theoretical curves ($q_0 = 5$, 1, 0.5, 0, -1) are superposed on the data. SS stands for the steady state model. (Based on J. Kristian, A. Sandage, & J. A. Westphal, The extension of the Hubble diagram-III. *Ap. J.*, **221**, 383 (1978).)

of luminosity evolution). For comparison, a number of theoretical curves of Figure 10.2 are superposed on the galaxy data. Although the Hubble diagram has reasonably small scatter, it is not tight enough to rule out (with a great deal of confidence) any of the theoretical curves. Further, since negative q_0 is also permitted, the accelerating universes of λ-cosmologies cannot be ruled out.

We conclude this section by briefly mentioning two nonstandard cosmologies. In the 1960s Sandage's claims of $q_0 \approx 1$ went against the steady state prediction of $q_0 = -1$. Now, however, the uncertainties of the $m-z$ relation are such that the $q_0 = -1$ curve is only 1σ away from the best value of q_0 (which is close to $q_0 = 0$). The steady state model cannot as a rule invoke luminosity evolution of galaxies unless the observations refer to a large group of galaxies formed simultaneously.

J. M. Barnothy and Tinsley argued from an early assessment of the data that the G-varying cosmology of Hoyle and the author (see Chapter 8) is disproved by the $m-z$ relation. However, V. Canuto and the author showed that correct computation of stellar evolution rates in the framework of varying G makes this cosmology consistent with the present range of the $m-z$ relation (see section 10.7.)

10.3 Number counts of extragalactic objects

So far, considerable work has been done on number counts of three types of extragalactic objects: (1) galaxies, (2) radio sources, and (3) quasars. Of these we will defer the discussion of quasars to section 10.6 and consider only galaxies and radio sources, although the latter contain quasars in some cases.

The basic idea behind these tests is to find out whether the number counts reveal the non-Euclidean nature of the spacetime geometry of the universe predicted by most models. Suppose we have a class of objects that (1) are uniformly distributed in space and (2) have the same luminosity L. If we further assume that (3) the universe is of Minkowski type, that is, with Euclidean spatial geometry, the number of sources up to a given distance R will go as

$$N \propto R^3, \tag{10.17}$$

while the flux from the faintest of the sources up to distance R goes as

$$\mathscr{F} \propto R^{-2}. \tag{10.18}$$

Eliminating R between these relations, we get

$$N^2 \mathscr{F}^3 = \text{constant}, \quad \text{that is,} \quad \frac{D \log N}{d \log \mathscr{F}} = -1.5. \quad (10.19)$$

Thus (10.19) tells us how N and $\mathscr{F}$ are related under our three assumptions, (1), (2), and (3). Under these assumptions N measures the volume and $\mathscr{F}^{-1/2}$ the radius of a spherical region centred on the observer, and (10.19) is simply the volume radius relation in Euclidean geometry.

In sections 3.9 and 4.6 we saw how the volume–radius relation differs from the cube law of Euclidean geometry when we consider Robertson–Walker models. We also saw in these sections how to work out the corresponding relations in non-Euclidean geometries. It is therefore possible, in principle, to test whether the observed relation agrees with one of the various cosmological models. Unfortunately, as with the $m-z$ test, various uncertainties prevent us from drawing a clearcut conclusion, as we shall see with the counts of galaxies and radio sources below.

10.3.1 Galaxies

In 1936 Hubble attempted number counts of galaxies in order to distinguish between model universes. However, he had to abandon the test because the number of galaxies to be counted is very large, and unless one goes fairly deep in space one cannot detect any significant departures from Euclidean geometry. Since the optical astronomer measures fluxes in magnitudes, the relation (10.19) may be re-expressed as a number magnitude relation:

$$\frac{d \log N}{dm} = 0.6. \quad (10.20)$$

Hubble's programme has been revived in recent years by a number of workers who now have at their disposal many electronic and solid state devices to facilitate galaxy counts to very faint magnitudes ($m \sim 24$). For example, in 1979 J. A. Tyson and J. F. Jarvis used techniques of automated detection and classification of galaxies on plates. The main problem at faint magnitudes is to be able to distinguish stars from galaxies. R. G. Kron uses a colour criterion to pick out the high-redshift galaxies. In the USSR, I. D. Karachentsev and A. I. Kopylov have done the exercise of counts down to 26^m. B. A. Peterson and others at the Anglo-Australian Telescope have also counted galaxies brighter than 24^m. We may add to this list the more recent work of R. Ciardullo and of H. K. D. Ye and R. F. Green and the CCD number counts by Tyson. Although problems still exist in normalizing these different surveys so that they agree on the number of galaxies per square degree at a given magnitude, there seems to be broad agreement that the slope of the $\log N - m$ curve is considerably flatter than the 0.6 of (10.20). The slope is actually close to 0.4 over the range 20^m to 24^m.

What does this result imply? If we use the formulae of sections 3.9 and 4.6 we will find the slope of the $\log N - m$ curve starts off with 0.6 in the nearby region and gradually flattens as we observe more and more distant regions. Thus the slope of ~ 0.4 may represent some kind of average. In the explicit calculation G. R. Burbidge and the author showed in 1981 that such a slope can be simulated by an empty Friedmann ($q_0 = 0$) model, provided we drop assumption (2) and assume instead that the galaxies have a luminosity function of the following form:

$$f(L) \propto \left(\frac{L^*}{L}\right)^{5/4} \exp\left(-\frac{L}{L^*}\right). \tag{10.21}$$

That is, galaxies do not have a fixed L, but a distribution of L. The constant L^* corresponds to an absolute magnitude of -20.6. Luminosity functions of this type are suggested by P. Schechter and others in surveys of galaxies in our local region.

It is necessary to take account of the K-correction mentioned in section 10.2 before a fit is made with a theoretical model. The observed magnitudes are J-magnitudes, and we need to know the intensity distribution $I(\lambda)$ for galaxies in the ultraviolet. The K-corrections are still uncertain, and it is therefore necessary to do the counts in different wave bands and also to look at the redshift distributions of faint galaxies.

On the basis of the data there have been claims of mild evolution over the look-back times. The evolution claimed is only statistical, and as Sandage has emphasized in a 1988 review, 'no check on the direct predictions of the standard model is available from this test ... unless *a priori* assumptions are made concerning the evolution.' It also seems that the notorious λ-term is likely to make a comeback. For example, M. Fukugita, F. Takehara, K. Yamashita, and Y. Yoshil have claimed that the counts demand a nonzero λ-term and can be best fitted for $\Omega_0 = 0.1$ and $\lambda c^2/3H_0^2 = 0.9$.

10.3.2 Radio sources: methods

By comparison with galaxy counts, counts of radio sources have the advantage that radio sources are not as numerous as galaxies. For this reason, after Hubble's galaxy-count programme came to nothing and radio astronomy became established during the 1950s, it was felt that time was ripe to have a go at the source-count test. M. Ryle at Cambridge, B. Mills at Sydney, and J. Bolton at Caltech did pioneering work on the source-count programme. Since the radio astronomer measures $\mathcal{F}$ over a specified bandwidth, he tends to plot $\log N$ against $\log S$, where S is the *flux density*, the flux $\mathcal{F}$ received over a frequency band divided by the bandwidth. The usual unit for S is the jansky (Jy), which equals $10^{-26}\,\mathrm{W\,m^{-2}\,Hz^{-1}}$. (Named after K. G. Jansky, who did pioneering work on radio astronomy in the 1930s.) Similarly, the *power* of the radio source is defined as luminosity over a unit frequency band per unit solid angle and is expressed in units of watts per hertz per steradian ($\mathrm{W\,Hz^{-1}\,Sr^{-1}}$).

There are several ways of plotting the radio source-count data. Since the astronomical literature contains references to all of them, they are only briefly outlined below.

Log N–log S

This is the form discussed earlier. The Euclidean geometry makes the prediction that

$$\frac{\mathrm{d}\log N}{\mathrm{d}\log S} = -1.5. \tag{10.22}$$

The Friedmann models and the steady model show a flattening tendency, as shown in Figure 10.4(a).

Log N/N₀–log S

Instead of plotting N against S, it is often convenient to plot N as a fraction of the number N_0 expected in Euclidean geometry against S. Figure 10.4(b) shows how such plots are expected to look in standard cosmology and the steady state theory. In this figure we have plotted not the ratio N/N_0 but the ratio $\Delta N/\Delta N_0$ of differential counts. That is, we denote by ΔN_0 the number of sources expected in Euclidean geometry in a given flux density range $(S, S + \Delta S)$, while ΔN denotes the actual numbers of sources found (or expected to be found in a given model). Thus $\Delta N_0 \propto S^{-5/2}\Delta S$, and we expect $\Delta N/\Delta N_0$ to decrease steadily from 1 as S decreases.

Luminosity–volume

Instead of plotting N against S, it is often more interesting to approach the source-count problem in the following way. Suppose S_m denotes the minimum flux density that can be picked up in a given survey. Let a source

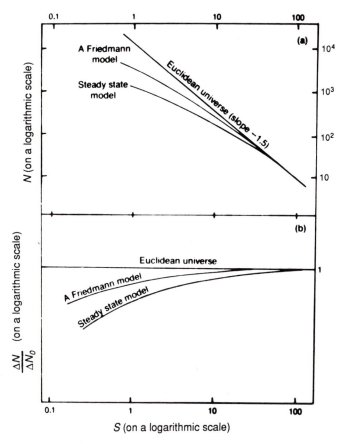

Fig. 10.4 Schematic plots of (a) $\log N$ against $\log S$ and (b) $\log (\Delta N/\Delta N_0)$ against $\log S$. The curves shown for a Friedmann model are representative of a number of curves that lie between the steady state and the Euclidean curves for various values of q_0.

with $S > S_m$ be found, and suppose we know its distance d (from its redshift, say). We then know its luminosity. We can then ask how much further the source could be moved for it to be barely detected in our survey.

Thus we have two volumes, V and V_m. The volume V is the volume of the spherical region centred on our location and with radius d at whose boundary the source actually exists. The volume V_m is that of the limiting sphere on whose boundary the source would be barely detectable in the survey. If the source were randomly distributed within the limiting sphere then the average value of the ratio V/V_m would be $\frac{1}{2}$. Thus in any cosmology with assumption (1) of uniform distribution the average value of the ratio V/V_m is expected to be $\frac{1}{2}$.

In practice, if we know the redshifts of the radio sources we can do a luminosity volume test and compute the average value

$$f = \left\langle \frac{V}{V_m} \right\rangle. \tag{10.23}$$

This computation will of course depend on the cosmological model chosen. If $f = \frac{1}{2}$ then our assumption of a uniform distribution is confirmed. If $f > \frac{1}{2}$, the survey implies that we are finding more distant than nearby sources. Since distant sources are seen at earlier epochs, this result implies a density evolution with more sources in the past than now. Similarly, $f < \frac{1}{2}$ indicates a density evolution of the opposite kind. M. Schmidt was the first to suggest this test in 1968 and to apply it to radio sources.

The maximum likelihood method

This method of analysing $N-S$ data was suggested by D. F. Crawford, D. L. Jauncey, and H. S. Murdoch in 1970, and it can be described in brief as follows.

Suppose the $N-S$ relation is of the form

$$N(S) = kS^{-\alpha}, \qquad S \geq S_m. \tag{10.24}$$

Write $\sigma = S/S_m$ so that $\sigma \geq 1$, and let σ_0 denote the maximum value of σ. Let $dp(\sigma)$ denote the probability that a source is found in the range $(\sigma, \sigma + d\sigma)$. Then from (10.24) we have

$$dp = \frac{\alpha \sigma^{-(\alpha+1)} \, d\sigma}{1 - \sigma_0^{-\alpha}}. \tag{10.25}$$

If we make ranges of σ small enough so that each range contains at most one source, we may have M such ranges, say. Denoting them by the label i ($i = 1, 2, \ldots, M$) and the corresponding probabilities by p_i, we write the likelihood function L as

$$\mathcal{L} = \sum_i \ln p_i. \tag{10.26}$$

The method consists of maximizing $\mathcal{L}$ with respect to α. Using (10.25) for p_i, a simple calculation (see Exercise 15) gives

$$\alpha = \frac{M}{\sum_i \ln \sigma_i}. \tag{10.27}$$

In the usual $\log N - \log S$ plot, the number N is built out of additions of numbers from successive flux intervals, and in this process errors tend to add up. Thus the different N-values in such a plot are not independent and the estimate of α based on them is not quite reliable. The maximum-likelihood method treats each observation independently, and the estimate of α given by (10.27) is therefore free of cumulative systematic errors.

10.3.3 Radio sources: the data

Before we come to the actual data the following points need to be made.

1. The number-count test is a test for the volume–distance relationship. The measure of the distance of a galaxy is its redshift. The radio astronomer is, however, unable to measure redshift directly. If the radio source is optically identified with a galaxy of known redshift, then only do we have a measure of its distance. The flux density S cannot be a reliable distance indicator unless we are sure that all sources have approximately the same luminosity. In practice the powers of radio sources vary over the range from 10^{23} to $10^{28}\,\mathrm{W\,sr^{-1}}$ from weak to strong sources. Thus it is possible to mistake a nearby weak source for a strong distant source.
2. Even if all redshifts in a survey were known, we would not have a complete sample to test the volume–redshift relation. This is because a sample that is complete with respect to a minimum flux density S_m is not necessarily complete with respect to a given maximum redshift z_m, and vice versa. In the former sample, very weak sources of moderate redshift are missed, while in the latter sample, very strong and very distant sources are missed. For practical reasons the radio astronomer has complete $S \geqslant S_m$-type samples and no complete $z \leqslant z_m$-type samples. The former samples suffer from the difficulty mentioned in point 1.
3. It has become clear, over the years, that simply counting objects as 'black blocks' can be misleading. We should know some basic features of what we are counting.

With this background we turn to observations of the radio source counts. The source counts have been obtained at various frequencies, and, as an illustrative example, Figure 10.5 gives the differential source counts (relative to Euclidean values) as in Figure 10.4(b) for four surveys at frequencies of 408 MHz, 1410 MHz, 2700 MHz, and 5000 MHz. Two important things are immediately noticeable from this diagram.

First, the surveys give different results at different frequencies. The (negative) curvature of the source count curves declines in magnitude as the survey frequency rises. The main reason for this discrepancy between curves is as follows. In a typical radio source the intensity frequency function (see section 3.6) has the form

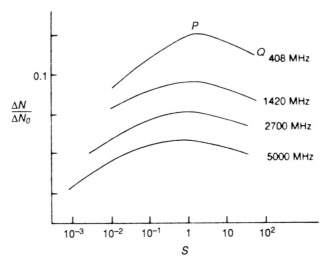

Fig. 10.5 Differential source counts at different frequencies. *S* is plotted on a logarithmic scale. The positioning of the four curves is arbitrary with respect to the $\Delta N/\Delta N_0$ axis. However, each marked interval shows an increase in $\Delta N/\Delta N_0$ by a factor of 10 as we move upwards along this axis. (After J. V. Wall & D. J. Cooke, Source counts at high spatial densities from pencil beam observations of background fluctuations. *MNRAS*, **171**, 9 (1975).)

$$J(v) \propto v^{-\alpha}, \qquad 0 \le \alpha \le 1, \tag{10.28}$$

where α is called the *spectral index*. Thus for a large α, the flux density falls more rapidly with frequency, with the result that a source is picked up more easily in a low-frequency survey than in a high-frequency one. For low α this effect is less noticeable. For this reason the ratio of steep-spectrum sources (high α) to flat-spectrum sources (low α) in a survey declines as the frequency of the survey increases. The predominance of flat spectrum sources in high-frequency surveys seems to make their $(\Delta N/\Delta N_0, S)$ relationship agree with the expected behaviour of Figure 10.4(b).

This second point relates to the preliminary rise of $\Delta N/\Delta N_0$ as S decreases from the highest flux value. This effect is most noticeable in the low-frequency survey at 480 MHz, and was the cause of a controversy between Ryle and Hoyle in the early 1960s. We describe it briefly for its historical interest.

If we compare the peak point P with the highest flux point Q in Figure 10.5 we can draw one of two possible conclusions:

1. There is a significant rise in the number ΔN compared to ΔN_0 as we go to lower flux densities from Q. If these sources are very distant and powerful ones, we are seeing an evolutionary effect, implying that the number density

of radio sources rose sharply in the past epochs compared with the present epoch.

2. Compared with P, the point Q shows a deficit of high-flux-density source in relation to the Euclidean value. If the sources are by and large not very strong, this deficit is a local one and simply indicates that we are in a 'hole', with fewer radio sources than average. Such effects could arise if there are inhomogeneities on the scale ~ 50 Mpc in the universe, as were expected, for example, in the 'hot universe' version of the steady state model (see Chapter 8).

Hoyle subscribed to the second viewpoint, while Ryle and his collaborators at Cambridge took the first possibility as correct. The Cambridge view, as it is now called, implied a strong evolution in number density and hence a disproof of the steady state theory. It is clear that the rise in number density cannot continue indefinitely, otherwise the radio background would be too high. So various functions describing the variation of the source number density with redshift z have been considered in order to fit the observed data at all flux levels. However, such a parameter-fitting exercise makes the test ineffective as a means of distinguishing between various q_0 cosmologies, since the geometrical differences are masked by the proposed evolutionary functions.

But to what extent is evolution really proved? The 3CR sample of radio sources has now been almost entirely optically identified and the redshifts of its members determined. If we assume that the redshifts are indicators of distance, it is now possible to re-examine the number counts of that sample. In 1989 P. DasGupta, G. Burbidge and the author carried out this exercise. They started with the hypothesis of 'no evolution' in the luminosity function. Based on the redshift data, this null hypothesis enables one to construct this radio luminosity function (RLF). Using this RLF it is then possible to carry out theoretical Monte Carlo samples of sources so that one obtains two-dimensional plots of number–flux-density distributions. These plots are then compared for deviations from the observed plot.

It turns out that the deviations are not large enough to be rejected on probability grounds. That is, the non-evolving model *cannot* be ruled out by the data. The deviations are smaller and the confidence levels (for retaining the null hypothesis) turn out to be higher for the steady state model. The agreement between 'no evolution' and observations improves further if we assume a 'local hole' of size ~ 50–100 Mpc for radio sources. Judging by the inhomogeneity (superclusters and voids) on this scale, the local hole hypothesis does not sound as outlandish today as it did in the 1960s.

10.4 The variation of angular sizes with distance

This test was discussed briefly in Chapters 3 and 4 (see sections 3.8 and 4.5), where we saw that the angular size of an object of fixed projected linear size does not steadily decrease with its spatial distance from us. Figure 4.6 shows how the angular size changes with the redshift of the object. In 1958 F. Hoyle first suggested that this property of non-Euclidean geometries could in principle be tested by astronomical observations.

Such a test could be performed for galaxies in the optical region. However, the redshifts of galaxies do not go far enough (that is, to $z \geqslant 1$) to make the predicted effects observable. For instance, in 1975 R. J. Dodd, D. H. Morgan, K. Nandy, V. C. Reddish, and H. Seddon examined the images of 3000 faint galaxies down to B-magnitude $= 23$ on the 48-inch UK Schmidt telescope in Australia. Instead of redshifts, which are not expected to exceed 0.5, they plotted the number N of galaxies larger than a specified angular size θ. The observed curve was, however, consistent with a broad range of Friedmann models ($0 \leqslant q_0 \leqslant 13$) as well as for $q_0 = -1$, corresponding to the steady state model. This result shows that galaxies are not likely to be useful as sensitive probes for selecting a narrow range of q_0.

For small redshifts the expected redshift dependence of angular size is $\theta \propto z^{-1}$. The eye estimates of angular diameters of first-ranked ellipticals in clusters, obtained by Sandage from the Palomar Schmidt and the 200-inch telescope plates in 1972, did show this dependence. At redshifts higher than ~ 0.5, one has to look at clusters of galaxies. Work by P. Hickson was along the following lines. He found the harmonic mean separation of the top 40 brightest galaxies within a specified radius of 3 Mpc of the centre of each cluster. Although Hickson and P. J. Adams claimed that their analysis of the data required evolution in cluster sizes towards progressively smaller sizes, there is doubt on the significance of this conclusion. For example, it is by no means clear that the clusters are relaxed and reached statistical equilibrium. Moreover, there is evidence for subclustering within a given cluster, showing that the mixing needed for virialization has not occurred.

Radio sources, as suggested originally by Hoyle, are likely to provide more useful information if the strongest of them can be seen at redshifts $z \geqslant 1$. Radio sources have by and large the double structures shown in Figure 10.6. The typical source has two radio-emitting blobs A, B, say, separated by a distance d, which ranges from a few kpc to ~ 1000 kpc.

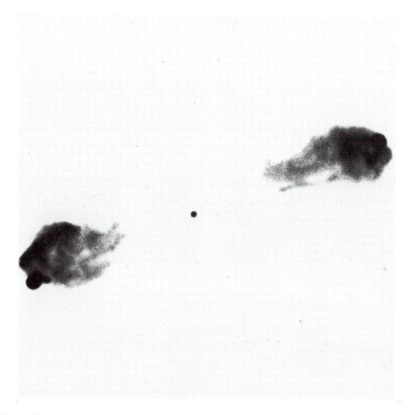

Fig. 10.6 The above image-processed radio picture of Cygnus A illustrates the structure of the most common type of extragalactic radio source with two radio-emitting blobs located symmetrically on the opposite sides of a central region. The central region is believed to be the source of activity that generates fast particles moving out in jetlike structures which create the two radio-emitting lobes after impinging on the intergalactic medium. (Source: National Radio Astronomy Observatory, USA.)

The typical angular size is ~ 20 arc second at the observer, angles that can be readily measured. However, unlike the angle subtended by a sphere, which does not depend on orientation, the angle subtended by a linear source AB, such as that shown in Figure 10.6, depends on the angle made by AB with the line of sight. Thus even the angles subtended by sources of the same linear size at the same distance will show a scatter.

The data published in 1974 by J. F. C. Wardle and G. K. Miley for quasars show enormous scatter, partly because of the above projection effects and partly because the linear size d is not fixed. Even so, the upper envelope of the plot shows a dependence of the form $\theta \propto z^{-1}$ for z up to 2.5 which is difficult to explain.

Figure 10.7 shows a plot of median angular size against redshift for radio sources, with the $q_0 = 0$ Friedmann curves superposed on it. The observed points are from the 1979 work of J. Katgert-Merkelijn, C. Lari, and L. Padrielli. The two theoretical curves are for median linear sizes of $125h_0^{-1}$ kpc and $165h_0^{-1}$ kpc. The agreement is not bad, although we cannot expect the data to single out a particular q_0 with any degree of confidence.

Because redshifts of radio sources are not directly measurable but have to be obtained by the process of optical identification, some radio astronomers have preferred to plot the angular size θ against the flux density S. Since radio sources have varying luminosities, this procedure adds a further source of scatter to the observations. However, R. D.

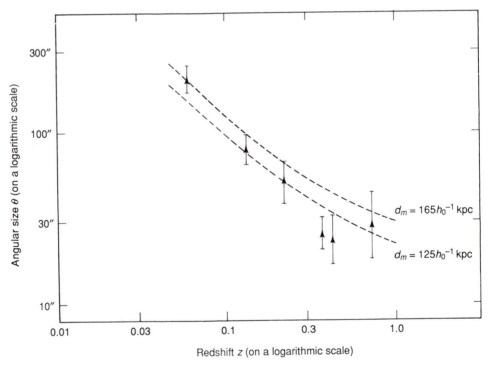

Fig. 10.7 Median angular size θ against the redshift z plotted for a number of radio sources, together with the theoretical curves for $q_0 = 0$ and $d_m = 165$ and $125h_0^{-1}$ kpc; the error bars seem to permit a wide range of values of q_0, although even a Euclidean result $\theta \propto z^{-1}$ cannot be ruled out. (Based on J. Katgert-Merkelijn, C. Lari, & L. Padrielli, Statistical properties of radio sources of intermediate strength. *Astron. & Astrophys. Suppl.*, **40**, 91 (1979).)

Ekers at Groningen and V. K. Kapahi and G. Swarup at Ootacamund did extensive work on this project from 1974 to 1975. In their work the median value of θ did not show the expected upturn at low flux densities, but instead tended to level off. This could imply one of the following interpretations:

1. Low S means large z. Since, at large z, θ should begin to increase according to Friedmann models with $q_0 > 0$, a flattening of θ implies evolution. In particular, the linear size d must decrease as z increases, implying that radio sources were smaller in the past than they are now.
2. Low S need not imply large z. We may simply be seeing sources of low luminosity. If there is a correlation between source size and source luminosity, the θ–S observations could be reproduced.

Kapahi preferred interpretation 1 and argued that a size evolution of the form

$$d \propto (1 + z)^\alpha, \qquad \alpha \approx 1 \tag{10.29}$$

can explain the θ–S observations. S. M. Chitre and the author took the opposite view, outlined above as interpretation 2. A better check on whether 1 or 2 is correct will come from the study of the structural properties of radio sources and from the measurements of their redshifts.

Subsequently, Kapahi considered four complete samples of radio galaxies with the following ranges of flux density and redshifts:

1. BFDL sample of A. H. Bridle, M. M. Davis, E. B. Fomalont and J. Lequeux with flux density at 1.4 GHz > 2 Jy and redshifts $0.075 < z < 0.2$;
2. GB/GB2 sample of J. Machalski and J. J. Condon, with flux density at 1.4 GHz > 0.55 Jy and redshifts $0.15 < z < 0.4$;
3. Same sample as (2) but with flux density at 1.4 GHz > 0.2 Jy and redshifts $0.25 < z < 0.6$;
4. Leiden Berkeley Deep Survey (LBDS) sample of R. A. Windhorst, G. M. van Heerde and P. Katgert with flux density at 1.4 GHz > 0.01 Jy and redshifts $z \geqslant 0.8$.

Spectroscopic redshifts are known for most of the sources only in the brightest BFDL sample. For the GB/GB2 samples they have been *estimated* from optical magnitudes by using the Hubble relation. The LBDS sample has sources that are either identified with galaxies of F magnitude > 22 or have no optical counterparts, implying that they have optical magnitudes fainter than the plate limit. Thus there is no direct information on redshifts but the expectation is that redshifts exceed 0.8 and are probably less than 2.

The median angular size versus redshift plot for these four samples shows again a steady decrease of θ_m as z^{-1}. To explain this, Kapahi invokes evolution of linear size with redshift, with sources at large redshifts being systematically smaller.

Although Kapahi has denied that Malmquist bias could creep into this result, we should keep the possibility open in view of the uncertain knowledge of redshifts. Suppose there is no evolution, but an anticorrelation between source luminosity and linear size. Then in a more distant (large z) sample the limits of flux density would admit only the relatively more luminous and hence more compact sources. This would bring down θ_m purely as a selection effect.

10.5 The surface brightness test

A test that combines the magnitude and angular sizes involves the measurement of surface brightness as a function of redshift. In the Robertson–Walker world models, formulae (3.44) and (3.53) in Chapter 3 give the apparent brightness of a source and its angular area respectively as:

$$\mathscr{F}_{bol} = \frac{L_{bol}}{4\pi r_1^2 S^2(t_0)(1 + z)^2} \tag{10.30}$$

$$A \equiv \frac{\pi}{4}(\Delta\theta_1)^2 = \frac{\pi d^2(1 + z)^2}{4r_1^2 S^2(t_0)}. \tag{10.31}$$

Dividing (10.30) by (10.31) gives the surface brightness of the source as

$$\sigma = \frac{L_{bol}}{\pi^2 d^2(1 + z)^4}. \tag{10.32}$$

Notice that σ does *not* depend on r_1; nor does it depend on q_0 – the parameter that labels different cosmological models. It depends only on $(1 + z)$ as its negative fourth power. Sandage has emphasized that this fourth power law is a signature of Hubble expansion, and as such it could be used to distinguish the expanding world models from other types of theories in which the redshift does not come from expansion.

Although we have not discussed such theories in Chapter 8, we should mention that there are such cosmologies. For example, there is the 'tired light' theory of J.-C. Pecker and J. P. Vigier, the chronometric cosmology of J. E. Segal, and so on.

Sandage has found that the $(1 + z)^{-4}$ law seems to be obeyed by first-ranked cluster numbers with a fairly narrow scatter, thus confirming the expanding universe picture for galaxy redshifts.

10.6 Quasars as probes of history of the universe

The tests described above assume that the redshifts of the objects used for the tests are of cosmological origin. This assumption is fairly sound in the case of galaxies for which, at least in the case of the first-ranked cluster numbers, the Hubble relationship is fairly tight.

By contrast, quasars have considerably larger redshifts. At the time of writing, 5000 quasars are listed in the catalogue of A. Hewitt and G. R. Burbidge, out of which over 1250 have redshifts between 1 and 2 and about 1200 have redshifts exceeding 2, the largest redshift listed being 4.897. These objects would therefore belong to region III of section 10.1 and should be considerably valuable as cosmological probes. We will assume here that the quasar redshifts are due to the expansion of the universe and so obey Hubble's law. To begin with, we discuss the evidence for this assumption which we shall refer to as the Cosmological Hypothesis (CH).

A plot of $\log z$ against m should, according to Hubble's law, give a slope $d \log z / dm = 5$ corrected for cosmological effects at large z. As early as 1966 G. R. Burbidge and F. Hoyle pointed out that the Hubble diagram for quasars is a scatter diagram with no apparent correlation between $\log z$ and m. This conclusion survived as more and more quasars were found, and it applies to the ~ 5000 quasars in the Hewitt–Burbidge list. Certainly Hubble, or for that matter any astronomer encountering these data in isolation, would not have concluded that any relationship exists between redshifts and magnitudes of quasars.

However, historically, quasars were discovered at a time when Hubble's law for galaxies was well established and none of the rival modes of redshifts – the Doppler or gravitational – were known to produce redshifts as high as the $z = 0.1$ common for galaxies. Thus it was natural to assume that quasar redshifts are also cosmological.

The conventional view when confronted with a scatter Hubble diagram has therefore been that the scatter is due to the vast spread in quasar luminosities. J. N. Bahcall and R. E. Hills argued in 1973 that a tight Hubble relationship for quasars is revealed when (1) corrections are made for various selection effects, (2) the quasar sample is divided into small redshift intervals (bins), and (3) the brightest quasar in each redshift bin is chosen. This conclusion has, however, been challenged by Burbidge and S. O'Dell, who find that their analysis along similar lines leads to much flatter slopes for $d \log z / dm$: slopes in the range from 2 to 3 instead of 5.

Whatever the outcome of such calculations, it is clear that the Hubble diagram cannot be taken as a proof of the correctness of the CH; at best, arguments of the Bachall–Hills type might make it compatible with the CH. Certainly, there seems no hope at present of using quasars to measure q_0 with the help of their Hubble diagram.

We next descuss other tests involving quasars.

10.6.1 Number counts

Using the luminosity volume test, M. Schmidt concluded that the average $\langle V/V_m \rangle$ for radio quasars in the 3CR catalogue was as high as 0.64 (compared with the Euclidean value of 0.50). Similar high values emerge in other surveys. On this basis it is usually argued that the quasar number density has been strongly evolving; that it was considerably higher in the past than now. Models of luminosity evolution as well as density evolution with enhancement factors like $(1 + z)^n$ $(n > 1)$ or exponential functions in the look-back time are used to fit the quasar number-count data. It is also argued that steep-spectrum radio quasars have stronger evolution than flat-spectrum radio quasars.

There also exist number counts of optical quasars which show a super-Euclidean slope of $\log N - m$ relation (a slope ~ 0.8 as opposed to 0.6 for the Euclidean universe) for the bright quasars. The $\log N - m$ curve flattens beyond the B-magnitude ~ 20. It is argued that luminosity evolution is responsible for this steepness. The numbers however, begin to fall off significantly beyond $z \approx 3$.

Curiously enough, the $\langle V/V_m \rangle$ for radio quasars turns out to be close to 0.5 if we assume that quasars are local (in region I) and uniformly distributed. This was found by R. Lynds and D. Wills in their examination of several complete samples of radio quasars.

In 1979 a new dimension was added to the source-count problem with the discovery (due largely to the Einstein Observatory) that X-ray emission is a characteristic feature of many quasars. Thus in principle it is possible to do a $\log N - \log S$ test for X-ray quasars. The early data suggested that the X-ray and optical luminosities of quasars are correlated. Hence, if optical number counts of quasars were to be taken as the basis of number counts of X-ray quasars also, then using formulae like (4.89) it would be possible to estimate the overall contribution to the X-ray background from quasars alone.

The optical number counts of quasars in 1979 suggested a steep rise in the number of faint quasars, and this led to the so-called X-ray background catastrophe. The quasars alone seemed to contribute over 100 per cent to the X-ray background. However, it is now realized that the quasar number density does not rise as earlier suspected. According to the later estimate of A. K. Kembhavi and A. C. Fabian, the contribution of X-ray quasars to the X-ray background should not exceed ~ 30 per cent of the total.

10.6.2 Angular-size–redshift relation

Compared with the chaotic situation in the case of the Hubble diagram, a clearer relationship between an observable (distance-dependent) property and the z of quasars emerges from a study of angular sizes. As pointed out in section 10.4, the largest angular size in a given redshift bin seems to decrease as z^{-1} for radio quasars. There is also a rough continuity between the θ–z plot of quasars and a similar plot for galaxies, suggesting that both objects probably belong to the same system. However, why should θ vary as z^{-1}, as no Friedmann model predicts? The curious thing is that if z were proportional to distance D then the observed result $\theta \propto D^{-1}$ is simply confirming Euclidean geometry! Attempts have been made to understand such a plot by making further assumptions such as evolution; but we are again forced to look upon such attempts as patchwork efforts. We certainly are not able to observe effects of non-Euclidean geometry, which we would have hoped for since the large z of a quasar means it is a distant object.

10.6.3 Absorption line systems

There are several quasars that show absorption lines as well as emission lines. The emission line redshift z_{em} of a quasar is usually the same for all lines. In some cases, however, more than one absorption line redshift z_{abs} is found. Also, mostly $z_{abs} < z_{em}$, although a few cases exist in which $z_{abs} > z_{em}$. Why do these redshift differences exist?

In principle, the difference between the emission line redshift and the absorption line redshift could be accounted for by (1) a relative motion between the emitting and the absorbing region, (2) a small contribution of gravitational redshift/blueshift between the two regions, or (3) the difference in the cosmological redshifts of the emitting and absorbing regions. Both (1) and (2) arise in the source, while (3) requires absorption to occur en route from the source to the observer. Note also that while (1) could be adjusted to have both $z_{abs} \geqslant z_{em}$, (2) and (3) require $z_{abs} < z_{em}$.

It is not clear from the work so far whether entirely satisfactory mechanisms exist in (1) to account for the various absorption line systems within the object. In the case of the quasar 3C 286, $z_{em} = 0.85$, while $z_{abs} = 0.69$; 21-cm observations of the source reveal a very small velocity difference ($\sim 3\,km\,s^{-1}$) across a distance of $\sim 300\,pc$ in the source. This result was quoted as a stumbling block to the theory, which seeks to explain the difference $z_{em} - z_{abs}$ as arising from high-speed gas driven outwards from within the quasar.

The more popular explanation of absorption line systems comes from (3), with the absorbers being intergalactic clouds or halos of galaxies situated *en route* from the QSO to the observer. Typically there are three types of absorption lines:

1. The broad absorption lines (BAL) or trough systems of CIV, SiIV, NV, OVI, etc. in addition to Ly α. The troughs are located on the blue side of the corresponding emission lines and have widths corresponding to a velocity $\leqslant 0.10c$.
2. The heavy-element systems containing sharp lines due to H and to heavier elements which may arise in a tenuous gas of near-solar composition. Here the difference between the emission and absorption redshifts corresponds to a velocity towards the quasar of $\leqslant 0.8c$.
3. The Ly α systems appearing with increased density on the blue side of the emission line.

Considerable work has been done to argue that the majority of the absorption lines arise from randomly distributed intervening cosmological objects. In some cases where an absorption line at a specific redshift has been found, a cluster of galaxies with the same redshift has been reported in the vicinity. As discussed in Chapter 9, however, the Gunn–Peterson test looking for substantial intergalactic neutral hydrogen *en route* to quasars did not yield a positive signal. This was interpreted more as evidence against the intergalactic hydrogen than the quasars being at their redshift distances.

10.6.4 Gravitational lens

In 1979 two quasars, $0957 + 561$ A, B, with the same redshift 1.4 and identical spectra were discovered. Their similarity led to the suspicion that they are two images of a single object produced by a gravitational lens. A cluster of galaxies with a redshift of 0.39 has since been identified as a probable candidate for such a lens.

Since the quasars are separated by only ~ 6 arc second, such an interpretation seems plausible and therefore provides support for CH, in the sense that the 'lensed' quasar is shown to be further away than the 'lensing' galaxy. However, there have been very few such clearcut cases of gravitational lensing.

10.6.5 Variability

Quasars show rapid variability in radio and optical wavelengths as well as X-rays, as indicated by recent data. A rule of the thumb is that if τ is the characteristic time scale of variation, the physical radius R of the object should not exceed $c\tau$. This leads to an energy generation problem that was first pointed out by Hoyle, Burbidge, and W. L. W. Sargent in 1966. The difficulty is briefly described as follows.

Since we measure the flux $\mathcal{F}$ from a quasar, its total luminosity is deduced from its distance. Under the CH, the distances are large and hence the luminosity is large. The quasar must therefore generate large quantities of energy in a small volume limited by the linear size $c\tau$. In the usual energy production scenario, the so-called synchrotron process, relativistic electrons radiate in magnetic fields. However, as Hoyle, Burbidge, and Sargent pointed out, this process results in the production of a very large density of photons. These photons collide with electrons, causing a very large Compton scattering, which degrades the energy of the fast electrons. Thus it is not possible to sustain energy production over distances even comparable to $R \simeq c\tau$; the electrons lose energy long before they travel a distance of this order.

The kinematic difficulty of whether quasars can manage to be confined to the radius $R \leqslant c\tau$ is partially alleviated by the following idea proposed by M. J. Rees. If an object expanding relativistically with a large Lorentz factor γ is viewed from a distance, it appears to increase its radial size at the rate γc. Thus the observed time scale of variation γ may be too small and the real inequality on R is $R \leqslant \gamma c\tau$ (see Exercise 20).

A compact size implied by the short time scale of variability is sometimes invoked in support of the idea that a quasar's energy is derived from a supermassive black hole. For example, the X-ray quasar OX 169 showed a significant drop in its X-ray luminosity within 100 minutes. The size limit implied by this time scale can accommodate a black hole of mass $\leqslant 10^8 M_\odot$.

10.6.6 Superluminal separation

Very-long-baseline interferometry (VLBI) observations have revealed a curious phenomenon in a number of quasars. In the central region of such a quasar, two radio components are observed to separate from each other very quickly. Since angular separations in milliseconds of arc are measurable by VLBI techniques, observations over a few months or a few years are sufficient to give a detectable effect. Thus it is found that the separation angle θ is changing (increasing) with time in such a way that the projected linear distance must change at speeds considerably faster than the speed of light – provided the quasars are at distances specified by the CH. Clearly, if the distances are much smaller than these, the speeds of separation become subluminal and the discrepancy with relativity disappears.

To retain the CH in spite of such data requires the conclusion that the observed separation is illusory. Various ways out have been suggested, and three of these are illustrated in Figure 10.8. Figure 10.8(a) shows the Christmas tree effect, which creates an illusion of motion by sequential lighting of stationary light bulbs. Figure 10.8(b) illustrates the so-called relativistic beaming, a variant of the idea proposed by Rees and described above under variability. The model illustrated in Figure 10.8(c) invokes a gravitational screen in the form of an intervening galaxy or a cluster of galaxies that bends the light rays (or radio waves) from the two components differentially so that their virtual images appear to separate at superluminal speed.

10.6.7 Morphology

It is argued that quasars and the nuclei of Seyfert and N-galaxies are basically similar objects and that in general we may think of a quasar as a galactic nucleus that is so bright that the rest of the galaxy either is not visible or is too faint to be seen. According to this argument, if the CH is correct, a large redshift implies a large distance and at that distance only the bright nucleus would be visible as a quasar. In some cases, such as the quasar Ton 256, it is argued that a fuzz surrounding the quasar has the luminosity distribution of an elliptical galaxy. To establish this line of evidence, which goes in favour of the CH, it would be necessary to show that a galaxy of stars indeed exists around the quasar. Absorption lines characteristic of stars in elliptical galaxies would be conclusive evidence for this purpose.

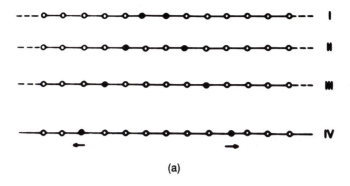

(a)

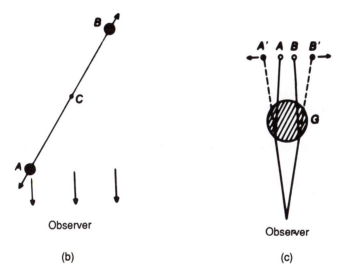

(b)   (c)

Fig. 10.8 Three ways of creating illusions of separation at speeds faster than the speed of light. (a) A row of lights is denoted by circles. The filled circles are lights that are lit. In stages I to IV the lighting is so contrived that a remote observer may think that two sources of light are moving outwards from the centre. (b) The observer sees the two actually separating components at different times; for light from the nearer component A leaves later than light from the further component B. C is the central nucleus at rest. An illusion of superluminal separation between A and B is created provided the line AB is almost aligned with the line of sight. This is very rare. Moreover, A is blueshifted and hence should be much brighter than B, which is redshifted with respect to the observer. It is usually presumed that B is at rest and A is beamed at the observer. (c) Here an intervening galaxy bends the rays from A and B so that the observer sees their images A', B', which could separate at superluminal speed even if A and B move apart with subluminal relative velocity. For this to happen the galaxy G must occupy a rather special position between the source and the observer.

10.6.8 Quasar–galaxy associations

One way of establishing that quasar redshifts are cosmological is to show that a galaxy and a quasar of the same redshift are physical neighbours. For we know that the galaxy redshift follows Hubble's law, and therefore the quasar redshift must also result from the CH.

During 1971 and 1972 J. Gunn, L. B. Robinson, and E. J. Wampler produced evidence of this kind. In 1978 a significant series of observations was reported by A. N. Stockton. Stockton chose all 27 known quasars with redshifts $\leqslant 0.45$ and visual magnitudes less than $19.2 + 5 \log z$ in the declination range $-15 < \delta < +55$. He then attempted to obtain spectra of all galaxies visible on the red Palomar Sky Survey plates lying within 45 arc seconds of any of the quasars. Of the 29 such galaxies he obtained the spectra of 25, of which 13 showed redshifts within $\sim 3 \times 10^{-3}$ of the redshift of the neighbouring quasar. Are these associations genuine or are they chance projections on the sky? Since the astronomer cannot measure radial distances of quasars, he has to use statistical arguments to settle the issue. Stockton's pairings would have come from chance projections with a probability less than 1.5×10^{-6}. Thus a statistician would be inclined to accept the associations as genuine and conclude that the quasars are near the galaxies and therefore at distances according to CH.

Taken in isolation this argument would be quite strong. Yet there is another side to this coin, which we shall discuss in the final chapter.

Broadly speaking, we may argue that there is considerable body of data *that is consistent with the quasar redshifts being cosmological*. It can also be argued that quasars and the nuclei of active galaxies which show emission lines form a continuous morphological sequence. Indeed, to push the argument further, one may argue that quasars *are* nuclei of galaxies which tend to outshine their galactic envelope, so that when they are seen from afar only the quasar is seen and not the galaxy. As a support to this argument, there are cases where a fuzz is found surrounding some low-redshift quasars.

However, we have as yet no direct proof that all the very-high-redshift quasars are in fact very far away, as the cosmological distance formula would have us believe. The nearest to direct evidence is the association between quasars and galaxies of the same redshift found by Stockton. Until a reliable distance indicator independent of Hubble's law becomes available for quasars, and it shows that the Hubble law is applicable to them, the CH must rest for its support largely on consistency arguments or just simple faith.

10.7 The variation of fundamental constants

Standard cosmology is based on the conservative assumption that physics as we know it here and now can be extrapolated to apply to the large-scale structure of the universe. As we mentioned in Chapter 8, such an assumption is justified on the basis of the economy of postulates, or Occam's razor. Among the nonstandard cosmologies, only the Perfect Cosmological Principle guarantees the validity of this assumption. On empirical grounds there is no reason to believe that the assumption must otherwise hold. Thus it is possible to have fundamental constants like c, $\hbar$, e, G, and particle masses varying with space and time. In Chapter 8 we encountered cosmologies that assume the last two items on the above list may vary with epoch. We will consider the evidence relevant to these issues in this section.

10.7.1 The variation of $\alpha = e^2/\hbar c$

We have noticed in the context of the Large Numbers Hypothesis (LNH) that $e^2/Gm_p m_e$ should vary with cosmic time t. Dirac assumed this to imply $G \propto t^{-1}$, with e, m_p, m_e constants. There is also the alternative conclusion to be drawn from the LNH that $e^2 \propto t$, with G, m_p, m_e constants. This was suggested by Gamow in 1967 because he believed that such a rapid decline in G as t^{-1} is ruled out by observations. If $\hbar$ and c are constants, then Gamow's interpretation leads to the conclusion that the fine-structure constant $\alpha \equiv e^2/\hbar c$ must vary with epoch as $\sim t$.

In 1967 J. N Bahcall and M. Schmidt measured the wavelengths of the O III multiplet line in the emission spectra of five radio galaxies with $z \sim 0.2$. If α is fixed then the wavelength difference $\delta\lambda$ between the observed multiplet lines as a fraction of the weighted mean wavelength λ of one of the lines must be the same in the observed spectra as in the laboratory spectra. If not, we have

$$\frac{\alpha(z)}{\alpha(0)} = \left(\frac{\delta\lambda}{\lambda}\right)^{1/2}_{\text{observed}} \times \left(\frac{\delta\lambda}{\lambda}\right)^{-1/2}_{\text{laboratory}}. \qquad (10.33)$$

Bahcall and Schmidt found that

$$\frac{\alpha(z = 0.2)}{\alpha(0)} = 1.001 \pm 0.002. \qquad (10.34)$$

If α was proportional to t, we should have got a value ~ 0.8 for the right-hand side of (10.34).

In 1977 M. S. Roberts compared the redshifts measured at optical wavelengths and at 21 cm in extragalactic sources to find that

$$\left|\frac{\dot{\alpha}}{\alpha}\right| \leq 4 \times 10^{-12} \text{ yr}^{-1}. \tag{10.35}$$

Again, this is an order of magnitude lower than the predicted variation rate $\alpha \propto t$.

10.7.2 The variation of G

This is an important observation, since the constancy of G is the basis of general relativity, on which standard cosmology is based. On the other hand, several nonstandard cosmologies predict $|\dot{G}/G| \sim H_0$ at the present epoch. We summarize below the direct and indirect evidence for the variation (or lack of variation) of G.

Radar observations.

In 1976 I. I. Shapiro and his colleagues reported the result of an analysis of several thousand observations of radar signals bounced off the inner planets between 1966 and 1975. Taking the other data from the Moon and the outer planets, the radar results give

$$\left|\frac{\dot{G}}{G}\right| < 10^{-10} \text{ yr}^{-1}.$$

P. M. Muller finds a value of $\dot{G}/G = (-6.9 \pm 3.0) \times 10^{-11} \text{ yr}^{-1}$.

Lunar mean motion

T. C. Van Flandern had examined Earth–Moon–Sun observations over several years using two time scales: atomic time, as measured by atomic clocks, and ephemeris time, derived from the Sun's motion around the Earth. The basis of these observations is as follows.

Suppose a body goes around another much more massive body in a circular orbit of radius r and mean angular velocity n. If M is the mass of the central body then Newtonian mechanics gives the following two relations:

$$GM = r^3 n^2, \qquad r^2 n = \text{constant} = h. \tag{10.36}$$

If we now introduce a slow variation of G with time, it is easy to deduce from the above two relations that

$$\frac{\dot{G}}{G} = \frac{\dot{n}}{2n}. \tag{10.37}$$

Thus the Earth's mean angular velocity around the Sun as measured by ephemeris time will slow down at the fractional rate of $2\dot{G}/G$ if G decreases with time. A similar equation should hold for the Moon, except that the Moon's motion is also affected by the tidal friction of the Earth–Moon system. Thus we have for the Moon

$$\frac{\dot{n}_M}{n_M} = \left(\frac{\dot{n}_M}{n_M}\right)_{\text{tidal}} + \frac{2\dot{G}}{G}. \tag{10.38}$$

If $\dot{n}_M/n_M$ is measured by atomic time, we should get (10.38). However, the same quantity measured by ephemeris time will take out the $2\dot{G}/G$ term arising from (10.37) and will only measure the tidal part. Thus the difference between the two observations should give us $2\dot{G}/G$.

The main uncertainty in this method has always been in obtaining a reliable estimate of the tidal effect. If the errors quoted in the various determinations of $\dot{n}$ are reliable, then there is a genuine contribution of the $\dot{G}/G$ term towards the Moon's motion. Early assessment of the data by Van Flandern suggested a value of the order $\dot{G}/G \sim (-6.9 \pm 2.4) \times 10^{-11}$ yr^{-1}, when considered within the framework of the Dirac cosmology. This rate is consistent with the Hoyle–Narlikar cosmology but is too high for the Brans–Dicke theory with $\omega \geqslant 30$.

The most accurate measurement to date of the rate of change of G was reported in 1983 from an analysis of the range data to the Viking landers on Mars. The experiment conducted by R. W. Hellings, P. J. Adams, J. D. Anderson, M. S. Keesey, E. L. Lau, E. M. Standish, V. M. Canuto and I. Goldman used the range measurements to the Viking landers and to the Mariner 9 spacecraft in orbit around Mars, the radar bounce range measurements from the surfaces of Mercury and Venus, the lunar laser range measurements and optical position measurements of the Sun and planets. A least-square fit of the parameters of the solar system model to the data shows that

$$\dot{G}/G = (0.2 \pm 0.4) \times 10^{-11}/\text{yr}.$$

Thus the result is certainly consistent with *zero variation* of G.

Stellar evolution

If G was greater in the past than now, stellar evolution would have proceeded at a faster rate, and this would lead to modifications of the *m−z* relation. In 1980 Canuto and the author showed that as far as the *G*-varying HN cosmology is concerned, the present data on *m−z* relation are consistent with the theoretical prediction. However, the uncertainties of the *m−z* relation are such that it cannot tell us definitely whether *G*-variation is taking place.

Biological evolution of the Earth

If *G* had been higher in the past, the Sun would have been brighter and the Earth closer to it than now. The Solar constant (the flux of radiation from the Sun outside the Earth's atmosphere, at present $\approx 1.388 \times 10^6$ erg cm^{-2} s^{-1}) therefore must have been considerably higher at the time of formation of the Earth than it is today. As estimated by Hoyle on the basis of the *G*-varying cosmology, at time when life began, say around 3×10^9 years ago, this constant may have been about three times its present value. Would life have been possible under such circumstances? Again, it can be shown that the *G*-variation in the HN cosmology is not inconsistent with the biological evolution of the Earth, although such evidence also cannot be used *to prove* that *G* does vary with epoch.

Exercises

1. Calculate the past light cone for Friedmann models by expressing $D(q_0, z)$ as a function of z. Plot these cones for $q_0 = 0, \frac{1}{2},$ and 1 as well as for the steady state model.

2. Discuss how the *m−z* curves for different cosmological models branch out for different values of q_0. Why does the uncertainty in the value of H_0 not hamper the test of the value of q_0?

3. What are the various issues that need to be considered before the *m−z* plot can lead to something of cosmological significance?

4. Discuss the aperture correction. In what way does it depend on q_0?

5. Show that
$$\frac{D(1, z)}{D(\frac{1}{2}, z)} = \frac{z}{2[(1 + z) - (1 + z)^{1/2}]},$$
and deduce that for $\alpha = 0.7$ the aperture correction introduces a magnitude difference of $\sim 0.09^m$ at $z = 0.7$.

6. Suppose $I(\lambda) \propto \lambda^2$ in the range 2500 Å $< \lambda <$ 5000 Å. A galaxy of redshift 0.5 is being observed in a wavelength band centred on

5000 Å. Another galaxy of redshift 0.7 is also observed at 5000 Å. Show that the K-terms for the two galaxies will differ by $\sim 0.41^m$.

7 Discuss the Stebbins–Whitford effect. Is it eliminated by taking due account of the K-correction?

8 Let $K(\lambda)$ denote the cross-section of absorption per unit mass of intergalactic dust at wavelength λ. Show that in a Friedmann model of given q_0, H_0, the apparent magnitude of a galaxy of redshift z_0, is increased owing to intergalactic absorption by an amount (at the measured wavelength λ_0)

$$\Delta m = 2.5\,(\log_{10} e)\,\frac{c\rho_0}{H_0}\int_0^{z_0} K\left(\frac{\lambda_0}{1+z}\right)\frac{(1+z)\,dz}{(1+2q_0 z)^{1/2}}\,,$$

where ρ_0 is the density of intergalactic matter.

Taking $K(\lambda) = (6400/\lambda_{\text{angstrom}}) \times 10^4\ \text{cm}^2\,\text{g}^{-1}$, $q_0 = \frac{1}{2}$, $\rho_0 \equiv 2.45 \times 10^{-33}\ \text{g cm}^{-3}$, and $h_0 = 0.5$, show that at $\lambda_0 = 6400$ Å, $\Delta m \approx 1^m$ for a galaxy of redshift unity.

9 Show how the luminosity evolution introduces uncertain corrections to the value of q_0. Using (10.16) with $x = 1.35$, compute the 'true' values of q_0 for the measured values $q_0 = 1$ and $q_0 = \frac{1}{2}$.

10 For the luminosity function of galaxies given by (10.21), show that the N–m relation in the $q_0 = 0$ Friedmann model is given by

$$N(<m) \propto \int_0^\infty \frac{x^{1/4} e^{-x}\,dx}{[x^{1/2} + \text{dex}\,(4.658 - m/5)]^3}\,,$$

where $\text{dex}\,y = 10^y$. Show that for a small m the above result becomes the same as for Euclidean geometry.

11 Show why the K-correction is necessary for the number counts of faint galaxies.

12 A radio galaxy of redshift $z = 0.1$ has a spectral index $\alpha = 1$ and luminosity of $10^{44}\ \text{erg s}^{-1}$ over the frequency range $150 \leqslant \nu \leqslant 1500$ MHz. For $h_0 = 1$ show that the flux density of the galaxy is ~ 350 Jy at 1000 MHz and ~ 1750 Jy at 200 MHz. (Neglect any cosmological effects.)

13 Express the radio power of the source in Exercise 12 in units of watts per MHz per steradian at the frequencies of 200 MHz and 1000 MHz, respectively.

14 Suppose that the probability that the ratio V/V_m lies in the range $(x, x + dx)$ for $0 \leqslant x \leqslant 1$ is proportional to $x^n\,dx$. Estimate n from the observed value of $\langle V/V_m \rangle = \frac{2}{3}$.

15 Suppose that for small enough intervals $d\sigma_i$ we have at most one source per interval. Writing (10.25) in the form

$$p_i = \frac{\alpha \sigma_i^{-(\alpha+1)} d\sigma_i}{1 - \sigma_0^{-\alpha}}$$

and maximizing with respect to α the expression

$$\mathscr{L} = \sum \ln p_i,$$

show that for $\sigma_0 \gg 1$ we get

$$\alpha = M \big/ \sum \ln \sigma_i.$$

16   Let $f(L)dL$ denote the number of radio sources per unit volume in the luminosity range $(L, L + dL)$. Suppose that for small redshifts the plot of $\log z$ against $\log L$ follows a straight line of slope $\frac{1}{2}$. Also assume that the number of points in equal intervals of $\log L$ is found to be constant. Using Euclidean geometry with distance $\propto z$, deduce from these observations that $f(L) \propto L^{-2.5}$.

17   Discuss why a sample of radio sources complete with respect to a minimum flux density is not necessarily complete with respect to a maximum redshift and vice versa.

18   A radio source survey gives $N = 10$ at $S = 12.5$ Jy, while $N = 93$ at $S = 5$ Jy. Show that in a Euclidean universe the above counts imply either a deficit of 13 sources at the high-flux end or an excess of 53 sources at the low-flux end. Use this example to comment on the Ryle–Hoyle controversy of the 1960s.

19   Show that in the Einstein–de Sitter model the number of galaxies intervening between a quasar of cosmological redshift z and the observer is given approximately by

$$N = 0.006 \left(\frac{R}{3 \text{ kpc}}\right)^2 \left(\frac{N_g}{0.1 \text{ Mpc}^{-3}}\right) [(1 + z)^{3/2} - 1],$$

where $h_0 = 1$, $R =$ the radius of the typical galaxy (assumed to be spherical), and $N_g =$ number density of galaxies. What can you say about the intervening galaxy interpretation of quasar absorption lines on the basis of this formula?

20   A spherical explosion leads to the expansion of an object with radial velocity V in the rest frame of a remote observer O. By considering the shape of the surface of simultaneity as seen by O. deduce that the object appears to expand laterally with speed $V(1 - V^2/c^2)^{-1/2}$.

21   Show that a supermassive black hole of mass $10^8 M_\odot$ has a

characteristic time scale of ~ 15 minutes. (In an accretion disc scenario, the disc may extend up to ~ 10 times the black hole radius, thus increasing the above time scale by a factor of 10.)

22 In the quasar 3C 345 an angular separation of central components was observed to increase from ~ 0.6 milliarc second in 1970 to ~ 1.6 milliarc second in 1975. The redshift of 3C 345 is 0.595. Show that if the redshift is cosmological the separation speed must be at least ~ 6.6c for $h_0 = 1$.

23 Suppose quasars are located at fixed distance D from their companion galaxies. Show that such an assumption leads to a $\log \theta - \log z_G$ plot with a mean slope of -1. Why is this assumption inconsistent with the CH?

24 Show that if the fine structure constant varies as t, then at a redshift of 0.2 the fine structure constant should be 77 per cent of its present value in the Einstein–de Sitter model.

25 Deduce that for a slow variation of G in Newtonian mechanics, the angular speed n of a particle going around a massive body in a circular orbit changes as follows:

$$\frac{\dot{n}}{n} = \frac{2\dot{G}}{G}.$$

26 Discuss the uncertainty introduced in the measurement of $\dot{G}/G$ by the tidal force between the Earth and the Moon.

11

A critical overview

11.1 Cosmology as a science

The preceding chapters describe the attempts of present-day cosmologists to study their subject within the discipline imposed by science. From the days when it was a subject of philosophical speculations and religious dogma, cosmology has now developed into a subject to which the scientific method of investigation can be applied. This change has resulted from improvements in techniques for observing the large-scale structure of the universe and from the wide applicability of the laws of physics.

Nevertheless, by claiming to describe the universe as a whole, cosmology transcends the realms of all other branches of science. Any conclusions about the universe are bound to be profound and hence must be drawn with caution. This caution is often found missing in statements about cosmology. All too often the investigator (whether a theoretician or an observer) is tempted to mistake the model of the universe for the real thing. Categorical remarks about the state of the universe are often found upon closer examination to be model-dependent. Firm claims about observations of the universe have had to be withdrawn later when a better assessment of the observational errors became possible.

To summarize the work of earlier chapters and to take stock of present cosmological studies, we have therefore adopted the following method. We begin with an enumeration of points in favour of the standard big bang models. Since these have been extensively set out in the text we shall be brief in describing them. Next we counter these arguments with those *against* the same models. Here we shall be more critical then we have been hitherto. Our reason for playing the devil's advocate is two-fold. First, we wish to correct (the deceptively simple) belief that the last word in cosmology has been said and that physics has reached a virtual end.

376

Second, we wish to highlight the fact that the universe is much more complex than conveyed by the standard hot big bang picture. On either count we should keep looking for alternative baskets for the cosmic egg.

11.2 The case for standard cosmology

An ardent supporter of standard cosmology will mention the following points in favour of the hot big bang models.

1. The models are based on Einstein's general theory of relativity. To the extent that it has been possible to test this theory, its predictions have always been borne out by observations. Thus we have confidence that the framework of our models is based on a sound theory.

2. The standard models are the simplest solutions of Einstein's equations. It is remarkable therefore that they are able to reproduce such a profound observation as Hubble's law. Moreover, these models *predicted* this law rather than its coming as an afterthought. This is clearly an indication that we are working on the right lines.

3. So far there is no satisfactory alternative to the theory of primordial nucleosynthesis for explaining the abundances of light nuclei, especially ^{4}He and ^{2}H. The agreement between the observed abundances and the theoretical values is good enough to generate confidence in the hot big bang.

4. The observation of the microwave background radiation and its Planckian spectrum is a striking confirmation of the early hot phase in the history of the universe. Again, as in (2), it goes to the credit of the picture that the observation had been *predicted* by the theory.

5. The recent successes of the Grand Unified Theories (GUTs) applied to the very early universe suggest that such a scenario must have some germ of truth in it. For example, the expectation based on primordial nucleosynthesis that there cannot be more than three neutrino species appears to be borne out by particle accelerators. In any case, the physical conditions under which the three basic forces of nature unite could have existed only in the very early universe. Since it is believed that redundant laws do not exist in nature, the situation leading to a GUT must have operated sometime; hence the very early universe is the logical choice.

6. A logical consequence of the GUT phase transition is 'inflation', which has turned out to be a fruitful input to the standard hot big bang cosmology and promises to resolve some of its outstanding problems.

7. The number counts and angular sizes of radio sources and quasars show evolution with epoch on the characteristic time scale of the

expansion of the universe. The evolutionary models demand increased density of quasars in the past, and this is consistent with the predictions of the standard models that the universe was denser in the past than at present.

11.3 The case against standard cosmology

The agnostic in the cosmological debate may use the following counter-arguments.

1. General relativity has been tested only in the weak-field approximation. We have no empirical evidence as to how the theory fares under the strong-field scenarios of cosmology. The standard models therefore are to be looked upon as nothing more than extropolations into speculative regions.

2. Relativistic cosmology in general and standard models in particular have the curious and unsatisfactory feature of a spacetime singularity. The appearance of infinities is considered disastrous in any physical theory. In general relativity it is worse, since the singularity refers to the spacetime structure and physical content of the universe itself. Moreover, it is sometimes sought to dignify this defect by elevating it out of the reach of physics. Thus one is not supposed to worry that the big bang violates all conservation laws of physics, such as the law of conservation of matter and energy.

3. There is a discrepancy between the astrophysical age estimates and the Hubble age of the standard models. The discrepancy is made worse if h_0 is close to 1 rather than $\frac{1}{2}$, $q_0 \geq \frac{1}{2}$ rather than $q_0 \sim 0$. This is particularly so if one goes by the inflationary scenario.

4. The photon/baryon ratio of $\sim 10^8$ is not explained by the standard models and the present temperature of $\sim 3\,\mathrm{K}$ of the radiation background remains to be derived from a purely theoretical argument. Although GUT provides one way of explaining N_γ/N_B, the present explanation still has the character of a post-dicting parameter-fitting exercise.

5. Despite numerous attempts by so many experts in the field, the formation of large-scale structure in the universe is ill-understood, especially in the context of the extraordinary smoothness of the microwave background.

On a somewhat epistemological issue, one feels uncomfortable at the way the research on the very early universe is being carried out. Because so much brainpower is currently being devoted to this field – and has been for the last decade or more – this issue needs to be stressed somewhat more forcefully. We will do so by comparison with other branches of physics.

In general, physics (or for that matter, science in general) progresses with a close interplay between theoretical ideas and observed facts. Sometimes theory is speculative and is checked by firm observations. On other occasions theory is well founded, but observations need to be further sharpened. In the work on the very early universe neither scenario holds: here we are dealing with theoretical speculations side by side with no direct observational evidence.

When the electroweak theory was formulated it was an exercise in theoretical speculations in gauge theories. It might or might not have worked. That it did work was eventually demonstrated by accelerator experiments. This is an example of how the scientific method worked in particle physics. In Gamow's work on the early universe, well-established physics was used in a cosmological scenario that was speculative (there are no astronomical observations of the universe when it was ~ 1 s old). However, the ultimate predictions of the work can be compared with hard facts: the elemental abundances and the radiation background.

Neither of these conditions hold *vis-à-vis* the very early universe. No one can deny that the work on GUTs is still highly speculative. Nor can the theories be dynamically tested with particle accelerators. To capture the full flavour of a grand unified theory one needs to attain particle energies $\sim 10^{15}$ GeV, which are far beyond the capabilities of present technology. On the cosmological side, the physics of the standard model with or without inflation at $t \sim 10^{-35}$ s is entirely speculative.

Thus we are matching one speculation with another. There is no harm in doing that provided we keep reminding ourselves that at best we can claim consistency of this matching with what we observe today. Instead, very definitive claims are often made about what the universe was like at these epochs.

Further, the requirement of 'repeatability of an experiment' is not satisfied in this picture. The GUT phase transition, inflation, etc. happened once only, and conditions conducive to them would not occur again. We may contrast this situation with nucleosynthesis in stars. This is an ongoing process with each star as an independant experiment.

The role of non-baryonic dark matter is highly reminiscent of the Emperor's new clothes in the story by Hans Christian Andersen. Except for neutrinos (whose massiveness is still open to question), no other form of such matter is experimentally established. Yet the various esoteric particles are taken for granted uncritically. Perhaps the cosmologists think that the physicists have established their existence on a firm footing, while the physicists think that such particles must exist because cosmologists tell them so. The hard fact is that there are no hard facts on either side!

So the sceptic may be permitted to ask, 'Is the work on the very early universe, inflation, dark matter, etc. real physics?'

11.4 The observational uncertainties

Here we wish to examine how firmly rooted are the main observations on which modern cosmology is founded. In particular, we will discuss the extent of validity of Hubble's law and the primordial (relic) interpretation of the microwave background.

11.4.1 How universal is Hubble's law?

Throughout the book we have taken it for granted that the redshift of an extragalactic object is cosmological in origin, i.e., that it is due to the expansion of the universe. In Chapter 10 we described this assumption as the Cosmological Hypothesis (CH). There we commented on the fact that while the Hubble diagram on which the CH is based shows a fairly tight $m-z$ relationship for first-ranked galaxies in clusters, a corresponding plot for quasars has enormous scatter. Although we discussed the cosmological tests on the basis of CH for quasars as well as galaxies, we found that in some cases special efforts are needed to make the CH consistent with data on quasars. These included, apart from the Hubble diagram, the superluminal motion in quasars, rapid variability, the absence of a Ly α absorption trough, etc.

To what extent is the CH valid for quasars? Let us begin with the type of data Stockton had collected in which quasars and galaxies were found in pairs or groups of close neighbours on the sky. The argument was that if a quasar and a galaxy are found to be within a small angular separation of one another, then very likely they are physical neighbours and according to the CH their redshifts must be nearly equal.

This argument is based on the fact that the quasar population is not a dense one, and if we consider an arbitrary galaxy then the probability of finding a quasar projected *by chance* within a small angular separation from it is very small. If the probability is < 0.01, say, then the null hypothesis of projection by chance is to be rejected. In that case the quasar may be physically close to the galaxy.

While Stockton found evidence that in such cases the redshifts of the galaxy and quasars, z_G and z_Q, say, were nearly the same, there have been data of the other kind also. In a book listed in the bibliography, H. C. Arp has described numerous examples where the chance projection hypothesis is rejected but $z_G \ll z_Q$. Over the years four types of such discrepant redshift cases have emerged:

1. There is growing evidence that large-redshift quasars are preferentially distributed closer to low-redshift bright galaxies (see Figure 11.1)
2. There are alignments and redshift similarities in quasars distributed across bright galaxies (see Figure 11.2).

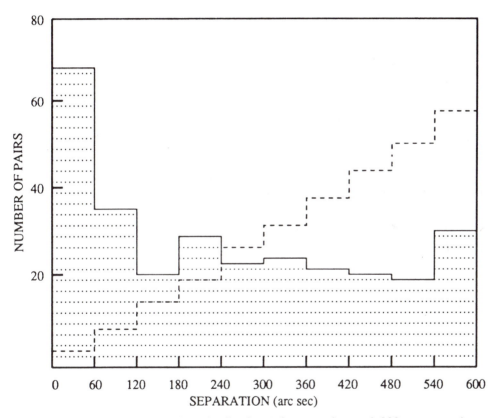

Fig. 11.1 A histogram of the distribution of separations of 300 quasar-galaxy pairs. If the quasars were randomly distributed with respect to bright galaxies then their numbers should have increased in proportion to the square of angular separation as shown by the broken line. Instead, there is a peak within 60 arcsec. The quasars are all of considerably higher redshifts than galaxies. (After G. Burbidge, A. Hewitt, J. V. Narlikar, & P. Das Gupta, Association between quasi-stellar objects and galaxies. *Ap. J. Suppl.*, **74**, 679 (1990).)

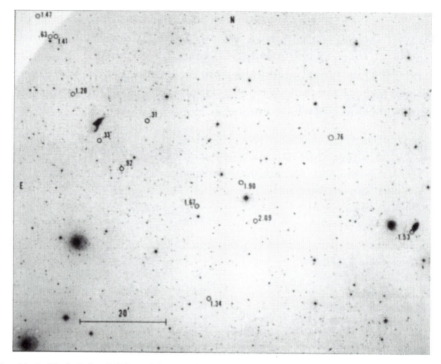

Fig. 11.2 Six bright QSOs with bunched redshifts as marked are seen here aligned across NGC 520. (From H. Arp & O. Duhalde, *PASP*, **97**, 1149 (1985).)

3. Close pairs or groups of quasars of discrepant redshifts are found more frequently than could have been due to chance projection (see Figure 11.3).
4. There are filaments connecting pairs of galaxies with discrepant redshifts (see Figures 11.4 (a) and (b).

The reader may find it interesting to go through the controversies surrounding these examples. The supporters of CH like to dismiss all such cases as either observational artefacts or selection effects. Or, they like to argue that the excess number density of quasars near bright galaxies could be due to gravitational lensing. While this criticism or resolution of discrepant data may be valid in some cases, it is hard to see why this should hold in *all* cases.

Another curious effect, first noticed by G. Burbidge in the late 1960s, concerns the apparent periodicity in the redshift distribution of quasars.

The periodicity of $\Delta z \approx 0.061$ first found by Burbidge for about seventy QSOs is still present with the population multiplied thirty fold (see Figure 11.5). What is the cause of this structure in the z-distribution? Another claim, first made by Karlsson in 1977, is that $\log(1 + z)$ is periodic with a period of 0.206.

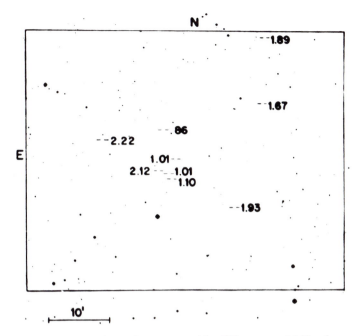

Fig. 11.3 The concentration of quasars with different redshifts (as marked in the figure) in the vicinity of the region of Right Ascension $11^h 46^m 14^s$ and declination $11° 11' 42''$ found by H. Arp and C. Hazard (*Ap. J.*, **240**, 726 (1980)).

On a finer scale, W Tifft has been discovering a redshift periodicity $c \Delta z = 72 \, \mathrm{km \, s^{-1}}$ for differential redshifts for double galaxies and for galaxies in groups. The data have been refined over the years with accurate 21-cm redshift measurements. If the effect was spurious, it would have disappeared. Instead it has grown stronger and has withstood fairly rigorous statistical analyses (see Figure 11.6).

For a universe regulated by Hubble's law, it is hard to fit in these results. The tendency on the part of the conventional cosmologist is to discount them with the hope that with more complete data they may disappear. At the time of writing this account there is no such tendency in the data!

It is possible that the effects are genuine and that our reluctance to ignore them also stems from the lack of any reasonable explanation. The explanation may bring in other noncosmological components in the observed redshift z. Thus we would write

$$1 + z = (1 + z_C)(1 + z_{NC}). \tag{11.1}$$

The cosmological component z_C obeys Hubble's law while the non-cosmological part z_{NC} exhibits the anomalous behaviour. What could z_{NC}

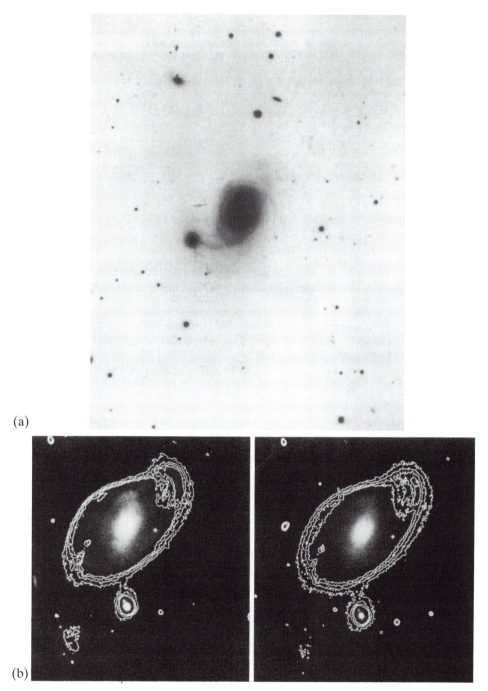

(a)

(b)

Fig. 11.4 (a) The large galaxy NGC 7603 ($cz = 8700\,\mathrm{km\,s^{-1}}$) appears connected to a compact companion ($cz = 16\,900\,\mathrm{km\,s^{-1}}$). (b) The luminous connection first found by Arp between NGC 4319 and Markarian 205 with redshifts $z = 0.0056$ and 0.07 respectively has been confirmed by J. Sulentic with CCD observations.

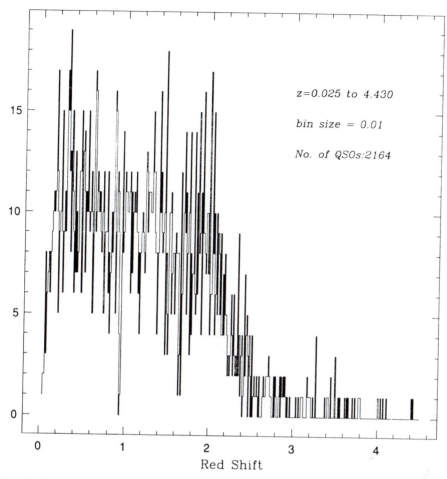

Fig. 11.5 A histogram of quasar redshifts showing peaks at approximate multiples of 0.06. The peaks are confirmed by power spectrum and other analyses carried out by D. Duari, P. Das Gupta, and J. V. Narlikar.

be due to? There are a few possibilities, none of which is thoroughly tested for full satisfaction:

1. Doppler effect arises from peculiar motions relative to the cosmological rest frame. It is a well-known phenomenon in physics.
2. Gravitational redshift arises from compact massive objects as discussed in Chapter 2.
3. Spectral coherence discussed by E. Wolf causes a frequency shift in propagation when light fluctuations in the source are correlated.
4. In the tired light theory a photon of nonzero rest loses energy while propagating through space.

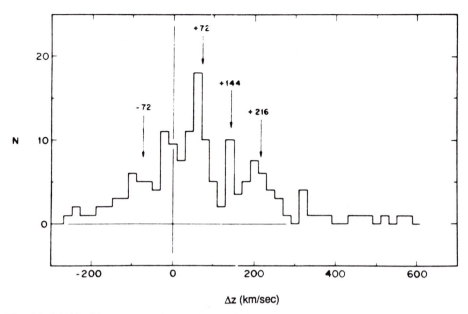

Fig. 11.6 This histogram of redshift differentials between dominant and companion galaxy redshifts shows peaks at multiples of $72 \, \mathrm{km \, s^{-1}}$. (Source: H. Arp, *Quasars, Redshifts and Controversies* (Berkeley, Calif.: Interstellar Media, 1987)

5. In the variable mass hypothesis arising from the Machian theory of F. Hoyle and the author, particles may be created in small and large explosions and those created more recently will have smaller mass and hence larger redshift.

To what extent can these alternatives provide explanations for the discrepant data? Would the discrepancies dwindle away as observations improve or would they grow in significance? Clearly these issues have enormous implications for Hubble's law in particular and for cosmology in general.

11.4.2 Is the microwave background primordial?

In 1968, three years after the discovery of the microwave background, F. Hoyle, N. C. Wickramasinghe and V. C. Reddish showed that if all the observed helium were generated in stars, the resulting starlight would have the energy density comparable to that in the microwave background. The essence of calculation is as follows.

Taking the observed baryonic density as $\sim 3 \times 10^{-31} \, \mathrm{g \, cm^{-3}}$ and $Y = 0.25$, the amount of helium per unit volume is 0.75×10^{-31} g. The production of 1 g of helium from hydrogen releases an energy

$0.007 \times c^2$ g $\approx 0.063 \times 10^{20}$ erg. Thus the energy density of starlight generated is $0.75 \times 0.063 \times 10^{-11}$ erg, i.e. $\sim 4 \times 10^{-13}$ erg cm^{-3}. If this energy could be thermalized it would produce a blackbody radiation of temperature 2.7 K.

This calculation contains the germ of an idea for creating the observed microwave background non-primordially.

If all helium could be produced nonprimordially, i.e., say mostly in supermassive stars, and if a mechanism could be found for thermalizing the resulting starlight, we would have the observed microwave background. The question is, how to achieve the thermalisation?

The answer to this question is now beginning to emerge. Hoyle and Wickramasinghe have suggested that dust grains in the form of graphite or iron needles can do the trick. If they are spread over the intergalactic space with a mean density no more than $\sim 10^{-35}$ g cm^{-3} they can absorb and re-emit starlight continuously until it is thermalized in time scales small compared with H_0^{-1}. A typical needle may be about 1 mm long and $\sim 1\,\mu$m in cross-section. The universal cosmic abundances of heavy elements certainly allow for a mean density of this order. Dust particles of this kind would be opaque to microwaves but exceedingly translucent to visible light.

How do such needles form? Laboratory experiments show that cooled metallic vapours condense into such whiskers. The formation is understood to occur in a sequence with first nucleation into liquid droplets of a few thousand atoms followed by sudden crystallization in the linear screw dislocation mode. Since such growths occur at an exponential rate with time, they outstrip growths which take place linearly with time.

In the interstellar space such whiskers would form with metals ejected by supernovae. Circumstantial evidence points to their presence near the Crab pulsar, whose radiation shows an apparent dip in the wavelength range 30 μm to 10 cm that could be ascribed to absorption by dust of this kind.

Two questions immediately occur with this scenario. First, will such a dust block our view of the distant parts of the universe, thus making it invisible in radio and optical wavelengths at even moderate red shifts of $z \leqslant 1$? The detailed calculations with the Friedmann and the steady state models set these doubts at rest. Discrete extragalactic sources at redshifts a upto $z \sim 2$ can be easily visible without difficulty, both in the radio and the optical.

The second question relates to the expected homogeneity of the thermalized radiation. The answer, as expected, is linked with the overall distribution of the cosmic dust. The absorptivity of iron whiskers is so large in the far infrared that they are pushed out by the strong outward radiation pressure from within a galaxy into the intergalactic space. Any inhomogeneity of the microwave radiation produces local pressure gradients that tend to redistribute the absorbers so as to reduce them. This process works so as to reduce $\Delta T/T$ to order $\leqslant 10^{-5}$.

This illustrates a scenario for making the microwave background of non-relic nature. This explanation works also in the steady state model, where the creation of matter, the formation of stars, their evolution and explosion, and the thermalization by ejected dust are all parts of an ongoing activity. It has the advantage that the observed radiation temperature is explained in terms of the ongoing astrophysical processes, rather than being ascribed to unobservable primordial processes. Any reader who finds the idea of 'whisker grains' strange should contrast the laboratory evidence for them with the pure speculations on non baryonic dark matter.

11.5 Outlook for the future

The above discussion should be sufficient to convince the reader that the subject of cosmology is still very much open. As discussed in Chapter 1, the majority of astronomers at the turn of the century held the view that the entire universe is contained in our Galaxy. Improved observing techniques soon demolished that view, and by the mid-1920s glimpses of the vast extragalactic world were beginning to enlarge the scope of cosmology.

Today we are similarly on the brink of another observational breakthrough, with many new-technology telescopes in the offing both on the ground and in space. It may well be that observations from these will confirm the standard big bang picture. It is equally likely that like our predecessors of eight decades ago we may be in for a radical revolution of ideas in cosmology. We end this book with a few speculations.

1. The most important cosmological parameter to be determined is H_0. The Hubble Space Telescope and other (ground-based) telescopes will no doubt attempt to narrow down the range of uncertainty in its measured value. The problem as we saw in Chapter 9, has become more complex, with the discovery of large scale streaming motion in our neighborhood. Nevertheless, problems of age of the universe, matter density, etc. need a clearcut value of H_0 to set limits on the allowable cosmological models.
2. Is $\Omega_0 = 1$? This question has assumed more importance since the advent of

the inflationary model. Independent measurements of q_0 and Ω_0 can, in principle, tell us if the Friedmann models are correct: for, if $\Omega_0 \neq 2q_0$, we may need the λ-term to make up the difference.

3. Will the microwave background reveal small-scale fluctuations that could be considered the signature of structure formation? Or will alternative evidence come to light implying that the background is not of primordial origin?

4. Will the dark matter turn out to be nonbaryonic? And, if so, will astronomers be able to identify it out of the zoo of such particles currently discussed by physicists? Or will the answer be baryonic after all?

5. On the theoretical side, the most significant advance will be made when there is a working theory of structure formation. Which came first, superclusters or galaxies? What happened to the relic radiation while the structures were forming? And, in general, does matter trace light?

6. Are the anomalous redshift cases referred to in the previous section significant? Are the periodicities in redshifts real? At what stage do we begin to worry about a theory to explain them? These issues go deeper than appears at first sight, since the very foundations of cosmology depend on the outcome.

7. With improvements in technology, it may be possible in the foreseeable future to put limits on $|\dot{G}/G|$ from pure laboratory experiments. This is a good test for Machian cosmologies and the LNH.

Finally, as in the long history of man's exploration of the universe, the bigger and better telescopes of the future will reveal unexpected new phenomena in the universe, phenomena that will provide the greatest intellectual challenges.

We conclude with this prognostication.

Table of constants

This table is taken from the following sources:

I. M. Ryshik and I. S. Gradstein, *Tables of Series, Products and Integrals* (Berlin: Veb Deutscher Verlag der Wissenschafte 1957).
Particle Data Group, 'Review of particle properties', *Rev. Mod. Phys. 52*, No. 2 (1980).
A. H. Wapstra and N. B. Gove, 'The 1971 Atomic Mass Evaluation', *Nuclear Data Tables 9A*, 265 (1971).
C. W. Allen, *Astrophysical Quantities* (London: Athlone Press, 1973).

Figures given in parentheses represent 1σ uncertainty in the last digits of the main numbers.

Mathematical constants

$\pi = 3.141\,59$ $e = 2.718\,28$ $\zeta(3) = 1.202\,06$
$\ln 2 = 0.69315$ $\ln 10 = 2.30259$ $\log e = 0.43429$
1 arc second $= 4.8481 \times 10^{-6}$ radians
1 steradian $= 3.2828 \times 10^3$ square degrees
Square degrees on a sphere $= 41\,252.961\,24$

Physical constants

Speed of light	$c = 2.997\,924\,58(1.2) \times 10^{10}$ cm s^{-1}
Planck's constant	$\hbar = 1.054\,588\,7(57) \times 10^{-27}$ erg s
	$= 6.582\,173(17) \times 10^{-16}$ eV s
	$h \equiv 2\pi\hbar = 6.626\,20 \times 10^{-27}$ erg s

Electronvolt	$1 \, eV = 1.602\,189\,2(46) \times 10^{-12}$ erg
Gravitational constant	$G = 6.6720(11) \times 10^{-8}$ dyn cm^2 g^{-2}
Charge of the electron	$e = 4.803\,242(14) \times 10^{-10}$ esu

Fine structure constant
$$\alpha \equiv \frac{e^2}{\hbar c} = [137.036\,04(11)]^{-1}$$

Planck length
$$\left(\frac{G\hbar}{c^3}\right)^{1/2} = 1.6 \times 10^{-33} \, cm$$

Planck time
$$\left(\frac{G\hbar}{c^5}\right)^{1/2} = 5.4 \times 10^{-44} \, s$$

Planck mass
$$\left(\frac{c\hbar}{G}\right)^{1/2} = 2.2 \times 10^{-5} \, g$$

Electron mass	$m_e = 9.109\,534(47) \times 10^{-28}$ g
Electron mass energy	$m_e c^2 = 0.511\,003\,4(14)$ MeV
Proton mass energy	$m_p c^2 = 938.2796(27)$ MeV
Neutron mass energy	$m_n c^2 = 938.5731(27)$ MeV

Planck energy
$$\left(\frac{c^5 \hbar}{G}\right)^{1/2} = 1.2 \times 10^{19} \, GeV$$

Thomson cross-section
$$\frac{8\pi e^4}{3m_e^2 c^4} = 0.66\,524\,48(33) \times 10^{-24} \, cm^2$$

Boltzmann constant
$$k = 1.380\,662(44) \times 10^{-16} \, erg \, K^{-1}$$
$$k^{-1} = 11\,604.50(36) \, K(eV)^{-1}$$

Radiation constant
$$a = \frac{8\pi^5 k^4}{15c^3 h^3}$$
$$= 7.5641 \times 10^{-15} \, erg \, m^{-3} \, K^{-4}$$

Number density of photons in a blackbody radiation of temperature T
$$\frac{2\zeta(3)}{\pi^2}\left(\frac{kT}{c\hbar}\right)^3 \cong 20.3 \, T^3 \, cm^{-3}$$

Weak interaction constant
$$\mathcal{G} = 1.02 \times 10^{-5}\left(\frac{\hbar^3}{m_p^2 c}\right)$$

Binding energy of deuterium	$= 2.224\,64(4)$ MeV
Binding energy of helium	$= 28.2969(4)$ MeV

Astronomical constants

Light-year	$1 \text{ ly} = 9.4605 \times 10^{17} \text{ cm}$
Parsec	$1 \text{ pc} = 3.0856(1) \times 10^{18} \text{ cm} \cong 3.26 \text{ ly}$
Radius of the Sun	$R_\odot = 6.959 \times 10^{10} \text{ cm}$
Mass of the Sun	$M_\odot = 1.989(1) \times 10^{33} \text{ g}$
Luminosity of the Sun	$L_\odot = 3.826(8) \times 10^{33} \text{ erg s}^{-1}$

Mass/light ratio for the sun

$$\frac{M_\odot}{L_\odot} \cong 0.51 \text{ g erg}^{-1} \text{ s}$$

Luminosity of a star of zero absolute magnitude ($M_{\text{bol}} = 0$)

$$L_0 = 2.97 \times 10^{35} \text{ erg s}^{-1}$$

Flux from a star of zero apparent magnitude

$$l_0 = 2.48 \times 10^{-5} \text{ erg cm}^{-2} \text{ s}^{-1}$$

Radio flux density	$1 \text{ Jy (jansky)} = 10^{-26} \text{ W m}^{-2} \text{ Hz}^{-1}$
Hubble constant	$H_0 = 100 \, h_0 \text{ km s}^{-1} \text{ Mpc}^{-1}, \, 0.5 \leqslant h_0 \leqslant 1$
Hubble age	$T_0 = H_0^{-1} \cong 9.8 h_0^{-1} \times 10^9 \text{ yr}$

Glossary of symbols and abbreviations

$3C, 4C, \ldots$	Cambridge Catalogue of radio sources
CDM	cold dark matter
CH	cosmological hypothesis (for quasar redshifts)
GA	great attractor
GeV	giga electron volt (unit for energy)
H	Hubble's constant
h_0	H measured in units of $100 \, \mathrm{km \, s^{-1} \, Mpc^{-1}}$
HDM	hot dark matter
HM	Hubble modulus
HN	Hoyle–Narlikar
IMF	initial mass function
L	luminosity
LG	Local Group
LNH	large numbers hypothesis
M	absolute magnitude
m	apparent magnitude
MeV	mega electron volt (unit for energy)
NGC	New General Catalogue
PCP	Perfect Cosmological Principle
PSR	pulsar catalogue label
q	deceleration parameter
$SU(n)$	special unitary group of n dimensions
$U(n)$	unitary group of n dimensions
VLBI	very-long-baseline interferometry

Bibliography

The reference material listed below is divided into three categories, (1) textbooks, (2) review articles and proceedings of scientific meetings and (3) lists of papers of a pioneering nature. In a rapidly growing subject like cosmology it is not possible to give an exhaustive list of references. It is hoped that this distribution of sources of information will meet the varying needs of the readers.

1 Textbooks

This is a representative (but by no means complete) list of books that cover part of the subject matter of the present book, with the numbers after each entry referring to the chapters of the present book where there is an overlap. Works preceded by an asterisk are less technical and mathematical than the present book.

I. J. R. Aithison & A. J. G. Hey, *Gauge Theories in Particle Physics* (Bristol: Adam Hilger, 1982). (6)

*H. C. Arp, *Quasars, Redshifts and Controversies* (Berkeley, Calif.: Interstellar Media, 1987). (11)

J. Binney & S. Tremaine, *Galactic Dynamics* (Princeton: Princeton University Press, 1987). (6, 7, 9)

G. R. Burbidge & E. M. Burbidge, *Quasi–Stellar Objects* (New York: W. H. Freeman, 1967). (10, 11)

E. M. Corson, *Introduction to Tensors, Spinors and Relativistic Wave Equations* (London: Blackie & Sons, 1953). (2, 6)

L. P. Eisenhart, *Riemannian Geometry* (Princeton: Princeton University Press, 1926). (2, 3)

*E. R. Harrison, *Cosmology: The Science of the Universe* (Cambridge: Cambridge University Press, 1988). (1, 3, 4, 9, 10)

S. W. Hawking & G. F. R. Ellis, *The Large Scale Structure of Space–time* (Cambridge: Cambridge University Press, 1973). (2)

S. W. Hawking & W. Israel, eds, *General Relativity, An Einstein Centenary Survey* (Cambridge: Cambridge University Press, 1979). (2, 3, 4, 5, 6, 7)

R. W. Hockney & J. W. Eastwood, *Computer Simulation Using Particles* (New York: McGraw-Hill, 1981). (7)

F. Hoyle & J. V. Narlikar, *Action at a Distance in Physics and Cosmology* (New York, W. H. Freeman, 1974). (8)

*F. Hoyle & J. V. Narlikar, *The Physics–Astronomy Frontier* (New York: W. H. Freeman, 1980). (3, 4, 8, 9, 10)

E. W. Kolb & M. S. Turner, *The Early Universe* (New York: Addison–Wesley, 1990). (5, 6, 7)

L. D. Landau & E. M. Lifshiftz, *Statistical Physics* (Oxford: Pergamon, 1970). (5)

L. D. Landau & E. M. Lifshitz, *The Classical Theory of Fields* (Oxford: Pergamon, 1975). (2)

*D. Layzer, *Constructing the Universe* (New York: Scientific American Library, 1984). (1, 3, 4)

C. W. Misner, K. S. Thorne, & J. A. Wheeler, *Gravitation* (New York: W. H. Freeman. (2, 3, 4)

*J. V. Narlikar, *The Primeval Universe* (Oxford: Oxford University Press, 1988). (3, 4, 5, 6)

P. J. E. Peebles, *Physical Cosmology* (Princeton: Princeton University Press, 1971). (4, 5, 7)

P. J. E. Peebles, *The Large Scale Structure of the Universe* (Princeton: Princeton University Press, 1980). (7, 9)

A. K. Raychaudhuri, *Theoretical Cosmology* (Cambridge: Cambridge University Press, 1979). (3, 4, 5, 6)

*J. Silk, *Big Bang* (New York: W. H. Freeman 1990). (3, 4, 5, 6, 7)

S. Weinberg, *Gravitation and Cosmology* (New York: John Wiley & Sons, 1972). (2, 3, 4, 5)

Y. B. Zeldovich & I. D. Novikov, *The Structure and Evolution of the Universe* (Chicago: University of Chicago Press, 1983). (2, 3, 4, 5, 9, 10)

2 Reviews and proceedings

The reader wishing to delve deeper into specific topics may wish to begin with the relevant entries from the following list. In most cases the titles are self-explanatory.

E. S. Abers & B. W. Lee, Gauge theories. *Physics Reports*, **9**, 1 (1973).

N. A. Bahcall: Large-scale structure in the universe indicated by galaxy clusters. *Ann. Rev. Astron. Astrophys.*, **26**, 631 (1988).

R. Balian, J. Audouze, & D. N. Schramm, eds, *Physical Cosmology*, Les Houches Lectures Session XXXII (Amsterdam: North Holland, 1979).

F. Bertola, J. W. Sulentic, & B. F. Madore, eds, *New Ideas in Astronomy* (Cambridge: Cambridge University Press, 1988).

R. H. Dicke Implications for cosmology of stellar and galactic evolution rates. *Rev. Mod. Phys.*, **34**, 110 (1962).

A. Dresler, The great attractor: do galaxies trace the large-scale mass distribution? *Nature*, **350**, 391 (1991).

S. M. Faber & J. S. Gallagher, Masses and mass to light ratios of galaxies. *Ann. Rev. Astron. Astrophys.*, **17**, 135 (1979).

P. N. Hodge, The Extragalactic Distance Scale. *Ann Rev. Astron. Astrophys.*, **19**, 357 (1981).

A. Hewitt, G. Burbidge, & L. Z. Fang eds, Observational cosmology. *IAU Symposium No. 124* (Dordrecht: D. Reidel, 1987).

J. V. Narlikar, Noncosmological redshifts. *Space Science Reviews*, **50**, 523 (1989).

J. V. Narlikar & A. K. Kembhavi, Non-standard cosmologies. In *Handbook of Astronomy, Astrophysics and Geophysics*, Vol. II, eds V. M. Canuto & B. G. Elmgreen (New York: Gordon & Breech, 1988).

J. V. Narlikar & T. Padmanabham, Inflation for astronomers. *Ann. Rev. Astron. Astrophys.*, **29**, 325, (1991).

T. Padmanabhan & K. Subramanian, Galaxy formation. *Bull. Astron. Soc. (India)*: to the published, 1992.

P. J. E. Peebles & J. Silk, A cosmic book of phenomena. *Nature*, **346**, 233 (1990).

A. Sandage, Observational tests of world models. *Ann. Rev. Astron. Astrophys.*, **26**, 561, (1988).

D. Schramm & R. V. Wagoner, Element production in the early universe. *Ann. Rev. Nucl. Sci.*, **27**, 37 (1977).

S. F. Shandarin & Y. B. Zeldovich, The large scale structure of the universe: turbulence, intermittency, structures in self gravitating medium. *Rev. Mod. Phys.*, **61**, 185 (1989).

G. Steigman, Observational tests of antimatter cosmologies. *Ann Rev. Astron. Astrophys.*, **14**, 339 (1976).

G. Swarup & V. K. Kapahi, eds, Quasars. *IAU Symposium No. 119* (Dordrecht: D. Reidel, 1986).

V. Trimble, The existence and nature of dark matter in the universe. *Ann. Rev. Astron. Astrophys.*, **25**, 425 (1987).

S. Weinberg, Beyond the first three minutes. *Physica Scripta*, **21**, 773 (1980).

S. Weinberg, The cosmological constant problem. *Rev. Mod. Phys.*, **61**, 1 (1989).

3 References to early work

Although the papers listed by chapter below are from a historical perspective, many of them are classics that continue to be relevant today.

Chapter 1

Classification of galaxy types

E. Hubble, Extragalactic nebulae. *Ap. J.*, **64**, 321 (1926).

W. W. Morgan, A preliminary classification of the forms of galaxies according to their stellar spectra. *Publ. Astron. Soc. Pac.*, **70**, 364 (1958).

S. van den Bergh, The luminosity classification of galaxies and stellar evolution. *Mem. Soc. Roy. Liège (Belgium)*, Sér. Cinquième, vol. III (1960).

Discovery and identification of Cygnus A

J. S. Hey, S. J. Parsons, & J. W. Phillips, Fluctuations in cosmic radiation at radio frequencies. *Nature*, **158**, 234 (1946).

W. Baade & R. Minkowski, Identification of radio sources in Cassiopeia, Cygnus A and Puppis A. *Ap. J.*, **119**, 206 (1954).

The first two quasars

C. Hazard, M. B. Mackey, and A. J. Shimmins, Investigation of the radio source 3C273 by the method of lunar occultations. *Nature*, **197**, 1037 (1963).

J. L. Greenstein & M. Schmidt, The quasi stellar radio sources 3C48 and 3C273. *Ap. J.*, **140**, 1 (1964).

Clustering and superclustering of galaxies

C. D. Shane & C. A. Wirtanen, A distribution of extragalactic nebulae. *A. J.*, **59**, 285 (1954).

F. Zwicky, *Morphological Astronomy* (Berlin: Springer-Verlag, 1957).

G. O. Abell, The distribution of rich clusters of galaxies. *Ap. J., Suppl.*, **3**, 211 (1958).

G. de Vaucouleurs, Recent studies of clusters and superclusters, *A. J.*, **66**, 629 (1961).

D. Lynden-Bell, S. M. Faber, D. Burstein, R. L. Davies, A. Dressler, R. J. Terlevich, & G. Wegner, Photometry and spectroscopy of elliptical galaxies. *Ap. J.*, **326**, 19 (1988).

Nebular redshifts

V. M. Slipher, Spectrographic observations of nebulae. Paper presented at the 17th meeting of the A. A. S., August, 1914.

E. Hubble, A relation between distance and radial velocity among extragalactic nebulae. *Proc. Natl. Acad. Sci. (USA)*, **15**, 168 (1929).

E. Hubble & M. Humason, The velocity–distance relation among extragalactic nebulae. *Ap. J.*, **74**, 35 (1931).

Chapter 2

Formulation of the field equations of general relativity

A. Einstein, Zür allgemeinen Relativitätstheorie. *Preuss. Akad. Wiss. Berlin, Sitzber.*, 778 (1915).

A. Einstein, Zür allgemeinen Relativitätstheorie (Nachtrag). *Preuss. Akad. Wiss. Berlin, Sitzber.*, 799 (1915).

A. Einstein, Die Feldgleichungen der Gravitation. *Preuss. Akad. Wiss. Berlin, Sitzber.*, 844 (1915).

General relativity from an action principle

D. Hilbert, Die Grundlagen der Physik. *Konigl. Gesell d. Wiss. Göttingen, Nachr. Math.-Phys. Kl.*, 395 (1915).

The Schwarzschild solution

K. Schwarzschild, Über das Gravitationsfeld eines Masspunktes nach der Einsteinschen Theorie. *Sitzber. Deut. Akad. Wiss. Berlin, Kl. Math-Phys. Tech.*, 189 (1916).

Birkhoff's theorem

G. D. Birkhoff, *Relativity and Modern Physics* (Cambridge, Mass.: Harvard University Press, 1923).

Gravitational redshift

R. V. Pound & G. A. Rebka, Apparent weight of photons (Lab. Expt). *Phys. Rev. Lett.*, **4**, 337 (1960).

W. L. Wiese & D. E. Kelleher, On the cause of redshifts of white dwarf spectra (40 Eridani B). *Ap. J.*, **166**, L59 (1971).

J. L. Greenstein, J. B. Oke, & H. L. Shipman, Effective temperature, radius and gravitational redshift of Sirius B. *Ap. J.*, **169**, 563 (1971).

Perihelion advance

A. Einstein, Erklarung der Perihelbewegung des Merkur aus der allgemeinen Relativitätstheorie. *Preuss. Akad. Wiss. Berlin, Sitzber.*, 831 (1915).
G. M. Clemence, The relativity effect in planetary motions. *Rev. Mod. Phys.*, **19**, 361 (1947).
C. W. Will, Periastron shifts in the binary system 1913 + 16: theoretical interpretation. *Ap. J.*, **196**, L3 (1975).
J. H. Taylor, L. A. Fowler, & R. M. McCullach, 'Measurements of general relativistic effects in the binary pulsar PSR 1913 + 16. *Nature*, **277**, 437 (1979).

Bending of light, radio waves, and microwaves

F. W. Dyson, A. S. Eddington, & C. Davidson, A determination of the deflection of light by the sun's gravitational field, from observations made at the total eclipse of May 29, 1919. *Phil. Trans. Roy. Soc. A*, **220**, 291 (1920).
C. C. Counselman III, S. M. Kent, C. A. Knight, I. I. Shapiro, & T. A. Clark. Solar gravitational deflection of radio waves measured by very long baseline interferometry. *Phys. Rev. Lett.*, **33**, 1621 (1974).
E. B. Fomalont & R. A. Sramek. A confirmation of Einstein's general theory of relativity by measuring the bending of microwave radiation in the gravitational field of the Sun. *Ap. J.*, **199**, 749 (1975).
K. W. Weiler, R. D. Ekers, E. Raimond, & K. S. Wellington. Dual frequency measurement of the solar gravitational microwave deflection. *Phys. Rev. Lett*, **35**, 134 (1975).

Radar echo delay

J. D. Anderson, P. B. Esposito, W. Martin, C. L. Thornton, & D. O. Muhleman. Experimental test of general relativity time-delay data from Mariner 6 and 7. *Ap. J.*, **200**, 221 (1975).
R. D. Reasenberg, I. I. Shapiro, P. E. MacNeil, R. B Goldstein, J. C. Breidenthal, J. P. Brenkle, D. L. Cain, T. M. Kaufman, T. A. Komarck, & A. I. Zygielbaum. Verification of signal retardation by Solar gravity. *Ap. J.*, **234**, L219 (1979).

Principle of equivalence

I. I. Shapiro, C. C. Counselman, & R. W. King. Verification of the principle of equivalence for massive bodies. *Phys. Rev. Lett.*, **36**, 555 (1976).
J. G. Williams, R. H. Dicke, P. L. Bender, C. O. Alley, W. E. Carter, D. G. Currie, D. H. Eckhardt, J. E. Faller, W. M. Kaula, J. D. Mulholland, H. H. Plotkin, S. K. Poultney, P. J. Shelus, E. C. Silverberg, W. C. Sinclair, M. A. Slade, & D. T. Wilkinson, New test of the equivalence principle from lunar laser ranging. *Phys. Rev. Lett.*, **36**, 551 (1976).

Precession of gyroscope

L. I. Schiff, Motion of a gyroscope according to Einstein's theory of gravitation. *Proc. Natl. Acad. Sci. (USA)*, **46**, 871 (1960).

Chapter 3

The Einstein universe

A. Einstein, Kosmologische betrachtungen zur allgemeinen Relativitätstheorie. *Preuss. Akad. Wiss. Berlin, Sitzber,* 142 (1917).

Newton's attempt to construct a model of the universe

I. Newton, 1692 and 1693 letters to Richard Bentley dated 10 December 1692 and 17 January 1693. In D. T. Whiteside, ed., *Mathematical Papers of Isaac Newton*, vol. 7, pp. 233, 238 (Cambridge: Cambridge University Press, 1976).

Newtonian cosmology

H. Seeliger, *Astr. Nachr.* **137**, 129 (1895) and *Münich. Ber. Math. Phys. Kl.*, 373 (1896).

C. Neumann, *Allgemeine Untersuchungen über das Newtonsche Prinzip der Fernwirkungen* (Leipzig 1896).

E. A. Milne & W. H. McCrea, Newtonian universes and the curvature of space. *Q. J. Maths.*, **5**, 73 (1934).

De Sitter's universe

W. de Sitter, On the relativity of inertia: remarks concerning Einstein's latest hypothesis. *Proc. Akad. Weteusch. Amsterdam*, **19**, 1217 (1917).

Weyl's postulate

H. Weyl, Zur allgemeinen Relativitätstheorie. *Z. Phys.*, **24**, 230 (1923).

The Robertson–Walker line element

H. P. Robertson, Kinematics and world structure. *Ap. J.*, **82**, 248 (1935).

A. G. Walker, On Milne's theory of world-structure. *Proc. Lond. Math. Soc. (2)*, **42**, 90 (1936).

Chapter 4

The Friedmann models

A. Friedmann, Über die krummung des reumes. *Z. Phys.*, **10**, 377 (1924) and *Z. Phys.*, **21**, 326 (1924).

The Einstein–de Sitter model

A. Einstein & W. de Sitter, On the relation between the expansion and the mean density of the universe. *Proc. Natl. Acad. Sci. (USA)*, **18**, 213 (1932).

Luminosity distance

W. Mattig, Über den Zusammenhang zwischen Rotverschiebung und Scheinbarer Helligkeit. *A. N.*, **284**, 109 (1958).

Variation of angular sizes

F. Hoyle, The relation of radioastronomy and cosmology. In R. N. Bracewell, ed., *Paris Symposium on Radio Astronomy*, p. 529 (Palo Alto, Calif: Stanford University Press, 1959).

Number counts of galaxies

E. P. Hubble & R. C. Tolman, Two methods of investigating the nature of the red-shift. *Ap. J.*, **82**, 302 (1935).

Olbers paradox

E. Halley, Of the infinity of the sphere of fixed stars. *Phil. Trans. R. Soc. Lond.*, **31**, 22 (1720). [The first known discussion of the paradox.]

H. W. M. Olbers. Über die Durchsichtigkeit des Weltraumes. In *Bode Jahrbuch*, **110** (1826).

The λ-cosmologies

A. S. Eddington, On the instability of Einstein's spherical world. *MNRAS*, **90**, 668 (1930).

Abbé G. Lemaitre, A homogeneous universe of constant mass and increasing radius accounting for the radial velocity of extragalactic nebulae. *MNRAS*, **91**, 483 (1931). [Translated from the original paper in *Annales de la Société Scientifique de Bruxelles*, *XLVII A*, 49 (1927).]

Chapter 5

The early work on primordial nucleosynthesis

G. Gamow, Expanding universe and the origin of elements. *Phys. Rev.*, **70**, 572 (1946).

R. A. Alpher & R. C. Herman, Evolution of the universe. *Nature*, **162**, 774 (1948). [The first prediction of the microwave background.]

R. A. Alpher, H. A. Bethe & G. Gamow, The origin of chemical elements. *Phys. Rev.*, **73**, 80 (1948) [This paper, with the sequence of authors Alpher/Bethe/Gamow, led to the name 'α/β/γ theory'.]

Stellar nucleosynthesis

G. R. Burbidge, E. M. Burbidge, W. A. Fowler, & F. Hoyle, Synthesis of the elements in stars. *Rev. Mod. Phys.*, **29**, 547 (1957).

Later work on primordial nucleosynthesis

C. Hayashi, Proton–neutron concentration ratio in the expanding universe at the stages preceding the formation of the elements. *Progr. Th. Phys. (Japan)*, **5**, 224 (1950).

F. Hoyle & R. J. Tayler The mystery of cosmic helium abundance. *Nature*, **203**, 1108 (1964).

P. J. E. Peebles, Primordial helium abundance and the primordial fireball. *Ap. J.*, **146**, 542 (1966).

Y. B. Zeldovich, The 'hot' model of the universe. *Usp. Fiz. Nauk*, **89**, 647 (1966).

R. V. Wagoner, W. A. Fowler, & F. Hoyle. On the synthesis of elements at very high temperatures. *Ap. J.*, **148**, 3 (1967).

Helium abundance and neutrino types

J-Yang, D. Schramm, G. Steigman, & R. T. Rood, Constraints on cosmology and neutrino physics from big bang nucleosynthesis. *Ap. J.*, **227**, 697 (1979).

Discovery of the microwave background

A. A. Penzias & R. W. Wilson, Measurement of excess antenna temperature at 4080 Mc/s. *Ap. J.*, **142**, 419 (1965).

Chapter 6

Steps towards a unified theory of basic interactions

J. C. Maxwell, A dynamical theory of the electromagnetic field. *Phil. Trans. R. Soc.*, **155** (1864) [Paper read on 8 December 1864.]

S. Weinberg, A model of leptons. *Phys. Rev. Lett.*, **19**, 1264 (1967) [The electro-weak interaction.]

A. Salam, Weak and electromagnetic interactions. In N. Swartholm, ed., *Elementary Particle Physics*, p. 367 (Stockholm: Almquist and Wiksells, 1968).

H. Georgi & S. L. Glashow, Unity of all elementary-particle forces. *Phys. Rev. Lett.*, **32**, 438 (1974). [The $SU(5)$ framework.]

Baryon excess in the early universe

G. Steigman, Observational tests of antimatter cosmologies. *Ann. Rev. Astron. Astrophys.*, **14**, 339 (1976).

M. Yoshimura, Unified gauge theories and the baryon number of the universe. *Phys. Rev. Lett.*, **41**, 281 (1978).

S. Weinberg, Baryon–lepton non-conserving processes. *Phys. Rev. Lett.*, **43**, 1566 (1979).

Inflationary universe

D. Kazanas, Dynamics of the universe and spontaneous symmetry breaking. *Ap. J.*, **241**, L59 (1980).

A. H. Guth, Inflationary universe: A possible solution to the horizon and flatness problems. *Phys. Rev.*, **D23**, 347 (1981).

K. Sato, First order phase transition of a vacuum and the expansion of the universe. *MNRAS*, **195**, 467 (1981).

New inflationary universe

A. Linde, A new inflationary universe scenario. *Phys. Lett.*, **B108**, 389 (1982).

A. Linde, Scalar field fluctuations in the expanding universe and the new inflationary scenario. *Phys. Letts.*, **B116**, 335 (1982).

A. Albrecht & P. J. Steinhardt, Cosmology for grand unified theories with radiatively induced symmetry breaking. *Phys. Rev. Lett.*, **48**, 1220 (1982).

Chaotic inflation

A. Linde, Chaotic inflation. *Phys. Letts.*, **B129**, 177 (1983).

Primordial black holes

S. W. Hawking, Black hole explosions? *Nature*, **248**, 30. (1974).

B. J. Carr, The primordial black hole mass spectrum. *Ap. J.*, **201**, 1 (1975).

Quantum cosmology

J. Hartle, Quantum cosmology. In *Highlights in Gravitation and Cosmology*, eds B. R. Iyer, A. Kembhavi, J. V. Narlikar, & C. V. Vishveshwara, p. 144. (Cambridge: Cambridge University Press, 1988).

Chapter 7

Jeans mass

J. H. Jeans, The stability of a spiral nebula. *Phil. Trans. R. Soc.*, **199A**, 49 (1902).

Growth of fluctuations

E. Lifshitz, On the gravitational instability of the expanding universe. *J. Phys. (USSR)*, **10**, 116 (1946).

R. H. Dicke & P. J. E. Peebles, Origin of the globular clusters. *Ap. J.*, **154**, 891 (1968).

Scale-invariant spectrum

E. R. Harrison, Fluctuations at the threshold of classical cosmology. *Phys. Rev.*, **D1**, 2726 (1970).

Y. B. Zeldovich, A hypothesis, unifying the structure and the entropy of the universe. *MNRAS*, **160**, 1P (1972).

J. M. Bardeen, P. J. Steinhardt, & M. S. Turner, Spontaneous creation of almost scale-free density perturbations in an inflationary universe. *Phys. Rev.*, **D28**, 679 (1983).

Dark matter and structure formation

J. R. Bond, G. Efstathiou, & J. Silk, Massive neutrinos and the large scale structure of the universe. *Phys. Rev. Lett.*, **45**, 1980 (1980).

P. J. E. Peebles, The peculiar velocity around a hole in the galaxy distribution. *Ap. J.*, **258**, 415, (1982).

Nonlinear evolution of structures

Y. B. Zeldovich, Gravitational instability: an approximate theory for large density perturbations. *Astron. Astrophys.*, **5**, 84 (1970).

S. D. M. White, C. S. Frenk, & M. Davis, Clustering in a neutrino dominated universe. *Ap. J.*, **274**, L1. (1983).

S. J. Aarseth, Direct N-body calculations. In *Dynamics of Star Clusters*, eds J. Goodman & P. Hut. *IAU Symposium No. 113*, p. 251 (Dordrecht: Reidel, 1985).

M. Davis, G. Efstathiou, C. S. Frenk & S. D. M. White, The evolution of large-scale structure in the universe dominated by cold dark matter. *Ap. J.*, **292**, 371 (1985).

Experimental data on massive neutrinos

V. Berger, K. Whisnant, & R. J. N. Phillips, Three neutrino oscillations and present experimental data. *Phys. Rev.*, **D22**, 1636 (1980).

F. Reines, H. W. Sobel, & E. Pasierb. Evidence for neutrino instability. *Phys. Rev. Lett.*, **45**, 1307 (1980).

V. A. Lyubimov, E. G. Novikov, V. Z. Nozik, E. F. Tretyakov, & V. S. Kozik, I T E F Preprint No. 62, Moscow (1980).

Massive neutrinos and cosmology

R. Cowsik & J. McClelland, An upper limit on the neutrino rest mass. *Phys. Rev. Lett.*, **29**, 669 (1972).

R. Cowsik & J. McClelland, Gravity of neutrinos of nonzero mass in astrophysics. *Ap. J.*, **180**, 7 (1973).

S. Tremain & J. E. Gunn, Dynamical role of light neutral leptons in cosmology. *Phys. Rev. Lett.*, **42**, 407 (1979).

Chapter 8

Steady state theory (first proposed)

H. Bondi & T. Gold, The steady state theory of the expanding universe. *MNRAS*, **108**, 252 (1948).

F. Hoyle, A new model for the expanding universe. *MNRAS*, **108**, 372 (1948).

C-field cosmology

F. Hoyle & J. V. Narlikar, Mach's principle and the creation of matter. *Proc. Roy. Soc.*, **A270**, 334 (1962) [Continuous creation.]

J. V. Narlikar, Singularity and matter creation in cosmological models. *Nature*, **242**, 135 (1973) [Explosive creation.]

Hot universe

T. Gold & F. Hoyle, Cosmic rays and radio waves as manifestations of a hot universe. In R. N. Bracewell, ed., *Paris Symposium on Radio Astronomy* p. 583 (Palo Alto, Calif: Stanford University Press, 1958).

The bubble universe–galaxy formation

F. Hoyle & J. V. Narlikar, A radical departure from the steady state concept in cosmology. *Proc. R. Soc.*, **A290**, 162 (1966).

F. Hoyle & J. V. Narlikar, On the formation of elliptical galaxies. *Proc. R. Soc.*, **A290**, 177 (1966).

Inertial forces and the absolute space

I. Newton, *Philosophiae Naturalis Principia Mathematica*, 1st edn (London: Streater 1687 [English translation by A. Motte (1729) revised by A. Cajori. Berkeley: University of California Press, (1934)].

E. Mach, *The Science of Mechanics* (Chicago: Open Court, 1893).

Sciama's theory of inertia

D. W. Sciama, On the origin of inertia. *MNRAS*, **113**, 34 (1953).

Brans–Dicke theory of gravity

C. Brans & R. H. Dicke, Mach's principle and a relativistic theory of gravitation. *Phys. Rev.*, **124**, 125 (1961).

R. H. Dicke, Mach's principle and invariance under transformation of units. *Phys. Rev.*, **125**, 2163 (1962).

Solar system tests of the Brans–Dicke theory

R. H. Dicke & H. M. Goldenberg, Solar oblateness and general relativity. *Phys. Rev. Lett.*, **18**, 313 (1967).

H. A. Hill & R. T. Stebbins, The intrinsic visual oblateness of the Sun. *Ap. J.*, **200**, 471 (1975).

Cosmological solutions on the Brans–Dicke theory

R. H. Dicke, Scalar tensor gravitation and the cosmic fireball. *Ap. J.*, **152**, 1 (1968).

G-variation in the Brans–Dicke theory

R. H. Dicke, Implications for cosmology of stellar and galactic evolution rates. *Rev. Mod. Phys.*, **34**, 110 (1962).

Hoyle–Narlikar theory of gravity

F. Hoyle & J. V. Narlikar, A new theory of gravitation. *Proc. R. Soc.*, **A282**, 191 (1964).
F. Hoyle & J. V. Narlikar, A conformal theory of gravitation. *Proc. Roy. Soc.*, **A294**, 138 (1966).

Electromagnetic response of the universe

J. A. Wheeler & R. P. Feynman, Interaction with the absorber as the mechanism of radiation. *Rev. Mod. Phys.*, **17**, 157 (1945).
J. E. Hogarth, Cosmological considerations of the absorber theory of radiation. *Proc. R. Soc.*, **A267**, 365 (1962).
F. Hoyle & J. V. Narlikar, Time symmetric electrodynamics and the arrow of time in cosmology. *Proc. R. Soc.*, **A277**, 1 (1963).

Spacetime singularity in HN cosmology

A. K. Kembhavi, Zero mass surfaces and cosmological singularities. *MNRAS*, **185**, 807 (1978).

G-variation in HN cosmology

F. Hoyle & J. V. Narlikar, Cosmological models in a conformally invariant gravitation theory I & II. *MNRAS*, **155**, 305, and **155**, 323 (1972).

Significance of large dimensionless numbers

P. A. M. Dirac, The cosmological constants, *Nature*, **139**, 323 (1937).

Large numbers hypothesis

P. A. M. Dirac, A new basis for cosmology, *Proc. R. Soc.*, **A165**, 199 (1938).

Dirac cosmology with two types of creation

P. A. M. Dirac, Long range forces and broken symmetries. *Proc. R. Soc.*, **A333**, 403 (1973).
P. A. M. Dirac, Cosmological models and the large numbers hypothesis. *Proc. R. Soc.*, **A333**, 439 (1974).

Chapter 9

Measurement of H_0

E. Hubble, A relation between distance and radial velocity among extragalactic nebulae. *Proc. Natl. Acad. Sci (USA)*, **15**, 168 (1929).
M. L. Humason, N. U. Mayall, & A. R. Sandage, Redshifts and magnitudes of extragalactic nebulae. *Ap. J.*, **61**, 97 (1956).

Period–luminosity relation of cepheids

H. Leavitt, Periods of twenty-five variable stars in the small Magellanic cloud. *Harvard College Observatory Circular* No. 173 (1912).

Measurement of distances using supernovae
R. P. Kirshner & J. Kwan, Distances to extragalactic supernovae. *Ap. J.* **193**, 27 (1974).

Anisotropy of Hubble flow
V. C. Rubin, N. Thonnard, & W. K. Ford, Jr, Motion of the Galaxy and the Local Group determined from the velocity anisotropy of distant Sc I galaxies, I & II. *A. J.*, **81**, 687 and 719 (1976).

Mass distributions in galaxies
I. King, The structure of round stellar systems: observation and theory. In A. Hayli, ed., *Dynamics of Stellar Systems*, p. 99 (Dordrecht, Holland: D. Reidel 1975).

Statistics of groups of galaxies
E. L. Scott, Distribution of galaxies on the sphere. In G. C. McVittie, ed., *Problems in Extragalactic Research*, p. 269 (New York: Macmillan 1961).
J. Neyman, Alternative stochastic models of the spatial distribution of galaxies. In G. C. McVittie, ed., *Problems in Extragalactic Research*, p. 294 (New York: Macmillan, 1961).
H. Totsuji & T. Kihara, The correlation function for the distribution of galaxies. *Publ. Astron. Soc. (Japan)*, **21**, 221 (1969).
P. J. E. Peebles, The gravitational instability picture and the nature of distribution of galaxies. *Ap. J.*, **189**, L51 (1974).

Missing mass in clusters of galaxies
F. Zwicky, Die Rotverschiebung von extragalaktischen Neblen. *Helv. Phys. Acta*, **6**, 10 (1933).

Local supercluster
G. de Vaucouleurs, Recent studies of clusters and superclusters. *A. J.*, **66**, 629 (1961).

q_0 from deceleration of nearby galaxies
A. Sandage, G. Tamman, & A. Yahil, The velocity field of bright nearby glaxies I–IV, *Ap. J.*, **232**, 352 (1979), and subsequent papers in later issues of *Ap. J.*

Age of the Galaxy
W. A. Fowler & F. Hoyle, Nuclear cosmochronology. *Ann. Phys. (NY)*, **10**, 280 (1960).
I. Iben, Jr, Post main-sequence evolution of single stars. *Ann. Rev. Astron. Astrophys.*, **12**, 215 (1974).

Abundance of light nuclei
R. V. Wagoner, Big bang nucleosynthesis revisited. *Ap. J.*, **179**, 343 (1973).

Microwave background measurements
See the references cited in Tables 9.4 and 9.5.

Dipole anisotropy of microwave background

G. F. Smoot, M. V. Gorenstein, & R. A. Muller, Detection of anisotropy in the cosmic black body radiation. *Phys. Rev. Lett.*, **39**, 898 (1977).
E. S. Cheng, P. R. Saulson, D. T. Wilkinson, & B. E. Corey, Large scale anisotropy in the 2.7 K radiation. *Ap. J.*, **232**, L139 (1979).

Zeldovich–Sunyaev effect

R. A. Sunyaev & Y. B. Zeldovich, Small-scale fluctuations of relic radiation. *Astrophys. and Sp. Sci*, **7**, 3 (1970).

Microwave background and high-energy cosmic rays

K. Greisen, An end to the cosmic ray spectrum? *Phys. Rev. Lett.*, **16**, 748 (1966).

Chapter 10

q_0 from the m–z relation

A. Sandage, The redshift distance relation, II. *Ap. J.*, **178**, 1 (1972).
J. Kristian, A. Sandage, & J. A. Westphal The extension of the Hubble diagram, II. *Ap. J.*, **221**, 383 (1978).

Corrections to the Hubble diagram

K. G. Malmquist, A study of stars of spectral type A. *Medd. Lunds. Obs. Ser II*, No. 22 (1920).
E. L. Scott, The brightest galaxy in a cluster as a distance indicator. *A. J.*, **62**, 248 (1957) [Scott effect.]
A. Sandage, The ability of the 200-inch telescope to distinguish between selected world models. *Ap J.*, **133**, 355 (1961).
J. B. Oke & A. Sandage, Energy distributions, K corrections and the Stebbins-Whitford effect for giant elliptical galaxies. *Ap. J.*, **154**, 21 (1968). [K-correction.]
J. E. Gunn & J. B. Oke, Spectrophotometry of faint cluster galaxies and the Hubble diagram: an approach to cosmology. *Ap. J.*, **195**, 255 (1974). [Aperture correction.]
B. M. Tinsley, Nucleochronology and chemical evolution. *Ap. J.*, **198**, 145 (1975). [Luminosity evolution.]
S. M. Chitre & J. V. Narlikar, The effect of intergalactic dust on the measurement of the cosmological deceleration parameter q_0. *Astrophys. and Sp. Sci.*, **44**, 101 (1976). [Interglactic absorption.]

G-varying cosmologies and q_0

J. M. Barnothy & B. M. Tinsley, A critique of Hoyle and Narlikar's new cosmology. *Ap. J.*, **182**, 243 (1973).
V. M. Canuto & J. V. Narlikar, Cosmological tests of the Hoyle–Narlikar conformal gravity. *Ap. J.*, **236**, 6 (1980).

Optical counts of galaxies

E. P. Hubble, Effects of redshifts on the distribution of nebulae. *Ap. J.*, **84**, 517 (1936).

Radio source surveys and counts

J. R. Shakeshaft, M. Ryle, J. E. Baldwin, B. Elsmore, & J. H. Thomson, A survey of radio sources between declinations −38° and +83°. *Mem. RAS* **67**, 97 (1955).

B. Y. Mills, O. B. Slee, & E. R. Hills, A catalogue of radio sources between declinations +10° and −20°. *Aus. J. Phys.*, **11**, 360 (1958).

D. O. Edge, J. R. Shakeshaft, W. B. McAdam, J. E. Baldwin, & S. Archer. A survey of radio sources at 159 Mc/s. *Mem. RAS*, **68**, 37 (1959).

J. G. Bolton, The discrete sources of cosmic radio emission. *Comptes Rendus de l'Assemblé Générale de l'URSI, Londres, 1960, Session V* (1960).

A. S. Bennett, The revised 3C-catalogue of radio sources. *Mem. RAS*, **68**, 163 (1962).

Luminosity volume test

M. Schmidt, Space distribution and luminosity functions of quasi-stellar radio sources. *Ap. J.*, **151**, 393 (1968).

Maximum likelihood method

D. F. Crawford, D. L. Jauncey, & H. S. Murdoch, Maximum likelihood estimation of the slope from number-flux density counts of radio sources. *Ap. J.*, **162**, 405 (1970).

Source-count interpretations for cosmology

P. F. Scott & M. Ryle, The number flux density relation for radio sources away from the galactic plane. *MNRAS*, **122**, 389 (1961).

F. Hoyle & J. V. Narlikar, On the counting of radio sources in steady state cosmology. *MNRAS*, **123**, 133 (1962).

D. L. Jauncey, Reexamination of the source counts for the 3C revised catalogue. *Nature*, **216**, 877 (1967).

F. Hoyle, Review of recent developments in cosmology. *Proc. R. Soc.*, **A308**, 1 (1968).

Variation of angular size and cosmology

F. Hoyle, The relation of radioastronomy and cosmology. In R. N. Bracewell, ed., *Paris Symposium on Radio Astronomy*, p. 529 (Palo Alto, Calif: Stanford University Press, 1959).

Surface brightness test

A. Sandage & J.-M. Perelmuter The surface brightness test for the expansion of the universe I & II. *Ap. J.*, **350**, 481, and **361**, 1 (1990).

Quasar catalogue

A. Hewitt & G. Burbidge. A new optical catalogue of quasi-stellar objects. *Ap. J. Suppl.*, **63**, 1 (1987).

A. Hewitt & G. Burbidge, The first addition to the new optical catalogue of quasi-stellar objects. *Ap. J. Suppl.*, **69**, 1 (1989).

Hubble diagram of quasars

G. R. Burbidge & F. Hoyle, Relation between the redshifts of quasi-stellar objects and their radio and optical magnitudes. *Nature*, **210**, 1346. (1966).

Number counts of quasars

M. Schmidt, Space distribution and luminosity functions of quasi-stellar radio sources. *Ap. J.*, **151**, 393 (1968).

A. K. Kembhavi & A. Fabian, X-ray quasars and the X-ray background. *MNRAS*, **198**, 921 (1982).

Gravitational lens

D. Walsh, R. F. Carswell, & R. J. Weymann, 0957 + 561 A, B: twin quasistellar objects or gravitational lens? *Nature*, **279**, 381 (1979).

Lyman-α absorption

J. E. Gunn & B. A. Peterson, On the density of neutral hydrogen in the intergalactic space. *Ap. J.*, **142**, 1633 (1965).

Energy production problem in quasars

F. Hoyle, G. R. Burbidge, & W. L. W. Sargent, On the nature of quasi-stellar sources. *Nature*, **209**, 751 (1966).

M. J. Rees, Studies in radio source structure, I. *MNRAS*, **135**, 345 (1967).

Superluminal separation in quasars

K. I. Kellermann & D. B. Shaffter, Superlight motion in radio sources and its implications for the distance scale problem. In C. Balkowski and B. E. Westerlund, eds., *Proceedings of the I.A.U./C.N.R.S. Colloquium*, held in Paris, September 6–9, 1976, p. 347 (Paris: C.N.R.S. 1977) [For observations and some theoretical models.]

J. V. Narlikar & S. M. Chitre, Gravitational screens and superluminal separation in quasars. *Ap. J.*, **235**, 335 (1980).

Variation of e^2/hc

J. N. Bahcall & M. Schmidt, Does the fine structure constant vary with time? *Phys. Rev. Lett.*, **19**, 1294 (1967).

M. S. Roberts, High redshift 21-cm lines. In C. Balkowski & B. E. Westerlund, eds., *Proceedings of the I.A.U./C.N.R.S. Colloquium*, held in Paris, September 6–9, 1976, p. 501 (Paris C.N.R.S., 1977).

A. D. Tubbs & A. M. Wolfe, Evidence for large-scale uniformity of physical laws. *Ap. J.*, **236**, L105 (1980).

Variation of G

F. Hoyle, The early history of the Earth. *QJRAS.*, *13*, 328 (1972).

I. I. Shapiro, Bounds on the secular variation of the gravitational constant. *BAAS*, **8**, 308 (1976).

Chapter 11

For standard cosmology

P. J. E. Peebles, D. N. Schramm, E. L. Turner, & R. G. Kron, The case for the hot relativistic big bang cosmology. *Nature*, **352**, 769 (1991).

Against standard cosmology

H. C. Arp, G. Burbidge, F. Hoyle, J. V. Narlikar, & N. C. Wickramansinghe, The extragalactic universe: an alternative view. *Nature*, **346**, 807 (1990).

INDEX

Page numbers in italics refer to the main discussion(s) of the topic in text.